Economics and Sustainability

Karl Bruckmeier

Economics and Sustainability

Social-Ecological Perspectives

palgrave
macmillan

Karl Bruckmeier
Berlin, Berlin, Germany

ISBN 978-3-030-56626-5 ISBN 978-3-030-56627-2 (eBook)
https://doi.org/10.1007/978-3-030-56627-2

Cover illustration: Fortgens Photography/gettyimages

This Palgrave Macmillan imprint is published by the registered company Springer Nature Switzerland AG.
The registered company address is: Gewerbestrasse 11, 6330 Cham, Switzerland

Preface

Economic and ecological knowledge need to be combined with other scientific and practical knowledge in the social-ecological transformation towards a sustainable society. This long process began three decades ago and has not significantly advanced since then. The reasons for the delays and difficulties are clear, but to date they have not been sufficiently addressed in analyses of the sustainability process. Knowledge- and power-related conflicts and controversies between scientists, decision-makers and actors in the transformation process accompany the research and the practice of sustainable development. But the interdisciplinary integration and synthesis of knowledge from different disciplines create methodological and practical problems. These problems guide the discussion of economics and sustainability in this book. For example, the book discusses how to deal with knowledge gaps around the multi-scale process of sustainable development, with the limits of disciplinary knowledge on the complex problems to be solved in the sustainability process, and with interdisciplinary cooperation in the production, integration, dissemination and application of knowledge. Instead of shifting these problems to policy and practice, where they are managed through "muddling through" and "clumsy solutions", they should be discussed where the creation of knowledge and learning of new practices take place, in research and education, in close connection with the development of new strategies and the practices of sustainability governance. An interdisciplinary textbook for sustainability studies at graduate and postgraduate levels seems more necessary in the sustainability discourse and process than the worn and overrated instruments of policy review and advice. Such a book needs to address the difficulties in the sustainability process: the creation of transformative literacy and agency in inter- and transdisciplinary practices of knowledge creation, in new interdisciplinary subjects such as sustainability science, ecological economics, social and political ecology and transformation research.

The interdisciplinary teaching of sustainability is part of the new knowledge practices that have developed in late modernity in research on the relations between nature and society. Interdisciplinary cooperation and knowledge exchange are not yet routine in the dominant academic cultures of specialised research and education, although they are practised in the sustainability discourse and policy process. The forms of conventional disciplinary and new inter- and transdisciplinary science, or "mode one" and "mode two", as they have been called at the beginning of the debate about new modes of knowledge production, coexist with tensions in the practices of environmental research and teaching. Disciplinary teaching is challenged through pluralistic knowledge practices, non-dogmatic forms of teaching economics and ecology, and cooperation with non-scientific actors and knowledge bearers developing in the sustainability process. In this process, of which interdisciplinary teaching is a part, it becomes necessary to deal with manifold and conflicting aspects of knowledge, power and interest between the many actors involved. Furthermore, it becomes necessary to leave the short-term perspectives of the policy processes, where the achievement of global sustainability is unrealistically reduced to a few decades, as in the United Nations' "Agenda 21" and "Agenda 2030" that guide the global sustainability process.

Following an enthusiastic start, with the global conference on environment and development in 1992 in Rio de Janeiro and the "Agenda 21" global action plan, the process has lost momentum and even slowed down, revealing the weaknesses of global action and cooperation. The decade from 2005 to 2014 focusing on education in sustainable development and the decade from 2021 to 2030 focusing on ecosystem restoration are examples of the UN's policy-based initiatives to anchor sustainable development in society at large and in people's social lifeworld (in the social practices and routines of their daily lives). These attempts to maintain the sustainability process are not enough to cope with the inherent difficulties of sustainable development. In sustainability research, in teaching and in governance practice, sustainable development needs to be renewed as a global process of sustainability transformation with a long-term perspective of many decades and generations. This book aims to become part of such a collective drive to build the knowledge cultures and governance practices for a long process that stretches into the distant future.

When the global discourse of sustainable development began in the late 1980s, the nature of the changes necessary to attain sustainability was less clear than it is now, when "another great transformation" involves how to deal with the sustainability process. Although the development of industrial society globally remains incomplete, the phasing out of the industrial society began towards the end of the twentieth century, coinciding with the sustainability movement. Regarding the sustainable and post-industrial economy and society of the future, little is known presently beyond than their names. The means of transformation, which will differ between countries, need to be discovered during the process following the signpost "sustainability". Knowledge about social, economic, political, cultural and ecological transformation processes flows together in the building of transformative agency to achieve sustainability. This approach to transformation requires collective action beyond the formal institutions and international regimes in environmental policy, with climate and sustainability regimes as the key processes. Global policy and governance depend from asymmetric international power relations and from other social and ecological processes that need to be taken into account and changed in the sustainability process. Addressing the problems facing sustainable development will require the dissemination of knowledge about modern society and its interaction with nature—about the modification of nature by humans in the industrial era of modernity, now called "the Anthropocene". Processes of global environmental change such as climate change, and of global social change such as economic globalisation, are clashing with the process of sustainable development. Social-ecological transformation and new forms of global governance are the primary challenges of the twenty-first century.

Berlin, Germany
June 2020

Karl Bruckmeier

Acknowledgements

The themes of this book developed in the "real-world laboratories" of teaching and research, during my work at the New University in Lisbon, Faculty of Social Sciences and Humanities, and the South Bohemian University in Ceske Budejovice, Economic Faculty. In seminars, workshops and conferences at these universities the themes of economics and sustainability have been discussed with many people. I am grateful for the discussions with the colleagues and doctoral students in the Department of Regional Management of the South Bohemian University: Roman Buchtele, Barbora Halírová, Josef Maxa, Nikola Sagapová, Jirí Sedlák, Iveta Sindelárová and Martin Slachta. A great debt of thanks I owe to the colleagues who read and commented drafts of the chapters: Iva Miranda Pires and Teresa Santos from the New University of Lisbon, and Miloslav Lapka, Eva Cudlínová and Jan Vávra from the South Bohemian University. Special thanks to the anonymous reviewers of the manuscript for their useful comments and the editorial team at Palgrave Macmillan.

Contents

Abbreviations

BRICS	Brazil, Russia, India, China, South Africa (newly industrialising countries)
EPI	Environmental Performance Index
ERO(E)I	Energy Return on (Energy) Investment
ESI	Environmental Sustainability Index
EU	European Union
EVI	Environmental Vulnerability Index
FAO	Food and Agriculture Organization of the United Nations
GDP	Gross Domestic Product
HANPP	Human Appropriation of Net Primary Production
IISD	International Institute for Sustainable Development
IPCC	Intergovernmental Panel on Climate Change
ISO	International Organization for Standardization
IUCN	International Union for Conservation of Nature
MEFA	Material and Energy Flow Accounting
OECD	Organisation for Economic Co-operation and Development
SPI	Sustainable Process Index
TEEB	The Economics of Ecosystems and Biodiversity
TRIPS	Trade-Related Aspects of Intellectual Property Rights
UN	United Nations
UNCED	United Nations Conference on Environment and Development (Rio 1992)
UNCHE	United Nations Conference on the Human Environment (Stockholm 1972)
UNEP	United Nations Environment Programme

UNESCO	United Nations Educational, Scientific and Cultural Organization
WBCSD	World Business Council for Sustainable Development
WBGU	German Advisory Council on Global Change

List of Figures

List of Tables

List of Boxes

Part I

The Sustainability Process: Context and Scope

1 The Policy Context of the Sustainability Discourse

The political framing of sustainability is reviewed in this chapter in four stages: (a) the genesis of the political and public sustainability discourse; (b) the description of global sustainability problems in the seminal document, the Brundtland Report (UN 1987); (c) the national and international sustainability policies developing after the Brundtland Report; and (d) the use of economic knowledge in the sustainability discourse (i.e. in the policy perspective that frames the discourse). The key concept of sustainability cannot be clarified with a single and simple definition, but only through discussion of variants from three complementary perspectives: the policy context, the historical context, and the knowledge context in which the terminology of *sustainability* and *sustainable development* (see Glossary) unfolds.

Chapter 2 reviews the history of the sustainability discourse within modern European society since the seventeenth century. Chapter 3 analyses the interdisciplinary knowledge contexts and the problems of knowledge integration in the sustainability discourse and policy process. These chapters provide a foundation for the detailed discussion of economic knowledge in the second part of the book, where the discourses of natural resource management, environmental and ecological economics are reviewed to show how economic knowledge can be integrated with other types of knowledge in the sustainability process.

The global sustainability discourse (Fig. 1.1) is multifaceted and complex; it faces myriad problems and has been developed through controversy, using knowledge from many disciplines, by many actors and stakeholders with conflicting interests, based on asymmetrical power relations, and facing many obstacles. Since the late 1980s few of the problems addressed in the Sustainable Development

K. Bruckmeier, *Economics and Sustainability*,
https://doi.org/10.1007/978-3-030-56627-2_1

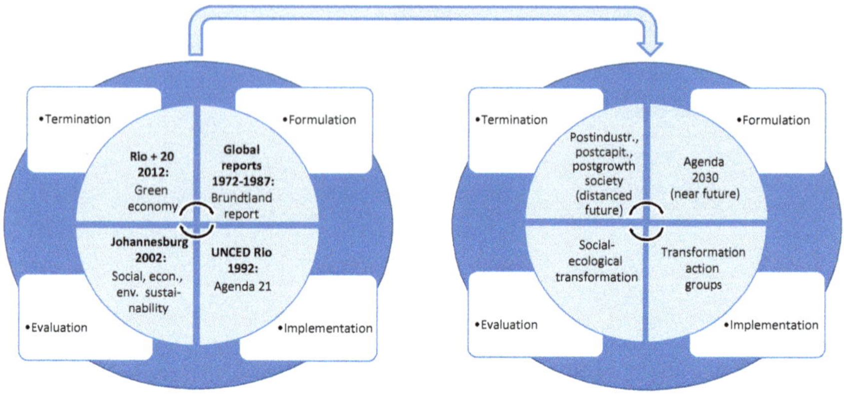

Fig. 1.1 The policy cycles of the sustainability process

Goals of the United Nations have been overcome. The majority of the problems will need to be dealt with in the future as part of a long process of transforming society and the economy.

The terminological problems associated with the discussion of sustainability cannot be resolved with a simple definition of sustainability as provided in the Brundtland Report (UN 1987). Several connected concepts need to be clarified: there are multiple definitions and interpretations of sustainable development and the concepts for describing modern society and the economy; no consensus about the terminology can be found in academic and political debates (Bruckmeier 2016: 130ff). To deal pragmatically with the abstract and inexact nature of terms such as sustainability and the nature of today's society and economy, global change and globalisation are used in all chapters as general terms without exact definitions. Short definitions of the terms are found in the glossary that follows Chap. 10. In each chapter the concepts are defined according to the theories and approaches that are discussed. The sustainability discourse cannot work with singular and simple definitions; open and pluralistic concepts are necessary, as with that of sustainability itself (see Box 1.1), or that of modern society, which in the social scientific and economic literature includes manifold descriptive facets: agricultural, industrial, capitalist society, post-industrial and post-capitalist society, civil society, world society and *world system* (see Glossary), knowledge or information society; the modern economy can be described with similar terms.

Box 1.1 "Sustainability": The Terminology

The notions of sustainability and sustainable development are used in different ways. In this book the discourse of sustainability is reviewed and discussed with the guiding concepts of intra- and intergenerational solidarity in resource use (Brundtland Report: UN 1987) and the three-dimensional concept of social, economic and environmental sustainability (the mainstream variant in use since the Johannesburg Summit of 2002). The concepts were created in the policy process for the purposes of initiating and guiding changes in economic resource use practices. Other definitions and scientific interpretations are discussed in the following chapters. The competing and changing interpretations of the abstract terms demonstrate the need for continuous reflection, discussion and review of the terms in the scientific and political sustainability discourses.

To structure the analysis of the multilayered sustainability theme, the concepts of discourse, policy (or governance) and process are used: *discourse* means the knowledge process in the scientific and public political debates of sustainable development; *policy* refers to the political decisions, programmes and regulations that direct the transitions to sustainability; *process* implies the broader social process of sustainable development, beyond policies, including activities of cultural, economic and environmental movements and organisations in the transition to sustainability, and the everyday processes and changes of the routines in social life.

The sustainability discourse proceeds without a single, specific, scientifically defined concept. Sustainable development is a *bridging concept*: of discursive and pluralistic nature, open to different interpretations and strategies of action. From an epistemological perspective sustainable development is an "essentially contested concept"; continuous controversies over the interpretation of the term arise depending on the scientific or political views and interests of the discourse participants. In the interdisciplinary sustainability discourse the process-quality of sustainable development as social, economic and ecological transition or transformation is highlighted. Sustainability is not a static state that can be achieved once and forever: a sustainable economy and society is continually developing and changing without exceeding its natural resource base.

To deal with essentially contested concepts such as sustainability, see the first learning exercise described at the end of the chapter.

Sources: Own text

Interdisciplinary knowledge practices create difficulties in the sustainability discourse when natural and social scientific concepts are used by scientists, decision-makers and stakeholders with different specialisations and interests. Within the scientific disciplines terminologies are not always standardised, and theoretical concepts and explanations often compete with those of other theories. Sustainability and sustainable development are examples of "essentially contested concepts" (Gallie 1956): abstract terms which can be interpreted in many ways by different people with different values and worldviews; they develop through collective learning and continuous improvement of arguments (see learning exercise one, described in the appendix of the chapter). As in everyday communication in the social lifeworld, the terms are not always clearly defined, as has been shown in ethnomethodological studies of language use in sociology (Garfinkel 1967). Nevertheless, communication and mutual understanding are possible when rules of communication are constructed spontaneously by participants, as part of their communication, not through explicit or scientific definition. The sustainability process is connected with the social lifeworld of the participants; it eventually becomes a process of changing ways of living and consumption—moving away from the imperial mode of living in industrialised countries: a mode of living where a minority of the global population consumes the largest part of global resources; this is possible through their unequal wealth and the asymmetric power relations between the industrialised Western countries and the non-industrialised countries (Brand and Wissen 2013).

1.1 Genesis of the Political and Public Sustainability Discourse

The global sustainability discourse began with the report "Our Common Future" of the World Commission on Environment and Development (UN 1987), known as the Brundtland Report. The notion of sustainable development came into use earlier, in the 1970s, after the United Nations Conference on the Human Environment (UNCHE), held in Stockholm in 1972 and organised by the United Nations. Earlier global reports had set the framework for the sustainability discourse. The report "Limits to Growth" (Meadows et al. 1972; updated: Meadows et al. 1992) mandated by the Club of Rome introduced the issues of economic growth, population growth, overuse of natural resources and pollution into public and political debates. In the Cocoyoc Declaration of 1974 (Ward 1975), scientists called for reform of the economic order, critically discussing global natural resource use, environmental degradation and unequal development. The Report "What Now?" (Dag Hammarskjöld Report 1975) reviewed global development and international cooperation, arguing for global sharing and redistribution of resources. The report of the North–South

Commission (Brandt 1980) "Securing the Survival" analysed the great chasm between the North (industrialised countries) and the South (developing countries) and argued for a transfer of resources from the North to the South. The diagnosis of problems found here anticipates ideas later developed in the Brundtland Report: global environmental problems are caused mainly by the growth of industrial economies, but also by global population growth; growth threatens the wellbeing and survival of future generations; global cooperation is necessary to protect the earth's atmosphere and other global commons and to prevent irreversible ecological damage. The report initiated a debate about the future global economy in terms of a sharing economy, relaunched in 2012 with the report "Financing the Global Sharing Economy" (STWR 2012). None of the global reports that preceded it were as influential as the Brundtland Report. The earlier reports were more critical in their analyses, their messages were often perceived as dystopian and they did not take the type of "soft diplomacy" approach seen in "Our Common Future". Its formulation of sustainable development, an idea that many could support, seemed to exude optimism, pushing the negative messages into the background.

The Brundtland Report initiated a global North–South policy to respond to problems of the uneven use, distribution and overuse of natural resources and global environmental pollution. The uneven development in the global economy (see Box 1.2) became the guiding theme in the sustainability discourse. Whereas the idea of sustainable development found global consensus among governmental and non-governmental organisations, the policy process became complicated, mainly because of the difficulties of changing the political and economic power relations in the modern world system.

Box 1.2 Genesis of the Political Sustainability Discourse: Global Reports About Environment and Development

1. "*Limits to Growth*" (Meadows et al. 1972): economic growth is connected with population growth, growth of natural resource use and of environmental pollution; global growth processes have become exponential, exceeding the limits of the natural resources on the earth, and could result in global economic and ecological collapse towards the end of the twenty-first century.
2. "*What Now?*" (Dag Hammarskjöld Report 1975): global development and economic growth are, in spite of international cooperation, not supporting the welfare and wellbeing of all humans; sharing and redistribution of resources will be necessary in future to limit economic growth and to create fairness in resource use.

(*continued*)

Box 1.2 (continued)

3. *North–South Commission—"Securing the Survival*" (Brandt 1980): global environmental problems are caused mainly by the growth of industrial economies, but also by global population growth; the growth processes threaten the wellbeing and survival of future generations; global cooperation is necessary to manage the earth's atmosphere and other global commons and to prevent irreversible ecological damage.
4. *World Conservation Strategy* (IUCN 1980): development, in order to be sustainable, should support conservation and protection of ecological processes and life-support systems, preservation of genetic diversity, and sustainable utilisation of species and ecosystems. "*Development is defined here as: the modification of the biosphere and the application of human, financial, living and not-living resources to satisfy human needs and improve the quality of human life. For development to be sustainable it must take account of social and ecological factors, as well as economic ones; of the living and non-living resource base; and of the long term as well as the short term advantages and disadvantages of alternative actions.*" (IUCN 1980: 18)
5. "*Our Common Future*" (UN 1987): sustainable development creates a common future for humankind—further economic growth is necessary, but it must maintain the functions and services of ecosystems; sustainable development results in the progressive transformation of the global economy and society.

The five global policy reports mark the genesis of a global environmental policy and movement involving governmental and non-governmental organisations. The reports are critical of the uneven economic development that divides the Global North (industrialised countries) and the Global South (former European colonies, developing countries) as well as its negative social, economic and environmental consequences, but no consensus was found about ways to solve these global problems. The most comprehensive and detailed definition of sustainable development is that of the World Conservation Strategy of IUCN.

Sources: mentioned in the text

The Brundtland Report is based on a critical review of the relations between the developed and the developing countries in the modern world system and of the global divide between the rich and poor parts of the world. To overcome the global divide, the report promotes (similar to the report of the North–South Commission [Brandt 1980]) the idea of global solidarity in resource use (sharing and redistributing resources within and between countries and generations). Whether this is achieved through another, environmentally friendly, phase of industrialisation or through transformation to a post-industrial society has remained a point of controversy in the sustainability discourse, centred around three points:

1. *The notion of sustainable development* requires clarification of the terms *development* (see Glossary) and *transformation* (see Glossary), both of which are used in the report. Yet development and transformation cannot be clarified through definition; they require discussion, research and the creation of flexible action strategies as part of the political sustainability process. Only after two decades of unsuccessful global sustainability policies did the need for system change and transformation become clear. The sustainability discourse arose when the "Second World", the former East European bloc of socialist countries, collapsed and their state-controlled economies were re-privatised. Sustainability became the main political discourse dealing with problems in the global society and economy. The conflict between East and West in the "Cold War" ended without freeing the world from global risks and dangers: the problems of the economic cleavage between the rich Global North and the poor Global South and increasing environmental damage remained. Indicators of the cleavage include the unequal global exchange and resource flows shown through material and energy flow accounting: the majority of resources flow from poor countries to rich countries. The consumption levels (in terms of ecological footprint) of countries vary widely, and large parts of the global population continue to suffer from poverty and hunger. Due to the relocation of industrial production to the late industrialising countries of the Global South, those countries now produce a significant amount of the negative environmental effects of industrialisation and will continue to do so in the future. How long the path to industrialisation can be maintained, and how many countries it can include, is unclear. According to global analyses of social or societal metabolism (see Glossary) and material energy flow accounting, the resources on the earth are not sufficient for all countries to reach the same level of resource use and wealth as the Western countries. Moreover, countries that began to industrialise only in the twenty-first century may not be able to complete the process and overcome poverty in this way. The systemic divide between the rich North and the poor South seems

to change and to be reproduced in the growth-based and globalising-based economy in new forms, for example as economic and digital divides (in rich countries as well, large parts of the population drop out of the income-generating economy and fall into poverty) or as the fragmentation of the national and global economy between local centres of wealth (global cities and villages) and poor hinterlands (*see learning exercise three for this chapter*).

2. *The integration of knowledge from different disciplines and the cooperation of local, national and international actors* is problematic, making outcomes uncertain. Global problems cannot be solved through specialised research and short-term political programmes and policy cycles. The achievement of more equal distribution of resources in terms of intra- and intergenerational solidarity is prevented by powerful political and economic players whose interests it would not serve. An ecological indicator for solidary sharing of resources is the fair earth share, discussed in the "State of the World Report" (Worldwatch Institute 2013: chapter four). Resource sharing has become a controversial issue in the discussion on the economic North–South divide: is this divide a temporary phenomenon, showing a dis-simultaneity, the delayed development of the Global South, which will end when all countries are modernised and industrialised, completing the global building of a modern society? Or is the divide permanent, as it has been in existence since the rise of the modern world system in the sixteenth century? Is the divide rooted in the incoherent structures and institutions of the world economic system, representing a global division of labour and power, which can only be changed through a transformation of the system? The sustainability discourse is torn between the choice of perfecting the growth-based economy or transforming it. The lack of a single social subject of sustainability transformation (group, class, institution) adds up to the uncertainties in this process; no single social class, no political or economic institution can achieve the sustainable development goals; so far networks, coalitions and action groups with different participants are the main actors.
3. *The state and functioning of the global economic system* and its potential to be transformed into an environmentally sustainable system (in which economic knowledge is not used in its disciplinary state, but only through integration with other knowledge) need to be discussed more critically. The heterogeneous components of social, economic and ecological sustainability cannot be analysed and described only in economic terms. Disciplinary knowledge from economics does not help to clarify controversial interpretations: it is provided from different schools, with arguments for and against system change, for and against economic growth. Given the multi-paradigmatic state of economics, the plurality of approaches and the mode of development of the discipline, it is difficult to provide a coherent analysis of the complex social, economic and ecological ramifications of the sustainability process. Additional knowledge and interdisciplinary

integration of knowledge from social and natural-scientific disciplines and research is required to deal with the problems addressed in the sustainability discourse. In the processes of knowledge transfer and application and political decision-making, new difficulties arise through the selective use of knowledge in the policy process. Sustainability policy and governance do not follow a scientific rationality, but one in which political and economic power and different interests determine what knowledge is sought, selected and accepted. The selectivity of knowledge practices, and the results and recommendations derived from research, generate continuous debates in the political sustainability process.

1.2 Sustainable Development in the Brundtland Report

The Brundtland Report is a global policy document, not a scientific report: it uses scientific knowledge and reasoning in selective, illustrative and exemplifying forms, but without in-depth analyses. The following sections summarise the guiding ideas of the report, its analysis of the modern society and economy, and the knowledge problems in the sustainability discourse. The *guiding ideas* of the Brundtland Report show the contours of the subsequent sustainability discourse in politics and science, which can be summarised as follows (Box 1.3).

Box 1.3 The Brundtland Report: View of Sustainable Development

1. *Definition: Sustainable development* is defined in the report as "*development that meets the needs of the present without compromising the ability of future generations to meet their own needs*" (UN 1987: 37, point 1); sustainable development implies limits—"*not absolute limits but limitations imposed by the present state of technology and social organization on environmental resources and by the ability of the biosphere to absorb the effects of human activities*" (UN 1987: 15, point 27).
2. *Guiding ideas: Connecting economy and ecology*: the global economy and ecology are locked together in new ways, accelerating economic and ecological interdependence among nations; sustainable development focuses on the connections between economic growth and protection of the environment, which requires changes in future economic development (UN 1987: 12, point 15).

(*continued*)

Box 1.3 (continued)

3. *Economic and non-economic variables*: for sustainable development "it is not enough to broaden the range of economic variables taken into account. Sustainability requires views of human needs and well-being that incorporate such non-economic variables as education and health enjoyed for their own sake, clean air and water, and the protection of natural beauty. It must also work to remove disabilities from disadvantaged groups, many of whom live in ecologically vulnerable areas" (UN 1987: 43, point 39).
4. *Common interests of humankind*: sustainable development is an attempt to formulate principles and preconditions for the common interests of humankind, of survival and wellbeing, and should help to find compromises between the interests of the industrialised and the developing countries; sustainability is not a fixed state of harmony, rather a process of change in which exploitation of resources, directions of investment, technological development and institutional change are made consistent with regard to present and future needs of humans (UN 1987: 15, point 30).
5. *Long-term development of the global society and economy*: the long-term perspective of sustainable development is formulated as a new form of development which aims at a *transformation of society and economy*: "development involves a progressive transformation of economy and society" (UN 1987: 37, point 3), referring to the problematic consequences of continuing economic growth driven by ambivalent technological innovations: "Emerging technologies offer the promise of higher productivity, increased efficiency, and decreased pollution, but many bring risks of new toxic chemicals and wastes and of major accidents of a type and scale beyond present coping mechanisms" (UN 1987: 19, point 69).

Sources: mentioned in the text

From the problems addressed in the report it can be concluded that new knowledge is required about the interacting and coupled social and ecological systems that must be changed in the sustainability process. This knowledge was not available at the time of the publication of the report. It has developed since then in the rapidly progressing interdisciplinary research in human, social and political ecology and sustainability science (see Chap. 3). The Brundtland Report discusses joint action, new policy and governance forms, and redistribution policies in combination with economic growth as the means to solve social and environmental problems (UN 1987: 41f). The principles of intra- and intergenerational solidarity are not translated into new forms of political action, international cooperation and citizen participation. In chap. 2 of the report, sustainable development is reformulated as *a process of change to make resource use, investments, technologies and institutional change coherent to meet the needs of present and future generations* (UN 1987: 38, point 15: a more concrete definition than the general one of the report). The arguments remain vague and abstract: distinguishing between economic growth and development as qualitative processes of change; the necessity of other forms of growth besides exponential growth; taking into account the connections between economic and ecological processes. With such reasoning the report creates the magic formula of sustainable growth as "a new era of growth—growth that is forceful and at the same time socially and environmentally sustainable" (Southwick 1996: 329). The report is also vague with regard to "human needs" and "limitations of resource use": technology and social organisation are seen as determining factors (UN 1987: 37, point 1).

The changes required for achieving sustainability can no longer be discussed in the framework of the outdated knowledge and insufficient reflections in the report. The following changes need to be discussed in the sustainability discourse and the policy process:

1. *scientific knowledge for the sustainability process* needs to be generated through interdisciplinary research, knowledge integration and synthesis; this is inexactly formulated among the guiding ideas of the report (Box 1.3): sustainable development requires a broad view of development that includes economic and non-economic factors and variables;
2. *political and collective action for sustainable development* requires a broadening of the policy process in the sense of governance, with participatory and civil society action;
3. *sustainable development requires a long-term perspective* and implies a transformation of the global economy and society: the idea of system transformation remains unclear in the report.

A coherent, scientifically and theoretically based, analysis of the global economy and its interaction with nature (in terms of material and energy flows) that could provide suggestions for how to achieve global sustainability is missing in the Brundtland Report. This is part of the explanation for the limited success of national and international sustainability policies. Furthermore, the report did not specify the scientific research and knowledge required to guide and maintain the sustainability process. The problems and obstacles to be dealt with in sustainable development that are insufficiently described in the report give an impression of its deficits (Box 1.4).

Box 1.4 The Brundtland Report: Anticipated Difficulties in the Sustainability Process

The report discusses the role of the international economy and the problems of population, human resources, food security, species and ecosystems, energy, industry, urban development, the global commons, peace and security, and finally proposals for institutional and legal change. The links between economic development and environmental consequences are discussed but not translated into strategies for sustainable development.

1. *A series of interconnected problems and obstacles* shows that the transition to sustainability is not a "win–win solution" from which everyone profits, but a process fraught with conflicts. *Interlocking crises and fragmentation of institutions block future development*: environmental, economic and social crises are no longer separate but connecting and interacting global crises; sustainable development is not a fixed state of harmony but implies painful choices and conflicts; limiting population growth and better use, distribution of and access to natural resources are necessary to achieve sustainable development; environmental and economic problems are connected with other social and political factors and problems; no single blueprint for sustainable development will be found, and no country can develop in isolation from others (UN 1987: 12, 32ff).
2. *Requirements of change—the report describes the preconditions and requirements for a sustainable economy and society in the future, but it does not specify the means to achieve sustainability beyond some preliminary ideas about institutional and legal change*: sustainable development requires "a political system that secures effective citizen participation in decision making"; "an economic system that is able to generate surpluses and technical knowledge on a self-reliant and sustained basis"; "a social system that provides for solutions for the tensions arising from disharmonious development"; "a production system that respects the obligation to preserve the ecological base for development"; "a technological system that can search continuously for new solutions"; "an international system that fosters sustainable patterns of trade and finance"; and "an administrative system that is flexible and has the capacity for self-correction" (UN 1987: 50, point 81).

(*continued*)

Box 1.4 (continued)

3. *Changes in international policy required to realise sustainable development*: searching for multilateral solutions, building a new system of international economic cooperation, global sharing and redistribution of resources—these guiding ideas give a vague impression of the obstacles to the global policy process of sustainable development but no guidance for building the necessary transformative capacity and action.

Source: UN 1987

Knowledge problems and selective knowledge use in the Brundtland Report: the report did not incorporate knowledge from critical analyses of the global economy and its interaction with nature that were already available when the report was prepared. Furthermore, the more critical analyses and conclusions that had appeared in earlier global reports issued since the 1970s (see Box 1.2) did not reappear in the report, nor did the critical scientific analyses of the global economy and the ecological earth system advanced, for example, in sustainability science, in social-ecological research, in world system analysis, in ecological economics, in analyses of unequal exchange and of complex interacting systems, and in global assessment studies (see Chap. 3). Political strategies and programmes remain weak without interdisciplinary knowledge from system analyses of the causes and the consequences of the problems being diagnosed and without strong support from governmental institutions and civil society.

The system concept applied in the theory of the modern world system describes an incoherent system: a global economy based on unequal economic and political power relations between rich (industrialised, Northern) and poor (developing, Southern) countries; politically and culturally the world system is fragmented in states and cultures between which manifold tensions and conflicts prevail. The modern world system is an imperfect system, when a system is understood as a coherent whole with different parts or subsystems, together providing a functioning system with emergent properties that do not exist at the level of subsystems or at the lower levels of the multi-scalar global social and ecological systems. The global economy is a system rife with conflict, instability, periodic crises, uneven development and continuous overuse of natural resources; it exists in a political context of fragmented political systems (states and regional power blocs) continually fighting for political, cultural and economic supremacy. In the context of the economic globalisation and deregulation of markets since the late 1960s, these power struggles have brought about a new line of conflict, that between public (political) and private (economic) organisations over the management and control of the earth's natural resources. This can be seen as an "invisible front", that is, one not visible in the public policy process. Sustainable development cannot succeed in overcoming the many problems and conflicts in the modern "system of inequality" by achieving social, economic and ecological solidarity without a profound analysis of the global economy and its interaction with the earth's ecological system from the perspective of social-ecological knowledge and research rather than economics.

Controversial messages of the report—economic growth, its forms and effects: the report is ambivalent in its attempts to bridge the contradictory ideas of economic, growth-oriented developmental thinking and of growth-critical environmental thinking (Mitcham 1995) and to extend the principles of scarcity in economics into areas such as ecology and the environment (Tijmes and Luijf 1995). This ambivalence was criticised as contradictory to the nature of the term sustainable development, but the debate ended in a deadlock. Sustainable development is less a conceptual problem of definition, ambivalence and unclear reflections, and more one of unclear ideas about the processes of change and transformation: how can sustainable development be achieved through politically guided transformation of the global economy; how can the contradiction between growth-based development and no-growth ideas be dealt with in the political practice of the sustainability process? Without clear aims and knowledge about possible pathways of transformation, sustainability policies experiment simultaneously with contradicting reform-oriented green growth strategies and ideas of social-ecological transformation towards future degrowth and post-growth economies (see Chaps. 4, 5 and 6).

The sustainability process cannot continue on the basis of the Brundtland Report, which provides an incomplete review of the problems and bears the characteristic limits of a policy document from the 1980s, leaving unclear the concept, the problems and the consequences of economic growth. The international development discourse refers to the 1980s as a "lost decade" for development, a time when politics and economics came to a standstill. With the changes in the global situation, with continuing global social and environmental change, and new knowledge and research, the sustainability process needs to be rethought and renewed, identifying the problems to solve on the way towards sustainability (Table 1.1).

1. *The first of two main problems insufficiently clarified in the Brundtland Report that need to be discussed further is: The idea of sustainable growth* reveals the dilemma faced by the report: it does not formulate a convincing critique of economic growth, it offers no convincing ideas about other types of growth, and it does not propose ideas about how forms of sustainable growth can be realised. The critique of modernisation, development and economic growth in the report should have been supported through analyses of the environmental consequences of economic growth. Up to now economic growth happens, ecologically seen, in primitive and polluting forms: the costs and damages, side effects, unintended

Table 1.1 Problems addressed in sustainable development

Type of problem	Environmental	Social	Economic
	Overuse of natural resources (globally), climate change, biodiversity reduction, land use change, pollution (of air, water, soils), environmental injustice, resource use conflicts	Hunger, health problems, education deficits, population growth, migration, social and economic discrimination and inequality	Scarcity of natural resources, unequal distribution of resources, North–South divide (unequal development), poverty, unemployment
Political dimensions of problem-solving	Environmental policy, development policy, health policy (public health)	Food security, gender policy, education, equality (citizenship, human rights)	Energy security (renewable energy sources), access and availability of resources (property rights and intellectual property rights)
Requirements of solution	Integration of local, national, global policies and governance (multi-scale governance), knowledge-based governance (inter- and transdisciplinary knowledge integration, knowledge exchange and sharing, scientific and practical knowledge)	Transformation of political systems (civil society, global governance, cosmopolitanism, strengthening of international cooperation), building of a world society	Transformation of the global economic system, post-industrial society (long-term perspective), reduction of global flows (energy and material resources), global regulation of economic processes
State of solution efforts	Slowly advancing (main problems: integration and coordination of national policies; intensifying conflicts and disasters; building of transformation action groups)	Slowly advancing (main problems: weakening of states; involvement, mobilisation and participation of citizens and development of civil society; changes in way of life and consumption)	Delayed (main controversy: forms of economic growth; main problems: deregulation of markets, commodification of nature, unequal development—economic and political dominance of the Global North)

Sources: UN-documents—Sustainable Development Goals, Agenda 21, Agenda 2030; own compilation

consequences and externalities of the use of natural resources and environmental destruction are not or not sufficiently accounted for in the political and economic processes supporting growth. The burdens of environmental pollution and maladaptive change are shifted to the future, to the poor in the Global South and to women, as argued by critical social and environmental movements. In the future, as the limits of natural resources available from ecosystems are reached, it will not be possible to develop economically by shifting the burdens of growth. The "limits to growth" report from 1972 made this argument more clearly than the Brundtland Report. The analysis of social limits to growth (Hirsch 1977), although less well known than the report of the Club of Rome, helped to broaden the debate about social and ecological limits to growth. The Brundtland Report's description of sustainable development waters down the limits to growth problems with an unclear notion of sustainable growth: growth is justified for developing countries as a means to improve the living conditions of the global poor, but no realistic proposal is given for how to change economic growth to become sustainable and to transform the global economy to sustainability. Both issues appeared, finally, on the agenda of sustainable development twenty years after the report was issued, when the global economic crisis from 2007 intensified the debate on limits to growth, degrowth and a post-growth economy.

2. *The rationality of the political sustainability discourse differs from those of scientific discourse and of civil society discours*e. These three discourses come together in the global sustainability process; they have no common form of rationality and knowledge production. Sustainable development requires a pluralistic knowledge process with inter- and transdisciplinary co-production and integration of knowledge in all three spheres: these knowledge processes are insufficiently reflected in epistemological and methodological terms. The interwoven knowledge processes in sustainable development change the relations between scientific and non-scientific knowledge production. Science has a dominant role, but with the use of scientific, managerial, local and practical knowledge in the governance processes it becomes necessary to work with other forms and epistemologies of knowledge production than those utilised in the social and natural sciences and within the scientific disciplines (see Chap. 8).

1.3 Sustainable Development in the Policy Process after the Brundtland Report

Following the Brundtland Report the discourse and political process of sustainable development developed quickly: from the local to the global level, involving many governmental and non-governmental actors, in many political projects, programmes and activities, in civil society and everyday life, and in a still increasing stream of interdisciplinary sustainability research. On the basis of the global framework programme "Agenda 21" for sustainable development, resulting from the United Nations Conference on Environment and Development (UNCED) in Rio de Janeiro in 1992, national and international environmental policies adopted sustainability as a guiding idea. Successes have been limited, however, due to the lack of advances towards sustainability in terms of reduction of environmental pollution and destruction as well as the more equal development and sharing of resources. The fair distribution of resources, the ecologically rational use of resources and significant changes in political and economic systems have not yet been realised and cannot be expected in the near future. The seventeen Sustainable Development Goals of the UN that guide the policy processes aim to reduce unequal development, environmental damage, poverty and hunger as endemic components of the modernisation process. The goals were meant to be achieved during the programme period 2016–2030; this short and unrealistic time frame for achieving the seventeen goals ignores the longer and more complicated transformations of political and economic systems required for sustainability.

Two problems blocking the transformation process are knowledge gaps and asymmetric power relations. The transformation process requires inter- and transdisciplinary knowledge production and integration, including system analyses of complex interacting systems to identify effects of interactions between modern society, the economy and nature. System analyses were neglected, remained controversial or suffered from insufficient data about the interaction between social and ecological systems. Asymmetric power relations and the supremacy of the Organisation for Economic Co-operation and Development (OECD) countries in the global economy and in politics resulted in insufficient support for system transformation. So far sustainability programmes were unable to change the unsustainable forms of growth in the global economy with the North–South divide and the lack of integration of programmes at local, regional, national, international and global levels. Experimentation with forms of a non-growing economy has occurred only in exceptional cases, in single enterprises, branches or ethically based forms of environmentally sound production. Policies have proceeded as "piecemeal engineering" and pragmatic strategies in the sense of internalisation of external effects or "wicked problems and clumsy solutions". Only limited improvements have been made to policy programmes by growing knowledge and understanding of the complexity of interconnected processes of social and ecological change. The lack of success has become increasingly evident with the availability of more knowledge and data about global environmental change and disruption of ecosystems, with the Millennium Ecosystem Assessment from 2005 and later global assessments, and with material and energy flow accounting.

Global climate policy aimed at reducing emissions of greenhouse gases and protecting the atmospheric ozone layer had a first success with the Montreal Protocol of 1987 for the reduction of emissions that destroy the ozone layer. Climate policy brought, furthermore, the first significant changes in natural resource use practices required for sustainable development: the search for and development of new energy forms and the transformation of energy systems through renewable energy sources. The state and development of climate research and policy is documented in the periodical assessment reports of the Intergovernmental Panel on Climate Change (IPCC). Research on climate change is globally coordinated and more intensive than for sustainable development, but as with sustainable development success has been limited and consensus building among governments has been difficult. Sustainability policies improved only after decades of inefficient programmes. The improvements cannot yet be seen in the achievement of goals, only in changes in the view and organisation of sustainability governance. Sustainability governance is improving through the adoption of transformation perspectives: transformation of the institutionalised forms of economic growth, transformation of the global economic system, and the adoption of inter- and transdisciplinary knowledge practices in research and policy.

1.3.1 The Global Sustainability Process

Global conferences on the environment and development guide the sustainability process. Prior to the Brundtland Report there occurred important political and public activities that influenced the later sustainability process: (a) the first global conference on the human environment, UNCHE, in Stockholm in 1972, working with a simpler concept of "ecodevelopment" that can be seen as a forerunner of sustainable development; (b) the publication of the report "Limits to Growth" of the Club of Rome in 1972, in which growth critique reached the global environmental discourse; and (c) the World Conservation Strategy of the International Union for Conservation of Nature (IUCN) from 1980, where the concept of sustainable development was spelled out from an ecological perspective.

With the UNCED conference in 1992 in Rio de Janeiro the idea of sustainable development spread globally; governmental and non-governmental actors supported this guiding idea, specified in a detailed global action programme, "Agenda 21", and in several legally binding agreements (the conventions on biodiversity, climate change and combatting desertification). An update of the "Limits to Growth" report, "Beyond the Limits" (Meadows et al. 1992), introduced an improved world model; it strengthened the sustainability discourse through the conclusions in terms of a conditional warning: the global decline of the economy can be avoided, and a sustainable society is technically and economically possible through changes in growth as discussed in the sustainability discourse.

With the Johannesburg Summit in 2002, controversies about the guiding idea of sustainable development became visible. At the occasion of the conference a sceptical message was communicated by a group of scientists who argued it was too late to realise sustainable development: the trend towards catastrophic global environmental change, especially climate change, was already too far advanced. With the summit came a view of sustainability as including three contrasting dimensions that need to be matched in the policy process—social, economic and environmental sustainability; this became the mainstream variant. In later sustainability policy the difficulties of organising a continuous transformation process became visible in two types of problems: (a) knowledge problems (in operationalisation of the three components and in the formulation of indicators to measure sustainability, of which the seventeen Sustainable Development Goals of the United Nations give an impression) and (b) power problems (the consensus about sustainable development is more symbolic than effective). The policy is dominated by powerful global players with vested interests in maintaining the global political and economic system, including the UN, OECD, European Union (EU), World Bank and Food and Agriculture Organization (FAO) as important actors.

With the conference in Rio de Janeiro 2012 opening the "Rio + 20 process", the concept of a green economy became a new mainstream idea for sustainable development. Since that time sustainability discourse and policy have continued to face controversies over strategies for the transformation to sustainability and the consequences of the currently limited success of sustainability policies. These controversies revolve around contradictory ideas about further processes:

1. to give up on the idea of sustainability as impossible, or to reduce it to resilience (suggested by some ecologists, for example, Benson and Craig 2014);
2. to continue with compromises and adaptations of sustainability policy and diplomacy (the mainstream variant of the UN-guided policy process, documented in the annual Progress Reports about the Sustainable Development Goals);
3. to renew the sustainability process with the concept of a new "great transformation" or a social-ecological transformation, as discussed intensively in the 2010s.

In the global policy process, the strong support for "Agenda 21" expressed at the 1992 conference has weakened; the difficulties of implementing sustainable development at local, regional, national and global levels and the different social, economic and ecological conditions for sustainable development in different countries became evident, requiring a renewal of the process with new knowledge and improved concepts. The UN formulated seventeen Sustainable Development Goals covering the social, economic and environmental aspects of sustainability; they are the core component of the action programme "Agenda 2030" for the years 2016–2030. The Global Sustainable Development Report (UN 2019) documents the complexity of the multi-scale process of sustainable development, with its many difficulties and the limited success of sustainability policies in the past decades, showing that "the world is not on track for achieving most of the 169 targets" of the seventeen Sustainable Development Goals. The report concludes that much more needs to be done to change policies that impede progress and to initiate "transformative changes". Four cross-cutting impacts of development are mentioned that counteract sustainability: rising inequalities, climate change, biodiversity loss and increasing amounts of waste from human activity. These negative trends are overwhelming the world's capacity to deal with them (UN 2019: XX).

Goodland et al. (1991) discussed critically economic growth, differentiating between quantitative economic growth and more complex forms of development required to prevent economic growth exceeding the carrying capacity of ecosystems. There is not yet a clear idea about the future of economic growth under conditions of sustainable development. The first step is to differentiate the two concepts and to conceptualise development as a complex process. Efforts to distinguish between different forms of economic growth have been refined since then, including ideas of green and blue growth, and the growth-critical ideas of the post-growth society and the "degrowth" discourse and movement. In order to reduce poverty worldwide through economic growth, a five- to tenfold expansion of the levels of resource use would be necessary—which is seen an ecological impossibility given the earth's available resources. As global limits to growth are approached, sustainable development requires other strategies in which poverty reduction is achieved through a more equal distribution and sharing of resources between the Global North and South. Overcoming unequal growth requires a reduction of the throughput of natural resources in the economies of the rich countries of the Global North. The term "throughput growth", referring to the growing use of natural resources in the global economy, requires changes in the global economy that have long been discussed by ecologists: the critique of Western lifestyles (Hardin 1968), and currently the change in the "imperial ways of life" in Western countries (Brand and Wissen 2013). The poor are not to blame for environmental destruction and overuse of the natural resource base—the rich are; this was the message with which the sustainability process was opened at the global UNCED conference in 1992.

On the occasion of the "Rio + 20" conference in 2012, some protagonists of the global sustainability discourse (Dodds et al. 2012) analysed the forty years of global sustainability diplomacy since the first global conference on the environment in Stockholm in 1972. The book, although not critically analysing the sustainability discourse and process, documents the dearth of success of global sustainability policy: the lack of coordination and coherence in the implementation of policy programmes, gaps in responsibility, a fragmented global environmental governance architecture, limited success in the implementation of "Agenda 21", the core document from the Rio conference in 1992, and in the years after 1992 the weakening of support for the global sustainability process. The authors suggest a series of activities to renew the process, including an Earth Charter, an International Court for the Environment and a Global Green Bond as an investment system to fund what will be a long process. With the critical discussion of the UN's "Agenda 2030", these modest suggestions were more critically discussed as well, opening a process of joint learning for sustainability transformation.

1.3.2 The Sustainability Discourse as a Joint Learning Process

The Brundtland Report left two issues undiscussed that created difficulties in the later sustainability process: (1) the complexity of the problems and processes that need to be discussed continuously on the basis of new and improved concepts and scientific knowledge from different disciplines; (2) the difficulty of creating transformation capacity and agency in sustainability policies. To solve the problems requires the building of a system of global environmental governance.

(1) *The discussion of definitions and interpretations of sustainable development* did not create consensus about the operationalisation, the knowledge requirements, the knowledge exchange and the need for cooperation between science, politics and civil society. Terms such as "bridging concepts" (Davoudi 2012) for communicating knowledge between natural and social sciences or science and practice, "floating" or "empty signifiers" (Szkudlarek 2007; Offe 2009), and "essentially contested concepts" (see above, Gallie 1956) are examples of how the diffuse, multifaceted idea of sustainable development was dealt with: as a linguistic, conceptual and communication problem. The first term highlights the necessity to create concepts that can be used in different disciplines and knowledge fields, by different actors and institutions, in science and policy, in theory and practice, for the purpose of reaching common views and cooperation in the sustainability process. The second and third terms highlight the difficulties in achieving such common understanding and cooperation with abstract, vague and normative notions such as sustainable development. These difficulties include classification, specification and

operationalisation, the normative and ethical implications of sustainable development, and the pathways towards global sustainability.

Beyond the conceptual discussion, the operationalisation and application of the terms in the sustainability process posed questions about the further use of the notion sustainable development: how to deal with scientific reasoning that sustainable development and sustainability are impossible because of the inherent complexity of the problems and processes in science and policy? How to deal with arguments from sceptical environmentalists that efforts to achieve sustainability are in vain, because of the probability of a global economic collapse and environmental catastrophes before sustainability is achieved? How to deal with the weakening support for sustainable development when transformation becomes a long process? To counteract scepticism and weakening support requires continuous creation, integration, discussion and transfer of knowledge to improve the sustainability process and practices of inter- and transdisciplinary cooperation in the scientific, political and civil society discourses.

The practical difficulties in communicating and defending the ideas of sustainability and sustainable development are that of dealing with economic growth: *how to connect the discourse of sustainable development with that of global "limits to growth"?* The two discourses of "limits to growth" guided by the Club of Rome and "sustainable development" guided by a broader epistemic community were not linked when they first arose; they became connected in the continuing environmental discourse of which both are part. The two discourses seemed to provide contrasting messages: the first about the possibility of a catastrophic future, the other about possible improvements to avoid catastrophes. The "limits to growth" debate is still perceived as being in the tradition of the "dismal science" of economics or ecology (Worster 1979: 113ff) that began with the Malthusian debate over population growth and economic subsistence in classical political economy towards the end of the eighteenth century. It continues in the Neo-Malthusian debate over global economic and population growth today as a discourse about poverty, misery and famine, about overshoot and collapse, complete with dystopic or apocalyptic views of the future of humankind and "doomsday prophecies". The discourse of sustainable development is based on a similar critique of the connected processes of economic growth and deterioration of the environment and scarcity of natural resources as the discourse of limits to growth. However, sustainable development seems to foresee better chances for achieving a socially and ecologically integrated future economy and society—an expectation not realised to date.

(2) Rather than terminological debates and definitions, it is necessary to *specify the economic and social changes and pathways towards sustainability*—to open up the sustainability debate to interdisciplinary and pluralistic knowledge practices.

The organisation and implementation of sustainable development is faced with unequal power relations in policy, economy and society, vested interests and difficulties in creating transformative capacity. Sustainable development as an idea created in the global policy process to deal with global social and environmental problems has lost much of its attraction as an inspiring and motivating idea that spread with the UNCED conference in Rio. The sustainability discourse remains controversial, but with new knowledge created since 1992 it has advanced from controversies about worldviews to controversies about what knowledge and empirical data should be applied. In the long and tough process of clarification of knowledge requirements, the discussion of transition paths to sustainability comes to the forefront. The powerful governmental actors engaged in the sustainability process, especially the UN, OECD, EU, World Bank and FAO, reformulated sustainable development in terms of "green growth" (or "blue growth" for marine resources). Critical scientists and environmental movements reformulated sustainable development as a new "great transformation", using a term from Polanyi (1944), or as social-ecological transformation, with growth-critical knowledge from ecological research (Daly 1996). The controversy between a transition to sustainability through sustainable growth or transformation of the global economy to a non-growing economy accompanies the future sustainability process.

Simultaneously sustainable development becomes more complex and complicated as an inter- and transdisciplinary process. Economic knowledge becomes part of interdisciplinary production, co-production and application of knowledge, in cooperation with science, policy and civil society, driven by collective learning processes. Sustainable development is a model process for a continually improving concept and strategy, where the knowledge available at a certain time is always insufficient; the strategy cannot simply be improved through research, but requires continuous processes of knowledge creation, integration and negotiation and improvements in different forms of knowledge practices in science, policy and civil society. Such practices include joint and social learning, as well as analysing and managing processes in complex and coupled adaptive systems, where development and transformation involve many interconnected changes.

The collective processes of knowledge creation, integration and learning in sustainable development are discussed in the newly developing interdisciplinary forms of research: ecological research (human, social, political ecology), sustainability research (sustainability science and transformative science) and interdisciplinary environmental research (especially about climate change and its consequences). In economics the interdisciplinary opening and knowledge integration happened mainly in heterodox approaches (institutional economics, development economics, interdisciplinary economics and ecological economics). Epistemic debates were the catalyst for the development of new inter- and transdisciplianry knowledge practices in science and politics: the epistemic debates about "citizen science", "new forms of knowledge production" and "*mode two*" (see Glossary), "*postnormal science*" (see Glossary) "*transdisciplinarity*" (see Glossary). In these debates scientific knowledge and its application is reflected on critically and connected with other knowledge forms influencing the sustainability process—practical, local, tacit, experiential and normative knowledge.

The debates about new forms of knowledge production have already brought changes in research, for example in ecology, in sustainability science and in environmental politics, where new ideas about knowledge creation, co-production and application have appeared: adaptive management or governance, policy as experiments, collective learning, cooperation between different social groups of knowledge bearers and pluralistic knowledge practices. The sustainability discourse and process mark the beginning of a new knowledge culture that broadens the horizons of specialised research, using scientific knowledge to create transformative capacity and agency for transitions to sustainability. This broadening of knowledge practices has improved understanding of the problems involved in the transition to sustainability: sustainable development, understood as the social-ecological transformation of the modern economy and society, can no longer be seen as just a political process. It includes interconnecting cultural, social, political, economic and ecological processes and changes; this is to some degree reflected in the creation of new action forms and perspectives with the newly adopted term governance, where sustainable development becomes the operational core process of global environmental governance (Biermann 2014).

1.4 Economic Knowledge for Sustainable Development

Since the Brundtland Report sustainable development has been discussed as a multidimensional process in which economic, social, political, cultural, technical and ecological processes interact. This evokes a series of questions about knowledge practices and interdisciplinary knowledge integration: what scientific knowledge is relevant for the sustainability process? Which economic and non-economic factors and processes influence or determine natural resource use practices and the processes of production, exchange and consumption? How can knowledge from other social and natural-scientific disciplines, especially about the complex processes of global social and environmental change, be connected with economic knowledge and knowledge about the policy and governance processes? Which epistemological and methodological questions come up in the integration of knowledge about complex interacting systems? How can controversies, competing and contradicting paradigms and theories be dealt with in the sustainability discourse? How to account for counter-current processes to sustainable development such as economic deregulation and globalisation? What are the limits of a purely economic view of environmental policy and sustainable development (the Stern review of climate policy, Stern 2007, can serve as a model case)? How to deal with theoretical knowledge in the sustainability process, especially for understanding the interaction between nature and society in modern society? How to deal with limits of knowledge and ignorance that appear continually in the analysis of complex and coupled social-ecological systems? How can the temporal horizons of the sustainability process—as a long-term process that stretches into the unknown distant future—be dealt with in sustainability governance?

Such questions, indicating the necessity of opening up economic science to inter- and transdisciplinary knowledge production and pluralistic knowledge cultures, are addressed in the following chapters (most systematically in Chap. 3). With the review of the Brundtland Report and the subsequent sustainability process, answers regarding the significance of economic knowledge can be attempted (Bruckmeier 2018: 12): sustainable development implies economic questions about the scarcity and the valorisation of natural resources, the consequences of economic growth, distribution and redistribution of resources, property rights and access to resources, social costs such as that of environmental pollution and economic instruments in environmental policy. Ideas for solving the problems on the way to sustainability include bio-economic approaches, ideas of a "green economy", of a post-growth economy, or ideas about degrowth and social-ecological transformation of the global economy and society. To solve the problems resulting from economic growth will require not only economic research but knowledge about other social and ecological systems interacting with economic systems. Scarcity of natural resources can, for example, be measured in ecological and economic terms, showing different forms of naturally or socially caused scarcity, or combinations of several forms. The discussion of social limits to growth by Hirsch (1977) provided first reflections about non-ecological forms, reasons and consequences of scarcity and resource limits. The systemic, social and ecological organisation of the economy in modern society needs to be studied through interdisciplinary and theoretical approaches to understand the intertwining of social and ecological factors in sustainable development (in economics this is practised systematically in ecological economics: see Chap. 6).

Economics is not equipped to explain environmental problems, their causes and consequences, and potential solutions: it cannot assess the functioning and the development of ecosystems, the consequences of environmental pollution for humans and nature, the environmental risks of overuse of natural resources; it cannot say much about the causes and consequences of global climate change or of biodiversity loss, of the limits of natural resource use or the planetary boundaries that cannot be exceeded by humans without ecological consequences. The coupling of ecological and social systems and the interaction of social and ecological forms of global change are mainly studied in interdisciplinary subjects such as cultural, human, social and political ecology. In orthodox academic economics, environmental destruction is reduced to costs and necessary action to governmental environmental policy. Calculating the costs of environmental damage and restoration is only a limited part of the knowledge required in the broader sustainability process; many costs of environmental damage, that of climate change and biodiversity loss, for example, cannot be compensated for in monetary forms. Strategies for sustainable development cannot be reduced to decisions between higher or lower economic costs or benefits; they require several assessment criteria, the construction of economic, social and ecological indicators and their combination. It seems impossible to find a single, "objective", economic criterion for the transition to sustainability. Sustainability, with its social, economic and ecological differentiations, is in the final analysis dependent on the maintenance of ecological functions and processes, as formulated in the concepts of sustainable development of the world conservation strategy (IUCN 1980).

1.5 Discussion and Conclusions

The formula of intra- and intergenerational solidarity in resource use through sustainable development is not intuitively clear and evident with regard to the requirements of change. The term solidarity obscures the necessary and conflict-provoking changes with a normative terminology. Inconvenient truths were openly discussed only after decades, when the "ecological shadows" of overuse of the natural base of the earth, of climate change, biodiversity loss, and changes in terrestrial and marine ecosystems became longer. The transition to sustainability is only possible with joint learning from successes and failures in the earlier process. The necessary learning and improvement of sustainability governance can only happen to a limited degree with the knowledge from monitoring and policy evaluation as routine processes in policy processes. A more difficult and controversial question is how to deal with inter- and transdisciplinary knowledge integration for improving sustainability research and policy or governance practices. Furthermore, the sustainability discourse needs to address the question of how the obstacles and conflicts on the way towards a sustainable future economy and society can be dealt with.

Three important themes of the future scientific and political sustainability process have become evident through the intensifying discourse and through interdisciplinary research, knowledge exchange and integration since the publication of the Brundtland Report:

1. *Sustainability transition*: More important than discussing the concept is to discuss possible forms and pathways of the transition or transformation to sustainability and the obstacles to this process. This is not only a question of policy evaluation and effective combinations of policy instruments. Policy research and economic research cannot provide sufficient knowledge about the broader changes required to achieve global sustainability. For the policy process the normative and political dimensions of the sustainability process need to be critically reviewed (van der Hel 2018). This includes questions of a new normative order, of environmental rights of citizens to support sustainable development and of changes to asymmetric power relations that distort and block the process.
2. *Uneven development*: The global economic system is based on uneven development and economically and ecologically unequal exchange; both components of the North–South divide need to be transformed in the sustainability process. The global economy is torn between globalisation with deregulated markets and attempts to develop a sustainable economy of the future (Cavanagh et al. 2002; Barnett 2004). The discussion is trapped in contradictory diagnoses of the problems of economic development, formulated in the positions of the World Business Council for Sustainable Development ("Denying poor people access to markets is planet-destroying as well as people-destroying"; Holliday et al. 2002: 41) and the International Forum on Globalization (the current market is unfair, distorted in favour of a small ruling elite, and "it is 'unfree', burdened by policies and conditions that hinder the poor from freely competing in it"; Cavanagh et al. 2002: 4, 53).
3. *Dilemmas and conflicts in the sustainability process* are analysed in the interdisciplinary research in social ecology, political ecology and ecological economics, and discussion of these issues became more intensive in the 2010s (Olsson and Gooch 2019). In a tentative form, to be developed in the following chapters, three contrasting processes with different logics, rationalities, interest bases and consequences need to be balanced: the global sustainability policy, the economic globalisation process and the anthropogenic disturbance of the earth's ecological system. In dealing with the conflicting trends, the sustainability process appears as both a *knowledge paradox* and a *development paradox*.

The *knowledge paradox* is related to the impossibility of predicting the future—organising a process of transformation, but not knowing what form the future economy and society should take as a result of the transformation. At present, at the beginning of the sustainability process, knowledge is insufficient and interests not sufficiently concerted to create agency and transformaiton capacity for this long process that stretches in the distant future (so far little is known about the building of transformation groups, the joint learning required from the actors, the regulative mechanisms of transformation). Large parts of the required knowledge need to be created on the path towards sustainability. Only through continuous research, knowledge integration, and the gaining of experience and insights through joint and social learning, as discussed, for example, in sustainability science, can progress in sustainable development be achieved. Joint learning is based on inter- and transdisciplinary knowledge production and integration, from which the actors can learn about the conditions for changing complex systems and processes in the transformation to sustainability.

The *development paradox* is related to the spatial expansion of societal development and change, its global dimensions and the temporal acceleration of processes of social and environmental change in modern society: the expansion and acceleration of global processes makes it difficult to understand the connections between local resource use and its global consequences. Fischer-Kowalski (2003) unfolds the theoretical arguments relevant for understanding the dependence of sustainability on the systemic structure of the economy and society, without which the sustainability discourse would be misleading: industrial society and the global economy obscure the connections between resource use and the local territory. Energy use in industrial societies is, through the global exchange of energy resources, practically independent of the domestic territory of the country and its size. CO_2 emissions do not have local or regional effects where they are produced, only in the sum of global emissions, which are a main cause of anthropogenic climate change that impedes the achievement of sustainability. The connection between industrial production, resource consumption and environmental pollution in industrial society is not visible at national levels and cannot be controlled at that level. The decoupling of resource use from local areas has shifted the environmental consequences to the global level, and here, finally, the connection between territory and (over-)use of resources is found to be out of balance, resulting in global environmental change. A new balance between production, consumption and pollution needs to be found in the sustainability process and on this planet: there is only one earth, and the import of life-supporting natural resources or the exodus of humans from the earth does not seem realistic in the foreseeable future.

Summary of Chap. 1: The Political Sustainability Discourse Since the Brundtland Report

1. The report "Our Common Future" from the UN (1987) created a new policy agenda with the term sustainable development and its interpretation as intra- and intergenerational solidarity. After the UNCED conference in Rio de Janeiro in 1992, sustainable development was established as an overarching policy worldwide with "Agenda 21". To date no significant progress towards global sustainability has been achieved, as the UN confirmed in their Global Sustainable Development Report from 2019.
2. The reasons for the lack of success in sustainability policy are manifold: the complexity of global social, economic and environmental problems and their interdependence; the difficulties in controlling and reducing economic growth and its negative environmental consequences; the asymmetric power relations in the global economy and society; the growing social and economic inequality; the realisation of the complexity of cooperative knowledge practices in science, politics and civil society; and the diminishing support for sustainability governance.
3. As a consequence of the lack of success, the sustainability process is changing. It is now seen as a social, economic and ecological transformation of the global economy and society, broader than policy processes and policy reforms. Lack of knowledge about the future sustainability process requires continuous knowledge creation and adaptation through research, experiments, trial-and-error processes and pluralistic knowledge practices in sustainability governance. Collective learning from earlier processes and failures is a core mechanism for continually maintaining and improving governance processes in a long-term transformation.

Appendices: Further Information and Material

1. Questions and Individual Exercises

Learning Exercise 1: Sustainability as an Essentially Contested Concept

Discuss (individually or in a working group) based on the following criteria the terms sustainability and sustainable development—to what extent can the criteria be applied to these terms?

Collier et al. (2006: 216ff) summarised the main criteria for essentially contested concepts from the writings of Gallie as follows:

1. Appraisiveness: this means that the concept designates something or assumes that it exists
2. Internal complexity: this means that the concept is complex and its definition has several components
3. Diverse describability: this means the concept can be described in several different ways
4. Openness: this means that the definitions of the concept can change
5. Reciprocal recognition: this means that the concept is seen by the discourse participants as controversial concept
6. Progressive competition: this means that with the continuing discourse the quality of arguments improves

To deepen your understanding of the problems with essentially contested concepts you can read the article of Collier et al. (2006).

Learning Exercise 2: Critical Text Analysis—The Brundtland Report—Aims and Assumptions

Analyse the following statements with regard to coherence, clarity, contradictions, forms of reasoning and value assumptions.

Guiding ideas:

1. sustainable development implies limits (regarding the state of technology, the social organisation of natural resource use, the ability of the biosphere to absorb the effects of human activities);
2. it can trigger a new era of economic growth (in chap. 2 of the report, ideas for reviving growth and changing the quality of growth are discussed);
3. continuing with the present forms of unequal development in a world of endemic poverty will lead to ecological and other catastrophes.

The ambivalent *role of technology* as creating innovation and growth: The report argues for the use and diffusion of environmentally sound technologies (UN 1987: 65ff) but refers also to new and uncontrollable risks resulting from innovative technologies. That technologies themselves require natural resources for their construction and may have unintended consequences, even in the case of "green technologies" or the use of renewable energy sources, was only critically discussed when the technologies were applied. The ecological and other limits that the report mentions are not specified.

Economic growth has so far not solved the problems of inequality and poverty: sustainable development requires changes of growth to make it less material- and energy-intensive and to create a more equitable impact; mutually reinforcing economic and social development and harmony between humanity and nature have not been achieved (UN 1987, chapter 2, points 35, 41, 81). The *idea of a sustainable world economy* remains unclear, as does that of growth, and does not demonstrate the changes required to make the global economy sustainable, leaving the impression that sustainability is possible within the current system by demanding "more rapid economic growth in both industrial and developing countries, freer market access for the products of developing countries, lower interest rates, greater technology transfer, and significantly larger capital flows, both concessional and commercial"; the idea is deadlocked in the formula "the international economy must speed up world growth while respecting the environmental constraints" (UN 1987, chapter 3, point 3, point 72, 74).

Possible catastrophes as a result of continued uneven development and growth: Unlike in the "Limits to Growth" report from 1972, these are downplayed in the Brundtland Report. Chapter 8 on industrial development shows the difficulties and dilemmas in seeking environment-friendly forms of industrialisation, following from the premise that many human needs can only be met through goods produced by industry. "Industry is central to the economies of modern societies and an indispensable motor of growth. It is essential to developing countries, to widen their development base and meet growing needs. And though industrialized countries are said to be moving into a post industrial, information-based era, this shift must be powered by a continuing flow of wealth from industry" (UN 1987: 144)—in contrast, however, the report states:

> It is becoming increasingly clear that the sources and causes of pollution are far more diffuse, complex, and interrelated—and the effects of pollution more widespread, cumulative, and chronic—than hitherto believed. Pollution problems that were once local are now regional or even global in scale. Contamination of soils, ground-water, and people by agrochemicals is widening and chemical pollution has spread to every corner of the planet. (UN 1987: 147)

Learning Exercise 3: Discussion About the Global North–South Divide (Group Work)

Organise a discussion about the Global North–South divide between two groups: one defending and one rejecting the existence of a permanent North–South divide between rich and poor countries. The group of defenders should prepare their arguments in the form of the "transformation hypothesis", based on analyses of the modern world system (Wallerstein): that centre–periphery relations are part of the system of modern capitalism and can only be dissolved through its social and economic transformation into a new economic and social system—a transformation that is currently under way. The other group should prepare their arguments for a dissolving North–South divide in the form of the "hypothesis of new industrialisation": industrialisation continues with late industrialisation (examples from BRICS [Brazil, Russia, India, China, South Africa] countries) and remains part of the new global economy. In the discussion (30–45 minutes) each group needs to respond to the arguments of the other group, not just defend their own positions. One or two participants will guide the discussion and document the main points (in a flipchart). Another person will write a protocol. In a follow-up discussion all participants will discuss their learning process (with the help of the protocol).

Introductory statement for the discussion:

With the advent of modern society and the European colonisation of the Global South, a global divide arose between the rich countries of the North and the poor (colonial, developing) countries of the South. This divide is not simply a geographical phenomenon but a social and economic divide based on the structure of the modern economy and society (some countries in the southern hemisphere, for example Australia and New Zealand, are part of the Global North). This divide can be assumed to be the main reason for the global social, economic and environmental problems to be solved through sustainable development. The controversy about this divide can be expressed in two questions: is this a divide deeply rooted in the modern world system that cannot be overcome without transforming the global economy and society? Or is it a temporary divide that will be dissolved finally with the late industrialisation of countries in the Global South (BRICS countries)? The dispute should also include discussion of adequate statistical indicators for the divide (for example GDP, income levels, human development index) and changing forms of the divide: its old forms (industrial/non-industrial countries: Stettner 1982, discussing the Brandt Report of the North South Commission: Brandt 1980) and its new forms (for example global digital/technological divide in the knowledge society based on internet-based information technology: Hilbert 2016).

Literature for additional reading on the North–South divide: Caporaso (1980), Thérien (1999), Litonjua (2012), Collier (2015), ITU (2019)

(A similar dispute is described as an exercise in Chap. 2, with different theoretical arguments)

2. Additional Reading (recommended to deepen your knowledge about the theme of the chapter):

Books about the policy and governance discourse of sustainable development: Worldwatch Institute (2013), Purdy (2015)

The scientific discourse of sustainable development: books in environmental sociology—Burns and Schaefer Caniglia (2016); political ecology—Peet et al. (2010); environmental economics—Harris and Roach (2018); ecological economics—Daly (1996)

Handbooks: sustainability and social science research—Leal Filho et al. (2018); history of sustainability (Caradonna 2018)

Important scientific journals with policy-related sustainability research (international, peer reviewed): "Environment, Development, Sustainability" (multidisciplinary); "Environmental Policy and Governance"; "Environmental Science and Policy" (interdisciplinary); "Environmental Sustainability" (multidisciplinary); "Green Technologies and Environmental Policy"; "Journal of Sustainable Development"; "Sustainability" (online journal); "Sustainable Environment Research"

References

Barnett, M. L. (2004). Are Globalization and Sustainability Compatible? *Organization & Environment, 17*(4), 523–532.

Benson, M. H., & Craig, R. K. (2014). The End of Sustainability. *Society and Natural Resources, 27*, 777–782.

Biermann, F. (2014). *Earth System Governance: World Politics in the Anthropocene*. Cambridge, MA: MIT Press.

Brand, U., & Wissen, M. (2013). Crisis and Continuity of Capitalist Society-Nature Relationships: The Imperial Mode of Living and the Limits of Environmental Governance. *Review of International Political Economy, 20*(4), 687–711.

Brandt, W. (1980). *Das Überleben sichern. Bericht der Nord-Süd-Kommission. Gemeinsame Interessen der Industrie- und Entwicklungsländer*. Köln: Kiepenheuer & Witsch.

Bruckmeier, K. (2016). *Social-Ecological Transformation: Reconnecting Society and Nature*. London: Palgrave Macmillan.

Bruckmeier, K. (2018). The Significance of Economic Knowledge in the Environmental Sustainability Discourse. In N. Soukupová & M. Matejcková (Eds.), *12th International Scientific Conference INPROFORUM: Innovations, Enterprises, Regions, and Management* (pp. 11–24). Faculty of Economics, University of South Bohemia in Ceske Budejovice.

Burns, T. J., & Schaefer Caniglia, B. (2016). *Environmental Sociology: The Ecology of Late Modernity* (2nd ed.). Norman, Oklahoma: Mercury Academic.

Caporaso, J. (1980). Dependency Theory: Continuities and Discontinuities in Development Studies. *International Organization, 34*(4), 605–628.

Caradonna, J. L. (Ed.). (2018). *Routledge Handbook of the History of Sustainability*. Abingdon, UK and New York, NY: Routledge.

Cavanagh, J., Mander, J., Anderson, S., Kimbrell, A., Barker, D., Korten, D., et al. (2002). *Alternatives to Economic Globalization: A Better World Is Possible*. San Francisco: Berrett-Koehler.

Collier, P. (2015). Development Economics in Retrospect and Prospect. *Oxford Review of Economic Policy, 31*(2), 242–258.

Collier, D., Hidalgo, F. D., & Maciuceanu, A. O. (2006). Essentially Contested Concepts: Debates and Applications. *Journal of Political Ideologies, 11*(3), 211–246.

Dag Hammarskjöld Report (1975). *The 1975 Dag Hammarskjöld Report on Development and International Cooperation*. Prepared on the Occasion of the Seventh Special Session of the United Nations General Assembly, New York, September 1–12.

Daly, H. (1996). *Beyond Growth: The Economics of Sustainable Development*. Boston: Beacon Press.

Davoudi, S. (2012). Resilience: A Bridging Concept or a Dead End? *Planning Theory and Practice, 13*(2), 399–407.

Dodds, F., Strauss, M., & Strong, M. (2012). *Only One Earth: The Long Road via Rio to Sustainable Development*. London: Routledge.

Fischer-Kowalski, M. (2003). Gesellschaftlicher Metabolismus, Territorium und Nachhaltigkeit. *GAIA, 12*(1), 44–45.

Gallie, W. B. (1956). Essentially Contested Concepts. *Proceedings of the Aristotelian Society, 56*, 167–198.

Garfinkel, H. (1967). *Studies in Ethnomethodolgy*. Englewood Cliffs, NJ: Prentice Hall.

Goodland, R., Daly, H., Serafy, S. E., & von Droste, B. (Eds.). (1991). *Environmentally Sustainable Economic Development: Building on Brundtland*. Paris: UNESCO.

Hardin, G. (1968). The Tragedy of the Commons. *Science, 162*, 1243–1248.

Harris, J., & Roach, B. (2018). *Environmental and Natural Resource Economics: A Contemporary Approach* (4th ed.). London: Routledge.

Hilbert, M. (2016). The Bad Thing Is That the Digital Access Divide Is Here to Stay. *Telecommunications Policy*. https://doi.org/10.1016/j.telpol.2016.01.006.

Hirsch, F. (1977 [1976]). *The Social Limits to Growth*. London: Routledge and Kegan Paul.

Holliday, C., Schmidheiny, S., & Watts, P. (2002). *Walking the Talk: The Business Case for Sustainable Development*. San Francisco: Berrett-Koehler.

ITU. (2019). *Measuring Digital Development. Facts and Figures 2019*. Geneva: ITU. Retrieved from https://www.itu.int/en/ITU-D/Statistics/Documents/facys/FactsFigures2019.pdf.

IUCN (Ed.). (1980). *World Conservation Strategy: Living Resource Conservation for Sustainable Development*. Gland, Switzerland: International Union for Conservation of Nature and Natural Resources.

Leal Filho, W., Marans, R. W., & Callewaert, R. (Eds.). (2018). *Handbook of Sustainability and Social Science Research*. Cham, Switzerland: Springer International.

Litonjua, M. D. (2012). Third World/Global South: From Modernization to Dependency/Liberation to Postdevelopment. *Journal of Third World Studies, 29*(1), 25–56.

Meadows, D. H., Meadows, D. L., Randers, J., & Behrens III, W. W. (1972). *The Limits to Growth. A Report for the Club of Rome's Project on the Predicament of Mankind*. New York: Universe Books.

Meadows, D. H., Meadows, D. L., & Randers, J. (1992). *Beyond the Limits*. London: Earthscan.

Mitcham, C. (1995). The Concept of Sustainable Development: Its Origins and Ambivalence. *Technology in Society, 17*(3), 311–326.

Offe, C. (2009). Governance: An "Empty Signifier". *Constellations, 16*(4), 550–562.

Olsson, G., & Gooch, P. (Eds.). (2019). *Natural Resource Conflicts and Sustainable Development*. Abingdon, UK: Routledge.

Peet, R., Robbins, P., & Watts, M. (Eds.). (2010). *Global Political Ecology*. New York and London: Routledge.

Polanyi, K. (1944). *The Great Transformation*. New York: Farrar & Rinehart.

Purdy, J. (2015). *After Nature: A Politics for the Anthropocene*. Cambridge, MA: Harvard University Process.

Southwick, C. H. (1996). *Global Ecology in Human Perspective*. New York and Oxford: Oxford University Press.

Stern, N. (2007). The Economics of Climate Change. *The Stern Review*.

STWR. (2012). *Financing the Global Sharing Economy*. Report. London: Share the World's Resources (STWR).

Szkudlarek, T. (2007). Empty Signifiers, Education and Politics. *Studies in Philosophy and Education, 26*, 237–252.

Thérien, J.-P. (1999). Beyond the North–South Divide: The Two Tales of World Poverty. *Third World Quarterly, 20*(4), 723–742.

Tijmes, P., & Luijf, R. (1995). The Sustainability of Our Common Future: An Inquiry of an Ideology. *Technology in Society, 17*(3), 327–336.

UN. (1987). *Report of the World Commission on Environment and Development: Our Common Future*. Transmitted the General Assembly as an Annex to Document A/4 42/427 – Development and International Cooperation: Environment. New York.

UN. (2019). *The Future Is Now. Science for Achieving Sustainable Development*. Global Sustainable Development Report 2019. New York: United Nations, Department of Economic and Social Affairs.

van der Hel, S. (2018). Science for Change: A Survey on the Normative and Political Dimensions of Global Sustainability Research. *Global Environmental Change, 52*, 248–258.

Ward, B. (1975). The Declaration of Cocoyoc: A Proclamation of Sustainable Development. *Bulletin of the Atomic Scientists, 31*(3), 6–10.

Worldwatch Institute. (2013). *Is Sustainability Still Possible?* State of the World Report 2013. Washington, DC.

Worster, D. (1979). *Nature's Economy. The Roots of Ecology*. Garden City, New York: Anchor Press/Doubleday.

The Historical Context: Sustainability in Modern Society 2

The chapter describes the history of the ecological discourse and of environmental policy and management in European countries, where modern society was born. The ideas of sustainability, limits to growth and maintaining the natural resource base through the protection of nature were developed in the political practices of governmental regulation of natural resource use. The main economic and environmental problems that arose in early modernity included the clearing of woodlands and air pollution in cities. During industrialisation, questions arose in the classical political economy about population growth, economic growth and industrial pollution of the environment. The sustainability discourse developed with interdisciplinary knowledge integration and with natural- and social-scientific knowledge created and applied in the practices of natural resource management.

Figure 2.1 illustrates the development, differentiation and integration of knowledge in the sustainability debates in the two phases of modern society in Europe, the preindustrial phase of agricultural society and the phase of industrial society.

2.1 Sustainability in Historical Perspective: Nature and Humans

The sustainability discourse and its encompassing critique of the global economy and industrial society, where overuse of natural resources, human modification of ecosystems and environmental pollution reached global levels, began in the 1970s. With the sustainability discourse the debate about a post-industrial society be-

K. Bruckmeier, *Economics and Sustainability*,
https://doi.org/10.1007/978-3-030-56627-2_2

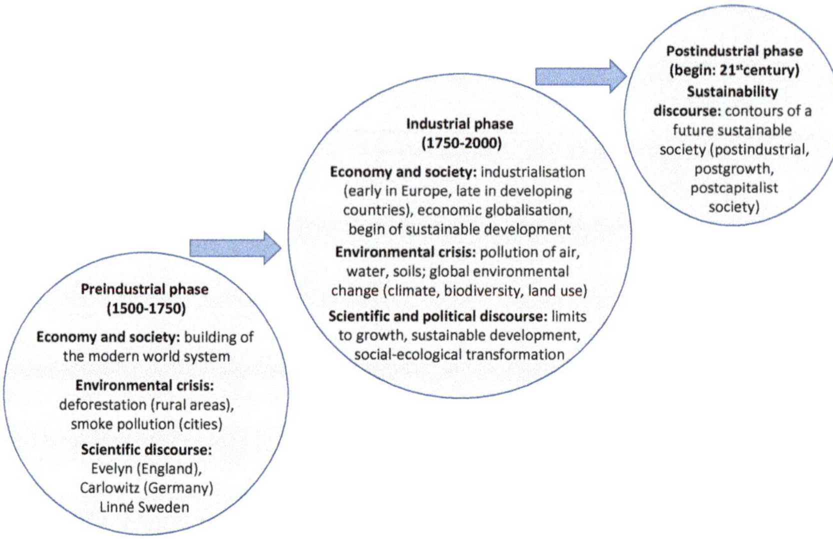

Fig. 2.1 History of the sustainability discourse in modern society

came global and critical, focusing on the scarcity of resources and global environmental problems. The *Routledge Handbook of the History of Sustainability* (Caradonna 2018) traces the ideas of sustainability in history through the practices of resource use and forms of life in ancient, indigenous and pre-industrial societies in different parts of the world. Earlier societies overused their natural resource base, for example through the ruin of fertile agricultural land in the Near East and the Mayan culture in Central America, which was one of the reasons for their collapse. The historical documentation of sustainability thinking, of ways of life and consumption, and of resource use problems in human history was seen by the authors of the handbook as useful for understanding resource use problems in human history and finding solutions for the current sustainability crisis (Tainter in Caradonna 2018: 40ff).

The modern term sustainability comes from modern Europe: it was coined in the early eighteenth century as *Nachhaltigkeit* in Germany in the discussion of timber crises (Caradonna 2018: 4; Grober in Caradonna 2018: 96ff; Grober 2012, describing the cultural history of sustainability). No single and common term, idea or principle exists for the history of sustainability; comparisons of historical periods and cultures need to work with several ideas and concepts: mode of production, forms of development, growth and natural resource use, protection of nature, overuse of resources and environmental destruction. More systematically, in theoretical terms, the sustainability of a society can be analysed in terms of (a) *accumulation regimes* as the ways in which the economic surplus in a society is generated and used (Labrousse and Michel in Jo et al. 2019); (b) forms of growth in population and resource use; (c) property and access rights for economic resources; and (d) *socio-metabolic regimes* (see Glossary). The social metabolism includes the organisation and regulation of natural resource use and the economic production forms with regard to the total flow of material and energy resources between nature and society.

The popular discussion of "learning from history" does not account for theoretically framed analyses. Costanza et al. (2007), discussing potential learning from integrating the history of human societies and the history of nature, show that the outcome remains controversial. Limiting the history of sustainability to that of modern society provides a clearer picture. A common thread in the modern, European history of sustainability is marked by the two connected concepts scarcity and growth, discussed intensively in the handbook (Dale and Schmelzer in Caradonna 2018: 71ff, 164ff, and further chapters by Heinberg, Muraca and Döring, Ehrenfeld). With these economic and ecological concepts and the industrial revolution as the culmination of the modern economy (Griffin in Caradonna 2018: 106ff), the sustainability discourse can be reconstructed, notwithstanding the manifold definitions and controversial discussion in the debates.

Overuse of natural resources and pollution of the environment have been occurring since early human societies were formed but intensified with the development of agricultural societies and once more with the development of modern industrial society, when it became a global problem (Hughes 2001, 2005; Williams 2006). With the concept of socio-metabolic regimes, sustainability problems can be described in theoretical terms to compare the resource use in historical and modern societies (Fischer-Kowalski and Haberl 1998: 578f; Fischer-Kowalski 1997: see Table 2.1). The societies differed widely in their cultural, social, political and economic systems, but most of them undermined their sustainability through overuse of natural resources and were not able to maintain their resource base in the long run (Box 2.1).

Box 2.1 Historical Studies and Theories of Natural Resource Use

The economic and environmental history of early societies and their resource use problems are analysed in a series of classic publications: in institutional economics (Wittfogel 1957; Polanyi et al. 1957), cultural anthropology (Sahlins 1972; Bennett 1976), interdisciplinary and historical ecology (Balée 1998, 2006) and environmental history (Mumford 1934; Crosby 1986; Stavrianos 1997; Krech et al. 2004; Hornborg et al. 2007; Caradonna 2018). Interdisciplinary research in environmental history (Hornborg et al. 2007), historical ecology (Balée 2006) and social ecology (Fischer-Kowalski and Haberl 1998) documents the complex interactions of social and ecological factors in the history of human societies, the human-made environmental catastrophes as well as successful management of natural resources over the long term.

In Europe and the Near East, overuse of natural resources and environmental pollution were already problematic in ancient societies and world systems. Following the transition to urban and rural settlements after the Neolithic revolution (about ten thousand years ago), the first large-scale political systems developed—empires with political and centralised regulation of natural resource use, distribution of resources and economic accumulation. In early empires and world systems—the Sumer, the Egyptian, the Roman Empire in Eurasia, the Mayan culture in Central America and others—deforestation, erosion, overuse of fertile land, salination of soils and water, and poor agricultural techniques contributed to their eventual collapse (Bruckmeier 2019: 19ff).

Paradigmatic cases of successful maintenance of the natural resource base of societies over an extended period of time are described for Asian societies in the historical theory of "oriental despotism" (Wittfogel 1957). The scarcity of water that determined the development of agriculture by early Asian societies was managed through its distribution in large channels and transport systems for maintaining agricultural production. The system was successfully managed through the state thanks to its centralised power. The history of Chinese peasant society is unique in that it maintained for more than four thousand years a socially and environmentally sustainable agricultural mode of production (King 1911; Needham and Bray 1984). This society experienced frequent periods of hunger due to crop failures as a result of climatic and ecological conditions that could not be managed, but it did not collapse, and it only came to an end in the late twentieth century with the delayed and politically enforced phase of industrialisation.

(*continued*)

Box 2.1 (continued)

The assumption of "population growth, overshoot and collapse" in classical political economy (Malthus) failed as a general explanation. Boserup (1965) developed a more complex framework of the interaction between population growth, technologies and changes of subsistence practices. Newer publications in the anthropological and historical sciences by Tainter (1988, 2006), or the popular science books written by Diamond with the simple hypothesis of "overshoot and collapse" (Diamond 1997, 2006), do not sufficiently explain societal collapse, nor do they speculate about the causes of collapse (thus the critique of Smith 2005 of Diamond). Fischer-Kowalski (2009) criticised the unclear use of the concepts civilisation and society in the historical analysis of the fall of the Roman Empire by Sieferle (2008). Historical ecology (Balée 2006) focuses on less complex analyses of local systems and landscape transformations.

Sources: mentioned in the text

The collective learning in terms of managing natural resources in the course of history yielded a series of successful practices and strategies to deal with resource limits: improvements of resource use practices and production, especially improvements of agricultural yields through new technologies, and control of population growth. Uneven development and economic divides arose among agricultural societies through the creation of social stratification and the redistribution of surplus production, which brought wealth to social, political and economic elites and dominant social classes. The dominant form of unequal appropriation of resources and means of production and consumption in history was the imperial form based on colonial conquest and expansion of the economic resource base. Ancient societies began the conquest and colonisation of new territories with the building of large economic systems maintained by states or empires. As long as large parts of the earth were not densely populated and not inhabited by Europeans, and most of the resources in those regions remained unexploited, expansionism in the form of economic and ecological imperialism created solutions for resource limits and overpopulation (Crosby 1986, analysing the practices of migration and spreading of Europeans in different parts of the world).

With the European conquest and colonisation of the Global South beginning in the fifteenth century, the economic and ecological processes were put in motion to build modern society: a global economic system with a world market and a capitalist mode of production. Global expansion resulted in the decoupling of economic development from the availability of local land and other resources through the use of fossil energy resources. The expansion brought exponential growth of the population, the economy and resource use, and it launched the global environmental change processes that are accelerating today.

2.2 Crises of Natural Resource Use in Modern Society

The environmental and resource use crises faced by modern society are not caused solely by overuse of the resource base and the extensive modification of ecosystems through agriculture, industry and urbanisation. Intra-societal causes, such as the unequal appropriation and distribution of resources and the periodic accumulation crises in the capitalist market economy, reinforce the crises. Environmental problems and crises have gone from being temporally and spatially limited to being global, permanent and interconnected crises of deforestation, land use, food and environmental destruction (Box 2.2).

Box 2.2 Interconnected Resource Crises in Modern Society

1. European *forest or wood crises* towards the end of the middle ages stimulated the search for new energy resources, which were found in the fossil energy sources that brought about industrial society (first coal, then oil and gas). Later, in the course of economic globalisation towards the end of the twentieth century, a new and rapid phase of deforestation occurred, especially in tropical rainforests and boreal forests in the arctic zone, destabilising the global climate and accelerating global warming (Runyan and D'Odorico 2016).
2. *Land use crises* began with the enclosures or privatisation of common land in early modernity as a precondition of the capitalist mode of production. Continuous social and environmental problems were created through the transformation of land use—expropriation, overuse, pollution and modification of agro-ecosystems. In late modernity the land use crises became global through global social and environmental change, with conflicts between competing forms of land use for urbanisation, food production, bio-energy production, land speculation and land-grabbing through the buying of agricultural land by governments and large corporations, especially in the countries of the Global South (Diergarten and Krieger 2015).

(*continued*)

Box 2.2 (continued)

3. *Food crises* developed from seasonal, local hunger crises to endogenous and permanent global hunger crises in late modernity, with a large part of the global population permanently suffering from malnutrition and hunger. This is a consequence of the uneven economic development in the modern global economic system: large parts of the global resources have been appropriated by the industrialised countries, first through the colonial economy and later through the unequal economic and ecological exchange between the developed and the developing countries in the modern global economic system (Rice 2007).
4. *Environmental crises* became permanent and global with the city-based industrialisation and the transboundary industrial pollution of air, water and soils in the twentieth century, resulting in global environmental change, especially climate change and extinction of species (Freedman 2014).

Sources: mentioned in the text

Environmental and resource problems were long ignored, assumed to be the unavoidable burdens of modernisation and industrialisation. Only the current global environmental crisis brought public awareness, intensified research, new social and environmental movements and a global sustainability policy. In the terminology of the theory of risk society (Beck 1996, 2009), the generalised crisis is described through the development of systemic risks that cannot be perceived by humans, only identified through scientific research. The economic forms of risk management developing in the modern economy, insurance and compensation for monetary losses of private enterprises, are not sufficient to deal with the environmental risks.

The natural sciences became increasingly important for the development of modern society: at first for developing industrial technologies, and later also for repairing the environmental damage through modern agricultural and industrial technologies. Research about agriculture played a decisive role in solving the food and health-related crises of population growth and urbanisation in the nineteenth century. The new agricultural chemistry (Liebig 1840, 1856) brought higher agricultural yields through plant nutrition with simple, inorganic matter, salts and acids, which opened the way to the industrial production of fertilisers. The hygienic revolution, made possible by medical research (especially by Virchow, the founder of modern pathology and social medicine), helped to prevent diseases among urban societies.

The unfolding of the sustainability discourse in European countries followed socially and economically specific forms of natural resource use in the pre-industrial agricultural and the industrial economy. The paradigmatic sustainability problem in the pre-industrial epoch of modern society was the wood crisis inherited from the feudal economy. The problems that appear today in the sustainability process—the maladaptive processes of global economic and ecological change—arose in the industrial era. The joint criterion for sustainability in both phases is the long-term perspective of resource use: how long can the natural resource base of aa society be maintained before it collapses through overshoot? This question can be translated into that of the secure future of the societal or economic system. All pre-industrial forms of society had low population numbers and much lower levels of consumption of material and energetic resources compared with industrial society. Through sustainable management of their natural resource base they could have existed for a long time (although this did not often happen, due to the manifold social and ecological reasons for change and collapse). Industrial society faces a triple burden that undermines its sustainability: it is part of (a) a global economic system with (b) exponential population growth and economic growth, (c) reaching global limits of resource use already after a very short historical period of about 250 years; it has no secure future.

2.2.1 Environmental History: The Forest Crisis in Modern Society

Local overuse of resources and deforestation have created environmental problems since the early agricultural societies in Europe and the Near East. From an ecological perspective the consequences of deforestation in all historical epochs are similar: biodiversity loss, erosion and loss of fertile soils. From the perspective of society, however, they differ: today deforestation is connected with the societal metabolism of industrial society and the anthropogenic global change processes of climate change, biodiversity loss, land use change and urbanisation. The transformation to a modern society was triggered by urbanisation in late medieval Europe and the wood crisis that sparked the search for new energy sources (Box 2.3).

Box 2.3 Environmental History of Modern European Societies

Radkau (2002, 2012) describes the world history of the environment within a framework of big epochs, including modern society as the modern era of ecology. As in most other historical, cultural anthropological and sociological analyses of modern society, the global development and modernisation that occurred with the European colonial conquest of the Global South in the sixteenth century marks the beginning. He differentiates between societies with subsistence economies dominated by agriculture and modern industrial societies. Until the nineteenth century agriculture and subsistence production were dominant, even in Europe, where most countries began the industrialisation during this century. Radkau's analysis focuses on governmental institutions more than economic institutions as mechanisms of transformation. In contrast to the culturally and religiously regulated relations between nature and society in subsistence societies, in the modern era new forms of power relations, including power over nature, were developed through political and legal regulation. The contradictory forms of exercising power over nature enable either environmental destruction (through industrialisation) or its prevention and control (through protection of nature). Radkau describes the incoherence of modern society and modernisation as supporting biological invasion and the spread of invasive species of plants and animals, and intensifying natural resource use, finally reaching global limits. The multifaceted process of modernisation appears in this view of the era of ecology as continuous ecological reflection and discussion of environmental problems, where finally the environmental movements of the twentieth century become the representatives of "green enlightenment". They demystify technological progress as way to solve environmental problems.

With his orientation to political power relations and processes, Radkau (2002) develops a critical analysis of forest crises in German history (Radkau and Schäfer 1987; Radkau 2018), which evoked controversial debates among German historians. He describes the history of the use of wood as a material and energy resource as being more complex than is indicated by empirical data about deforestation: it was not a general forest crisis everywhere and similar for all, but more severe in certain countries and threatening social classes differently. Against dogmatic views of forest history, his interpretations of the public debates about real or assumed forest crises and shortages of wood show that the forest debates were used by governmental institutions to introduce new regulations and laws for forests and forestry, supporting modernisation and privatisation and suppressing older forms of forest use by peasants and as commons.

(*continued*)

Box 2.3 (continued)

The system-related analyses of societal and economic systems, modes of production and the societal metabolism of societies are not sufficiently addressed in Radkau's history of the era of ecology: theoretically based interdisciplinary analyses in historical and theoretical studies are required, where the interaction between cultural, social, political, economic and ecological processes and systems are analysed more systematically than the political dynamics.

Sources: mentioned in the text

The birthplaces of modern society were the growing cities in the medieval feudal societies of Europe. Urbanisation, accelerating a millennium ago, brought successive innovations that made modern society possible over the course of several centuries (markets, long-distance and oversea trade, financial economy, manufacturing, development of new technologies). Cities as local forms of society were autonomous social and economic systems in the rural territories of feudal societies, although they were ecologically dependent on the continuous flow of resources from rural to urban areas. Some cities succeeded in building independent political systems, city states and republics within the feudal empires. Paradigmatic cases include the city republics in Southern Europe (Northern Italy) and the international federation of the "Hanse" cities in Northern European countries. These medieval cities experimented with a new mercantile economy based on international and overseas trade as the source of economic development and wealth. From the twelfth to the sixteenth century, the modern economy and society developed in the cities as islands of modernity within medieval feudalism. In the new urban economy, the professions (merchants, craftsmen, engineers and scientific professions) that later made up the new bourgeois social class in modern society arose. The urban economies initiated the development of the market-based economy of modern capitalism within a stationary, non-developing feudal economy. The feudal system was based on rural and agricultural production and land as the main factor of production. The new economy was differentiated by the modern economic trinity of the combined factors of capital, labour and land. The conquests, colonisation and overseas trade of Portugal and Spain in the fifteenth and sixteenth century initiated the development of the modern economic system at global scale, producing, along with industrialisation, the dynamics of exponential population growth and economic growth that remain in effect today.

(1) The forest crisis as a form of energy crisis: The resource and energy crises in Europe that resulted in the transition from the feudal to the modern capitalist economy were caused by deforestation and increasing scarcity of wood as the main energy resource in the late phase of the agricultural society. The political and scientific debates about wood scarcity showed similarities with the discourse that later unfolded about sustainability. As in all historical forms of human societies, deforestation was in complicated ways connected with the social organisation of economic systems. In earlier societies and in the feudal European societies the ecological problems caused by deforestation could not be solved, but the impact of the scarcity of wood could be reduced through the search for new material and energy resources to replace wood, for example, using turf instead of firewood, or through long distance trade of wood. The wood crises faced by feudal societies were finally resolved by modern society with the transformation of the energy system through industrialisation and the use of fossil energy resources. New environmental crises related to deforestation as a cause of global climate change arose in the twentieth century.

(2) Sustainable management of forest resources in early modern society: Ideas about environment-friendly and sustainable use of natural resources were first developed in the seventeenth century, when concepts and practices of reforestation spread across several European countries, along with the elaboration of scientific ideas about how to manage forests sustainably. The forest crisis is not just about the loss of wood through forest clearing, but also maintaining the stocks of natural resources in forests, including all forest-based living and non-living resources, in a productive state: trees, bushes, wild fruits, rivers and lakes, game and fish. The food resources found in the forest had an important function for the local population in feudal society and in pre-industrial modern society: the forests were "the poor man's overcoat", that is, in years of crop failure and hunger, the rural population had to survive until the next season on the resources in the woods. Even in late modernity the forests are an important resource base for many poor people in developing countries (Hobley 2007). Until early modernity wood was the main energy resource for society (Box 2.4). The scientific sustainability discourse began from there, visible in the pioneering works of Evelyn in England and Carlowitz in Germany. Evelyn (1972, 2018) and Carlowitz (1713) developed ideas about a solution to the forest crisis in interdisciplinary analyses connecting ecological, economic and political or managerial ideas. In early modern ecology or natural history, embedded in the philosophical discourse of enlightenment in the seventeenth and eighteenth centuries, many scientists had similar ideas, including Derham and his physico-theology (1713) in England and Linné (2010) in Sweden (for a detailed description of the history of ecology in early modernity see Trepl 1994: 81ff).

Box 2.4 Wood Crises in Medieval and Modern European Societies

Ecological ideas for solving the wood scarcity problem locally by reducing deforestation have been documented since the Middle Ages in European countries, for example: the use of technically less efficient instruments (axes instead of saws), the ban on wood clearing in certain areas (as in the sixteenth century in the Elizabethan Acts in England, prohibiting the cutting of trees within a distance of twenty-two miles from London) and ideas about reforestation. However, none of the responses to local resource problems solved the wood crises until industrialisation began, when the "subterranean forests" of coal, oil and gas replaced wood as the dominant energy resource. Until the Napoleonic wars reappeared forest crises in Europe, also in the UK, where the last oak forests vanished through the building of the big military and commercial fleets.

After the industrial revolution that began in England and Scotland, the dependence on wood, both as a material and an energy resource, was reduced, but the wood crises were not resolved: new demands for wood were created with the development of the industrial society and its intensification of resource use. In spite of the use of fossil energy sources in the industrialised countries (until today not all countries in the Global South are industrialised, whereas in the Global North some countries are already approaching a post-industrial economy) the amount of wood used did not decline: from a global perspective it is increasing, as are other indicators of intensification of resource use, population growth and economic growth. The change that has occurred in forest management in European countries since the industrial revolution has been that reforestation has become more systematically practiced. In some countries the forest cover is growing, and the amount of wood harvested is outpaced by the level of regrowth.

Globally, however, the loss of forest continues, especially in the countries of the Global South, where there has been a dramatic loss of tropical rainforests. The forest cover on the earth has shrunk from about 40% to 20% in the last thousand years and is expected to shrink further during the twenty-first century. Furthermore, the difficulties of maintaining forests in a healthy state have increased since the twentieth century as a result of air pollution and climate change. The death of forests as a consequence of acid rain was among the problems intensively discussed in the 1970s and 1980s in Europe, but interest in the discussion later waned. Meanwhile the risks for forests are growing again, in Europe and elsewhere—industrial emissions have only

(continued)

Box 2.4 (continued)
been reduced in some countries, and climate change continues, with large-scale losses of forest through heavy storms, damage of trees due to changing climate conditions, air pollution, forest fires and invasion of species.

Sources: for deforestation and forest crises in history see the website of "Environmental History Resources"; for the current forest crisis see the website of "Rainforest Action Network" and Chakravarty et al. (2012).

Evelyn wrote in 1661 about the bad air quality produced by the burning of coal that damaged buildings and the health of the urban inhabitants of London. This did not prevent the extensive use of coal as an energy source for industrialisation, which was made possible by the coal regime. Damages were only reduced technologically, for example, through "end of the pipe" technologies (according to the terminology from today): higher chimneys and improvements to fireplaces that brought better indoor air quality and improved air quality in the city through the wider dispersion of smoke. In 1664 Evelyn described deforestation and discussed the emerging ideas about the protection and maintenance of forests, including reforestation. The growing production of iron and glass was identified as the main reason for the reduction of forests in Great Britain and was seen as a risk for the British fleet, which was dependent on wood for shipbuilding. The author argued for systematic reforestation and for the founding of new forests and parks. The discussion about reforestation in Great Britain was later taken up again by environmentalists such as Wallace (1878) and became part of the broader theme of the restoration of nature (Clark and York 2007).

Carlowitz wrote the first textbook about forest economy in Germany, based on his expertise and experience as governmental administrator of the mining industry in Saxony. His analysis of the wood crisis explicates the economic and ecological causes of wood scarcity through silver mining in the pre-industrial Saxonian economy. The technologies of silver mining and smelting required high quantities of wood and were a main cause of deforestation in the region. Furthermore, he also criticised the clearing of woodland for agricultural production (see Box 2.5). He was aware of the forest reforms of Colbert from 1669 that were designed to reduce forest loss, especially from the building of military and commercial fleets.

Box 2.5 Forestry Economics According to Carlowitz

In Carlowitz's description of forestry economics ("sylvicultura oeconomiae", 1713: 24) he differentiates the modes of tree production in forests in two basic forms: (1) supernatural, meaning the original creation of forests through the creator of the world, and (2) the succeeding natural production, according to physical principles. The second form is divided into two specific forms: (1) the purely natural mode, or the way in which wild forests grow and develop without human intervention, and (2) the artificial mode, or the way in which forests grow and develop when human art and science help nature to produce more resources for the well-being of humans (Carlowitz 1713: 24). He argues that sustainable forestry can be practised by reformulating the principle of the "eternal forest" as a principle of artificial forestry by humans: to obey the limits of natural forest growth—not to cut more wood from the forest than is growing every year. The means to restore the loss of wood through deforestation is the replanting of woods. This practical knowledge existed previously in forestry, but without clear and measurable criteria for the maintenance of forests. Carlowitz used the same term to describe this approach as is used today: sustainable use of forests.

His forestry economics is embedded in a philosophy and worldview where humans must not ruin forests, because it is against their nature and the divine order: forests should be improved through work. Deforestation, seen in the context of enlightenment thinking and physico-theology, appears as human failure, destruction and misuse of forests. He is aware of the risks of deforestation and analyses the reasons for the scarcity of wood in detail in its many man-made forms in detail in chap. 4 of the book, where he names various reasons for the loss of forest in European countries: the large number and size of buildings, the big rooms that must be heated, the wood-intensive technologies of cooking and heating in stoves and chimneys, the cutting of the best trees, the introduction of animal husbandry that increases deforestation to create pastures and the high demand for wood for mining, which he sees as greater than the demand for ship-building (Carlowitz 1713: 40ff). He searches out scientific arguments for the cultivation and maintenance of forests from the natural sciences, physics and what he calls the natural view of forests: observing what nature does and developing an art and a science that imitate nature to prevent further damage to forests. He argues that improving human capacities through science and technology helps to maintain the forests(Carlowitz 1713: 39f). The technologies he suggested include better

Box 2.5 (continued)
insulation of houses, better furnaces, stoves and heaters to save wood, planned reforestation and substituting wood with other resources such as turf. Carlowitz wrote about sustainable forestry shortly before the industrial revolution that introduced the fossil resource coal to end the dependence on wood. Yet one can find in his thinking an idea that was later developed in economic sustainability theory: sustainability as a non-declining utility of natural resources to maintain human welfare in the long run (Solow 1993; Howarth 2007).

Source: Carlowitz 1713

The term "economy" in Carlowitz's analysis comes from the German variant of mercantilism (Wallerstein 1980), called "cameralism" (Wakefield 2009): an administrative science and practice to manage the economy, the natural resources and the finances of the state. Modern economics as an academic discipline developed later with the differentiation of classical political economy from moral philosophy (Ferguson, Smith) in the second half of the eighteenth century, parallel to the unfolding of the new industrial mode of production in Great Britain. For Carlowitz, the economic principles of forestry are part of ecological knowledge, as in the writings of Linné (2010), who argued some decades after Carlowitz for the "economy of nature" in the sense of a sustainable economy, but without applying the term. In the economy, humans imitate nature; when the processes in the human economy are synchronised with those in the economy of nature, the resources are used in a way that does not cause damage to nature. For Linné as for Carlowitz, sustainability in the sense of following nature's lead is not primarily a question of scientific knowledge (this is only a means to achieve an end), but one of belief. Their worldview was influenced by the physico-theology of the enlightenment epoch, which is a synthesis of theological thinking and natural science: God can be found in his works of creation, in nature. According to this worldview humans cannot act against nature, only with it.

The principles of the forest economy that Carlowitz describes are based on biological and ecological knowledge about trees and how to use that knowledge for sustainable forestry. Carlowitz did not analyse systematically the changing forms of the economy with the capitalist mode of production, although he was aware of the changes towards a commercialised and commodified economy, as is evident from his critique of land use. The extended reconstruction of the economy after the disastrous Thirty Years War in Germany during the seventeenth century included the building of a centralised economic system with state-owned manufacturing, improved agricultural production and support for population growth in the areas depopulated during the war. Carlowitz's analysis focuses on the forest economy in a simplified sense, on the introduction of ecological principles or "imitating nature" in forestry, with no analysis of the modern economy or the changes in the mode of production and in property rights that had already occurred at the time. Only a few decades after his seminal publication, the industrial revolution was launched, changing the conditions for developing a sustainable economy. The main difference is that the sustainability of the forestry was no longer assessed in ecological terms or in natural form, as in the forest economy of Carlowitz, but rather in monetary form, as profit. The simplification of resource use by humans as imitating an "economy of nature" is the ecologically rational core of the idea of sustainability in the current discourse as well, but it needs to be reformulated according to a different terminology and for another economic system: to maintain the functioning of ecosystems and of the earth system, and not to exceed the global limits of resource use. Daly (1990: 2) formulated the principles of a steady state economy in a similar way: the rates of use of renewable resources should not exceed the rates of regrowth; the rates of use of non-renewable resources should not exceed the rate of creating sustainably renewable resources to replace them; and the rates of emission of pollutants should not exceed the rate of receptivity of ecosystems for these substances.

(3) Sustainable management of forest resources today: The current sustainability discourse and process is compared to the historical examples of sustainable resource use and pollution, working with a broader concept of sustainability where more complex relations between economic and ecological systems are discussed. In science and in environmental politics the problem perceived as dominant, overshadowing all others, is climate change resulting from fossil energy sources and conversion technologies in industrial society. The forest discourse is a part of this through the new forms of deforestation outside Europe: deforestation appears now as a global problem contributing to climate change.

The late modern forest crisis is not happening in European countries, where deforestation has slowed down. European forests are damaged by air pollution and climate change, including damage from storms. Large-scale deforestation is happening in other parts of the world, mainly in tropical rainforests in Asia, Africa and South America (Rudel 2005). Deforestation is also occurring in the boreal forests in the northern periphery of the continents adjacent to the forestless polar zone, especially in Russia. Large-scale deforestation is one of the causes of the acceleration of global climate change. Other problems in the forest areas include the commercialisation of forest products beyond trees, especially pharmaceutical plants, such as in the Amazon, where the forests were seen as commons by the indigenous inhabitants. The use of such plants by pharmaceutical firms happened also in forms of biopiracy: the firms did not observe the traditional property rights and commons; only after international protests began gradually a change to less imperial forms of resource appropriation, by compensating indigenous people or sharing the benefits from resource use with them.

In the early modern energy crisis, the depleted forests could be restored through reforestation. Reforestation of tropical rainforests to their prior state, however, is more difficult and may be impossible. The solution to the late modern energy crisis will not be reforestation, although it can be part of the process to reduce climate change and its consequences. Sustainable resource use becomes a question of phasing out the industrial energy system in favour of one based on renewable energy sources and reduced environmental pollution. This will become part of the larger transformation of the industrial society through sustainable development that includes the solution of multiple and connected problems of social, cultural political, economic and ecological origin (Bruckmeier 2019: 93ff). This complexity is indicated in the United Nations' seventeen Sustainable Development Goals.

2.2.2 Environmental History: Interdisciplinary Research

To understand and solve the interconnected problems related to sustainable development, historical analyses of resource use problems and epistemological and theoretical framing in interdisciplinary studies are required. How can knowledge from several disciplines in broad analyses of social-ecological transformations of historically specific large-scale societal, economic and social-ecological systems be brought together? Historical science plays a specific role insofar as the transformation to sustainability is a historical process and can be compared with earlier social-ecological transformations in human history, both in similarities and differences. The epistemological problems that arise in historical studies begin with the conceptualisation of history and time, of the relations between the past, the present and the future that appear in interdisciplinary analyses of "the history of the future". For the current sustainability discourse and its historical contextualisation, two historical studies, neither originating from the historical sciences, have enhanced the historical reflection:

1. Polanyi's (1944) analysis of the "great transformation" to the modern market economy, where earlier economic systems and societies are characterised as integrated and embedded in cultural systems, became the paradigmatic model

for further transformation research. In this research the term transformation becomes an interdisciplinary theoretical concept. The socially and ecologically disintegrated modern society with deregulated markets is a consequence of modern capitalism; a new transformation is required to prevent the social and ecological self-destruction of modern society, as Polanyi and other economists in institutional economics argued (Kapp 1976; Zinn 1980). This discourse addresses the following question: *What makes transformation a specific form of social change to be differentiated from other forms of social and economic change?*

2. In cultural anthropology, Moscovici (1968) applies a similar differentiation to that of Radkau between subsistence-based agricultural and industrial societies, rejecting for ecological research the (economic and political) differentiation of capitalist and socialist modes of production. For Moscovici the ecologically decisive differences are those between agricultural and industrial societies. In his analysis, human history and society appear as part of the history of nature; after the fruitless attempt to dissociate society from nature in earlier modernity, in late modernity society will again be integrated with nature. His classification of modes of production introduces another controversy to social and economic science: *Is the industrial organisation of production or its capitalist (and socialist) form the greater problem to be dealt with in environmental policy and sustainable development?* The collapse of socialism in Eastern European rendered the question of a socialist mode of production less significant. But in the sustainability discourse, the question of a transformation of the global capitalist system comes up again: the natural-scientific descriptions of global environmental problems and limited resources cannot be translated into societal analyses and political practices of transformation without analysing the systemic constitution of the modern economy and society.

Interdisciplinary historical analyses of society and societal resource use (Box 2.6) touch on questions relevant to the analysis of coupled societal and ecological systems. They create the preconditions for "scientific learning from history": the systems involved in natural resource use, not the single events and decisions in the policy process, need to be studied to understand the systemic principles of nature–society interaction that vary in each type of society and economy.

Box 2.6 Societal History and Resource Use Processes: Interdisciplinary Studies

Among the historical analyses of the forms and dynamics of modernisation, several theoretical variants are relevant to understand the complexity of nature–society interactions in modern and industrial society:

1. Braudel (1949)—*temporal structures in social change.* Braudel's seminal work on Mediterranean society brought significant changes in the view of human and societal history. He differentiates temporal structures in historical processes in three components: the *geographical time*, the *long historical time* ("longue durée") and the *short historical time* (of events and individuals: this is the least important type, the surface of history). The first two temporal perspectives and their interaction are important for interdisciplinary environmental and social-ecological research: the geographical time is the slow, repetitive and cyclic processes in nature, and the long historical time is the much faster development and change of societies, economies and culture.
2. Wolf (1982)—*the local and neglected history of the modern global capitalist system.* Wolf connected the world system theory of Wallerstein with analyses of the local history of transformation and modernisation of the economy and society from cultural anthropological research (Wallerstein emphasised the necessity of analysing local processes in the perspective of the world system, to show the complicated building of a multi-scale system: Wallerstein 1974, 1980). Both macroscopic system analysis and microscopic analysis of local development processes that differ culturally, politically and economically are necessary to understand the dynamics of capitalist modernisation. As an economic system with a global economy and world market as its core, modern capitalism was never a homogeneous system and coherent mode of production; it remained imperfect and changing, temporarily integrating other and older forms of production. Local peasant cultures and economies endured for a long time, as shown by the existence of a subsistence-based, non-market-dependent small-scale peasant economy until the twentieth century in large parts of the world, including in European countries.

(*continued*)

Box 2.6 (continued)

3. Debeir et al. (1991)—*history of energy systems*. The complexity of energy use and the development of different societal energy systems in the course of human history cannot be understood from the empirical data alone. These show the continuous increase and intensification of natural resource use in terms of per capita consumption since the early societies of hunters and gatherers. The data cannot explain the dynamics of interacting social and ecological processes in different forms and phases of human society: theory-based analyses of the complexity of social and ecological factors are necessary. This is done in analyses of changes in modes of production. Political economic theories are used by the authors to explain changes in energy regimes. They do not systematically analyse the societal metabolism in modern society, which would require measuring the total material and energy flows between nature and society.
4. Hornborg et al. (2007)—*environmental history as world system history*. The authors systematically and theoretically analyse the social and ecological complexity of forest and other natural resource use crises (also addressed in Radkau's European environmental history), as complexity resulting from the interaction of social and ecological systems in the modern global economic system. With critical theory as the basis for the historical analysis, a different view of resource use and environmental problems is presented—that of the modern economy and society as systems in continuous change, in long ecological transition processes triggered by resource use crises. The current transformation problems in the global sustainability process appear as combinations of different types of social, political, economic and ecological crises.

Sources: mentioned in the text

Interdisciplinary historical studies are part of the theoretical analyses of nature–society interactions, social metabolism and societal processes of natural resource use (Table 2.1). With such research the complexity of processes of development and change of social-ecological systems is gradually understood and explained: the forms of social-ecological changes and transformations, the connections between discourses about nature and sustainability, the different forms of sustainability problems in the pre-industrial and industrial phases of modern society, the

Table 2.1 Socio-metabolic systems of societies in human history

	Hunter and gatherer society	Agricultural society	Industrial society
Population density	Very low	Low	Very high
Territorial dimensions	Small (local)	Large (historical world systems and empires)	Global (modern world economic system)
Mode of production in ecological terms	Subsistence production—hunting, gathering, fishing	Subsistence production—agriculture, forestry	Industrial production, agriculture, forestry, services
Mode of production in economic terms	Subsistence production	Subsistence production, reciprocity, embedded markets, simple commodity production	Global exchange and trade, world market, capitalist commodity production
Agricultural population	0%	>80%	<10%
Settlement	No permanent settlement, mobility	Permanent settlement, villages (mainly) and cities	Permanent settlement, cities (mainly) and villages
Dominant energy	Biomass	Biomass	Various fossil and renewable energy sources
Energy input in gigajoule per capita and year	10–20: food, wood (basic human metabolism: 3.5)	ca. 65: 50 fodder, 3 vegetarian food, 12 wood (basic human metabolism: 3.5)	250: 170 fossil energy, 5 hydropower, 14 nuclear energy, 61 biomass (basic human metabolism: 3.5)
Material input in tons per capita and year	ca. 1: food, wood (basic human metabolism: 1)	ca. 4: 0.5 vegetarian food, 2.7 fodder, 0.8 wood (basic human metabolism: 1)	19.5: 4.7 biomass, 5.2 oil, coal, gas; 9.7 minerals, metals, others (basic human metabolism: 1)
Dominant forms of resources	Living resources—food (plants, animals)	Living resources—food (plants, animals), agricultural land	Non-living physical resources (minerals, metals, coal, oil, gas)
Dominant forms of changing and damaging nature/environment	Local (overhunting)	Local (deforestation, erosion of agricultural land)	Global (climate change, biodiversity loss, deforestation, urbanisation, pollution of air, water, soils)
Historical time	Since ca. 1.8 million years ago	Since ca. 10,000 years ago	Since ca. 250 years ago

Sources: Fischer-Kowalski and Haberl 1998; own compilation

ecological critique of modern society, the building of the modern world economic system, and cultural differences within the modern global society and economy and the consequences for the sustainability process.

2.3 Connected Discourses About Nature and Sustainability in Modern Society

The European conquest and colonisation of the Global South since early modernity was the last form of colonisation of new territories in human history, leaving only the oceans and the Arctic and Antarctic areas as spaces and resource bases for future expansion and growth of the global economy. This expansion began recently, simultaneously with the discourses of global limits to growth and sustainability, through economic globalisation and with risky technologies such as deep sea oil drilling, to create future economic growth based on controversial ideas of sustainable or "green" growth for land-based resources and "blue" growth for the maritime economy. When these resources are exploited, no further territorial expansion can help to stimulate economic growth, only the opening up of non-material resources: ecosystem services, genetic manipulation of plants, animals and humans, and the use of information and knowledge as economic resources. These processes are already under way as further commercialisation of hitherto non-valorised resources. Concerns about the "commercialisation of everything" and the disregard for social and ecological limits of growth are growing among ecologists and economists.

Since the twentieth century the human use of natural, living and mineral resources has been approaching global limits and the hitherto effective means of controlling and limiting natural resource use, population growth and environmental damage are gradually becoming ineffective. For the first time it has become necessary to act globally in response to the maladaptive changes that came with economic globalisation. New strategies for the ecological transition to sustainability and experimentation with different forms of limiting global resource use are required. It has become possible to calculate the earth's global limits or "planetary boundaries" with knowledge from modern physics, where the concepts of energy and matter are defined, operationalised and measured in terms of quantitative and qualitative changes through human use of resources. The sustainability discourse began as one about economic and ecological transition or transformation, where the question of limits to growth is the core, and integrated analyses of societal, ecological and cultural change become necessary.

1. *Societal processes of transition* are not sufficiently reflected and analysed in economics, neither the historical changes leading to the modern world economic system nor the future changes necessary for the sustainability transition. In orthodox economics social, technological and ecological change appear only as conditions of economic growth, affecting it positively or negatively. Privatisation of natural resources (especially land) in the previous accumulation processes, industrialisation, urbanisation, the building of a world market, and the globalisation and commodification of exchange, production and consumption became the critical issues in economic and ecological research.
2. *Ecological processes of environmental and land use change were part of economic activities of natural resource use*: reduction of the forest cover of the earth, land use change through agriculture and urbanisation, environmental pollution of soil, water and air, and the new processes of anthropogenic global change, climate change, biodiversity loss and the creation of novel ecosystems where nature is greatly modified by humans and their technologies. These changes are not sufficiently analysed in economics. In the dominant view of the modern economy as driven by growth, the changes appear in the form of externalities that cannot, or can only imperfectly, be internalised in the economic system: as the ecological costs of progress.
3. *Socio-cultural changes in modern society accompanying ecological and economic transition* include modernisation processes that are based on the development of modern science: modernising societal communication systems, education systems, the state and the economy, use of new science-based technologies in industrialisation and urbanisation, and medical knowledge for the "hygienic revolution". Modernisation processes are part of the building of a new society with new social classes and political democracy in the industrialising countries in Europe. The modernisation processes accompanying the changes in natural resource use, industrialisation and economic growth brought new cultural, philosophical and scientific views and interpretations of nature and changing relations between humans and nature (Box 2.7). In environmental sociology the conventional economic view of rationality and economic growth appears as a form of the "human exceptionalism paradigm" or the dominant Western worldview, where the ecological consequences of economic action are ignored.

Box 2.7 Changing Views of Nature and Humans in Modern Society

The view of nature and of humans began to change in the transition from medieval feudal to modern European society in the fifteenth and sixteenth century with the development of the European philosophical discourse of humanism, focusing on humans and supporting anthropocentric views of nature. The subsequent enlightenment philosophy brought two contrasting epistemological perspectives, rationalism (mind and reason as sources of cognition) and empiricism (cognition and knowledge resulting from sensual perception), which can be seen as the epistemological origins of the modern sciences differentiated into natural, social and humanitarian sciences. The emergence of a modern natural-scientific view of nature happened slowly, in different and contradicting forms that show the cognitive struggle for emancipation from medieval theological dogmatism. Two paradigmatic examples are the philosophy of Spinoza in the seventeenth century and the ecological thinking of Carlowitz (1713) in his analysis of forestry. Spinoza's ethical and other philosophical writings merge the ideas of god and nature in the formula "god or nature", not easily interpreted, but compatible with the view that god is synonymous with the creation of nature. Carlowitz formulated the principles of sustainable forestry as a synthesis of ecological and economic thinking. Although he described the damage done to nature by humans empirically and concretely, his worldview (rooted in physico-theological thinking) did not allow him to think of humans as destroyers of nature. The fluctuation between the contrasting ideas of humans as masters of nature and humans as stewards of nature, both found in Christian ethics, are visible in many variants of enlightenment thinking and philosophy of nature in the sixteenth and seventeenth centuries.

The history and philosophy of nature, with its changing ideas of creation and development, includes early forms of ecological thinking such as natural or physico-theology (Derham 2009) and its de-theologised variants in the "economy of nature" (Linné 2010), based on older ideas of a harmonious natural order. The rise of modern economics with classical political economy (developing from moral philosophy in the works of Ferguson and Smith in the eighteenth century), and the development of modern biology and ecology in the nineteenth century (including the Darwinian theory of evolution and as its critical counterpart, the holistic philosophy of the cosmos by Humboldt), mark changes in scientifically based views of nature.

(continued)

Box 2.7 (continued)

The ideas of an active and productive nature date back to older sources reinterpreted in modern philosophy. In ancient Greek philosophy Aristotle attributed the capacity of production ("poiesis") to nature and to humans. In the course of modernisation the dominant view became that nature can be appropriated, managed and modified by humans for their needs and interests with knowledge from the natural sciences. This implies that the productive capacity of nature can be manipulated by humans. This view is critically reflected in the environmental discourse, for example, by Merchant (1980, 2005) as the "death of nature" and the transition from an "organic worldview" to a "mechanistic worldview" in early modernity, under the influence of physics. When the modern natural and later the social sciences participated in the discourse of nature, the concepts inherited from philosophy were analytically dissected, deprived of their philosophical meanings and transformed into positive and empirical knowledge. This supported the anthropocentric view of humans as masters of nature (originally a Christian concept) that was dominant in modern Western societies until the emergence of a new environmentalism and environmental movements in the second half of the twentieth century.

Sources: mentioned in the text

With regard to the environmental consequences of modernisation, the socio-cultural changes can be summarised in three points:

1. *The socio-cultural changes seen in modern society indicate a plurality of conflicting modernisation processes* in asynchronous development, with competing worldviews, value- and interest-based controversies, with political, economic and cultural conflicts, where philosophical, religious and political ideas and cognitive interests clash and interact. The changing views of nature in modern science and society are reflected in philosophical, political, social-scientific and cultural discourses in the arts, humanities and religion that accompany the transition from a medieval, feudal and religion-dominated society tomodern society in Europe. This transition began with a scientific conflict, the change from a geocentric to a heliocentric worldview in early modern science, based on the work of Kepler, Copernicus, Galileo and Newton. In the further course of modernisation views of nature and humans changed from religious to scientific-

based views that spread in politics and in the lifeworld. Enlightenment philosophy opened the way to the future and accelerated the demystification of nature, which no longer appeared as a transcendental or metaphysical order of divine creation. Nature became the explanatory concept for the creation of life, of a generative capacity, and of the continually reproducing living nature: such ideas led to the modern ecological and evolutionary view of nature. The modification of nature by humans was first seen as improving the productive capacity of nature with the help of science and technology. Only in late modernity, after industrialisation and the damage it caused to the environment, did critical views of the human modification of nature spread in Western countries. The socio-cultural development of modern society thus appears as a self-correction of the modernisation process.

2. *Several concepts were used in the modern social sciences to describe retrospectively the development of new views of nature, or of human relations to nature, in the epoch of modernity*: nature was "demystified" or "disenchanted", "subjectivised" and "rationalised" through new knowledge practices in the sciences and the lifeworld. The multifaceted concept of "rationalisation" included the forms in which natural resources became subjected to human use in economic and political processes, or "rationally" organised. In paradigmatic form this view of modernisation as disenchantment and rationalisation can be found in the sociology of Weber (2019), where the modern form of society is described using the concept (called "ideal type") of "bureaucratic authority" and formal organisation that characterised all forms of organisation and cooperation in the cultural and religious, political and economic spheres of modern society. This "rational" form of political and economic organisation and control of natural resource use coincided with the ideas of unlimited growth and resource use. It is a historically and socio-culturally specific form of thinking and organisation, as can be seen today, when the rationality of hierarchical bureaucratic organisation is changing under the influence of new information technologies and the experience of environmental problems that require different rationality criteria.
3. *The socially relevant views of nature changed in late modernity, in complicated processes* that cannot be sufficiently understood based on the abstract terms rationalisation and disenchantment. These terms offer a simplified understanding of the historical process in modern society that connotes linear and unidirectional development. Views of nature develop and change through continuous dialogue and controversy with other views of man and nature. The "modern worldviews", including physical and ecological views of nature, split into anthropo-, physico- and bio-centric views (for example, in environmental ethics: DesJardins 2006). Sometimes they are simplified as two contrasting views: humans as autonomous or independent from nature, versus humans as depen-

dent on or integrated in nature, that is, as anthropocentric versus ecocentric worldviews. In U.S. environmental sociology, these views of nature were formulated as a contrast between an anthropocentric "human exce(m)ptionalism paradigm" and an ecocentric "new ecological paradigm" (Catton and Dunlap 1978) that structured the controversies in science, politics and environmental movements. In recent decades the Western world has seen a shift from anthropocentric to ecocentric views that have spread alongside environmental awareness, in empirical social research articulated in different forms, and as the value change from material to post-material values (Inglehart 1977, 1990). Beyond the worldviews and guiding ideas about nature–society relations that suffer from abstractness, more concrete experiences of economic and ecological crises in modernity direct the ecological discourse: it becomes less one of ontological beliefs and more one of applying empirical knowledge and experiences to environmental action and sustainability governance. Parallel to the sociocultural and value changes in the late twentieth century developed the interdisciplinary ecological sciences (human, cultural, social, political ecology: see Chap. 3). Moreover, the social sciences, especially cultural anthropology, sociology, economics and political science, adopted the theme of nature in new subdisciplines of environmental sociology (Burns and Schaefer Caniglia 2016), environmental and ecological economics, and environmental history.

2.3.1 Ecological and Social Critique of Industrial Society in Europe

Social movements have arisen to fight against the social and economic changes to agriculture and the economy and for the rights of peasants and the rural poor in European countries since the beginning of modern society, long before industrialisation (see Box 2.8). The industrial society that began to develop in Europe around 1750 was rooted in an energy regime based on fossil energy sources, which had two consequences for the further development of a modern economy in European countries. The economic processes became independent from the local limits of land, and the scarcity of wood was no longer a development problem as it had been in the pre-industrial agricultural economy. The development of the industrial society resulted in large-scale pollution of the environment, exponential economic growth, exponential growth of natural resource use and exponential growth of the global population. These themes were taken up in the ecological and social critique of industrial society by social and environmental movements. The critique was often ethically or religiously motivated, in utopian and non-utopian forms, all of which were oriented toward ideas of social justice, equality, solidarity the "good society".

Box 2.8 Social Movements in the History of Modernity

Long before the environmental movements of the twentieth century arose, there were frequent critiques of industrialisation and protests against privatisation of common lands. Early environmental movements in Europe, as with the "environmentalism of the poor" in the Global South today (Martinez-Alier 2002), were movements of the poor for social justice and access to land and natural resources, as well as utopian and religious movements.

The industrial revolution in Europe happened in the framework of the capitalist mode of production, initiated through a process referred to in political economy as "previous accumulation", which with the privatisation of land created a new social class of landowners and entrepreneurs who expropriated the land of the agricultural producers. This process created new markets beyond that for consumer goods: markets for land and labour and the commodification of land and labour. The previous accumulation, for example in England, was accompanied by protests and revolts by utopian movements of the landless fighting for fair and equal access to common land and against its privatisation: the "diggers" and "levellers" (Kennedy 2008) that can be seen as forerunners of modern socialist movements. Peasant revolts motivated by economic or religious goals occurred in European countries from the late Middle Ages (Blickle 1981; Bercé 1987). The French Revolution of 1789, in a country with a pre-industrial feudal society, aimed to create a just society according to the model of nature, but brought the breakthrough of industrialisation in the country. The revolution was strongly influenced by Rousseau and his critique of modern society based on an analysis of human and societal relations to nature, private property, social inequality and human alienation from nature (Bertram 2018).

The peasant movements in European countries during the Renaissance, the diggers and levellers in pre-industrial England, the industrialisation-critical romantic movements in Germany and France, the urban movements in high modernity, the French Revolution, and the socialist and working-class movements in the nineteenth century were not only movements of victims of modernisation and uneven development, of lower classes, plebeians, expropriated peasants, proletarians, the poor, suppressed and marginalised people. Many of the movements of the rural and urban poor beginning in early modernity had environment-related interests and motives, whatever form the movements took: rural or urban; as religious, heretical, utopian, millenarian or sectarian; as cultural, political or economic movements.

(*continued*)

Box 2.8 (continued)
Their environmental motives and goals were articulated in similar forms as today in the sustainability discourse: as redistribution and the fair sharing of resources—fair and equal access to food, land and other life-supporting resources, healthy living conditions and housing, a clean environment and access to education.

Sources: mentioned in the text

Romanticism is an inexact term used to describe various cultural movements (from the late eighteenth century until the end of the nineteenth century in Europe and the United Sates) in literature, painting and music, and critical intellectual movements with different ideas, motives and goals, which varied widely between the countries. One of its main themes was the critique of modern industrial society, which is often ecologically motivated and articulated as "back to nature", influenced by the thinking of Rousseau (Cooper 1999). What made it part of the anti-industrial and environmental movements can be formulated as the "tragedy of ugliness", a critique of industrial pollution of the environment, less in political and predominantly in cultural forms, and by artists. In late modernity a similar cultural movement for reflecting and studying ecological critiques of environmental destruction in literary forms and its connections with environmental movements, eco-criticism, was developed (Glotfelty and Fromm 1996). The movement, which arose in the United States and spread to many countries, involves the study of literary texts on environmental themes from an interdisciplinary perspective, as well as other approaches which view literature and poetry as specific forms of cultural ecology (Zapf 2008).

The French Revolution brought about the development of modern bourgeois society, political democracy and the modern national state. It accelerated the transition to industrial society in France, but it was still a revolution that took place in the political context of the late feudal agricultural society. The sustainable society that was the aim of this revolution was a society based on an agricultural model of nature, as the names of the months in the new calendar introduced after the revolution showed. The revolutionary movement was influenced by the ideas of Rousseau (especially Robespierre: Israel 2014), whose theory of the social contract became a key theme of the revolution.

Philosophical and religious motives of environmentalism in modern society have their origins in Christianity and an ethic of stewardship. The philosophical movement of New England transcendentalism (Thoreau and Emerson) in the early nineteenth century created a religious ethic rooted in the agricultural society and connected with utopian agricultural movements (Gura 2007). Early political ecologists include Marsh (1965 [1864]) in the United States, who criticised the transformation of the earth by humans and the degradation of nature, and Ruskin in Great Britain, who lamented the decline of craftsmanship as a result of industrial capitalism. Non-Western religions and ethics of natural resource use include Buddhism, in which the harmony between humans, animals and plants is portrayed in ideal forms. Until the early twentieth century most environmental movements were of a utopian type, based on non-scientific ideas and functioning as part of the routine of everyday life, in education and professional practices. This is in stark contrast to the environmental movements of the twenty-first century: they are no longer limited to pioneering minorities of environmentally aware activists, as they were for most of the twentieth century. Environmental activism has become a mass phenomenon and part of everyday life. New movements are developing, for example the youth movement "Fridays for future" (Hagedorn et al. 2019). Sustainability-related movements cannot be separated from other social and environmental movements and cannot be reduced to the activists in environmental policy and governance: they exist in manifold and different movements, networks and forms of activity throughout society.

The different types of environmentalism are analysed in a comparison of environmental movements in the Global South, or the "environmentalism of the poor" (Martinez-Alier 2002; Guha and Martinez-Alier 1998), with the environmentalism of the rich in the Global North. The movements in the South are movements of rural poor living directly from the resources available in local ecosystems. The movements of the middle class in industrialised countries offer a more philosophically and scientifically based critique of industrial society. The environmentalism of the poor is fighting against poverty and hunger and for ecological and economic redistribution, whereas the environmentalism of the rich relies more often on the idea that economic growth is good for the environment (Guha and Martinez-Alier 1998: 75f). The contrast between the two environmentalisms is reappearing in the global sustainability discourse. An early critical analysis of the progress of sustainable development (Lee et al. 2000: 62ff) directed attention to the differences between ideas about sustainable development in the Global South that centre around questions of redistribution, solidarity and "environmental democracy", and the sustainability discourse in the Global North. The changing environmentalism, its illusions (Dauvergne 2008: 209ff) and the paradoxes of the transition of environmentalism (Mulvihill 2009) have been a topic of discussion in recent years.

Dauvergne (2008: 215ff) discusses the failure of environmentalism to create sustainable patterns of global consumption, instead becoming part of the policies in the Global North that deflect the consequences and risks onto ecosystems and the powerless, mainly in the South. Environmentalism in all its forms is increasingly confronted with the consequences of economic growth—the "ecological shadows" of global consumption. Changes in consumption behaviour have been too slow to mitigate the negative impacts of global consumption on the environment—not only sustainability policies, but also environmentalism, need to be transformed. Mulvihill (2009: 504f) saw the essentialist, uncompromising, ideological and solidarity-oriented environmentalism of the first decades of the new movements fading away. This diagnosis is in line with the description of social movements as becoming integrated into the system they are criticising in ecological modernisation theory (Mol et al. 2009), or in broader research on social movements. The newly emerging forms of environmentalism are described as more anti-utopian, more informed by the growing environmental research, more diversified and decentralised, less insular, less Western and more global.

2.3.2 Sustainability in Pre-Industrial and Industrial Modern Society

In terms of natural resource use, modern society in Europe developed from an agricultural to an industrial mode of production. This transformation process is currently under way in the late industrialising countries in the Global South. The European economic modernisation process happened in two phases: a pre-industrial and an industrial phase (Bruckmeier 2019: 24ff).

1. *The first phase (1500–1750)—pre-industrial society*: The ecological discussion began with the forest crisis in early modernity and included the development of forest economics. In this phase of the development of modern society the dominant mode of production was agricultural, although it operated within a global economic system of modern capitalism. The agricultural and the industrial modes of production differ significantly in the social organisation of their resource use, in their societal metabolism, and in the social and class structures which determine the sustainability of the respective economic systems (Sieferle et al. 2006). Economic growth was very low in the pre-industrial phase; the sustainability problems were created by the agricultural energy systems. The dominant material and energy resource was wood, and the deforestation required to gain agricultural land led to a scarcity of wood and environmental damage in the

form of loss of forests and destruction of fertile agricultural land through erosion.

2. *The second phase (1750–today)—industrial society*: Industrialisation began in the eighteenth century in England and Scotland, and in most other European countries in the nineteenth century. The ecological discussion of the environmental consequences of industrialisation was part of the newly developing field of political economy. The idea of sustainability led to the critical discussion of the growth paradigm in classical political economy. Freeing the economy from the energetic limits of the agricultural metabolism unleashed economic growth and gave rise to two questions: how long could it continue, and what would be the environmental consequences? In this phase the sustainability discourse was broadening from a discourse of local and national consequences of environmental change caused by industrialisation to one about the consequences of global environmental change such as climate change, biodiversity loss, land use change and urbanisation. The building of the industrial society happened at several levels, from the local to the global: the result was a multi-scale society with an autonomous market economy that was not dependent on global political or cultural systems as the early historical world systems were. The industrial society developed, beginning in Europe, in one country after the other, not simultaneously. Industrialisation spread to the United States and Japan in the nineteenth century, then to Eastern European countries and Australia, and it is currently spreading to some larger countries in the Global South, including Brazil, Russia, India, China and South Africa (BRICS), but it has not yet spread globally. Applying the term society to this system creates several difficulties, which cannot be clarified through general definitions of society as modern. Analytically the term can be differentiated for local, national and global or world society. Predecessors of the modern world system were societies in the political form of empires with large populations of up to millions of individuals (see Chap. 3).

The industrial phase of a modern global economy and society includes the processes of globalisation and global environmental change. The development of modern society is described more concretely in the environmental and historical research of Hughes, Richards, Bess, McNeill. The process of economic globalisation began with the global colonial expansion in early modernity. Hughes (2005) sees this process, the building of the world market economy and supranational institutions, as threatening the conservation of nature, even under the guiding idea of sustainable development that is torn between promoting and limiting growth. Richards (2003), in his global environmental history of the early modern world, identifies the key social-ecological processes that dominated in the early modernisation phase. These include the appropriation and cultivation of land through continued shifting of frontiers of settlement; biological invasions where many plant and animal species, but also humans, left their original places and ecosystems, invading ecosystems in other countries and continents; the commercialisation of hunting, including fishery as the hunting of wild fish; and increased problems of scarce energy resources. All these transformation processes are connected to the shifting spatial frontiers of natural resource use during the colonisation of the globe. "The unending frontier", as in the title of Richard's book, is a mechanism of modernisation that comes to an end only in the twenty-first century, when all habitable space on earth is occupied by humans and integrated in the global economy.

The second phase of modernisation under the auspices of industrialisation is one where technologies change the natural world significantly. Technological innovations and their application have accelerated since the 1700s (Bess 2005) with the beginning of industrialisation. These innovations became powerful instruments of "artificialisation" of society and nature, polluting the environment and creating artificial ecosystems; the process began to accelerate after 1945 as "the gale-force of artifice". McNeill (2000) describes the forms of industrialisation in the twentieth century and the consequences for the environment and environmental changes as a huge uncontrolled experiment with the earth and its natural resources: the increase and intensification of natural resource use and growth unleashed the "engines of change"—population growth, growth of resource use, economic growth, urbanisation, technological change and innovation, and cultural and value shifts that supported such changes. The related exponential growth processes are a late phenomenon in human society: ecological and economic research has shown that they did not exist in pre-industrial society. Economic growth (in the sense of per capita growth of gross domestic product, GDP), according to the analyses of Maddison (1995), was very slow prior to 1800: before that, resource use and wealth in all parts of the world remained rather low and fairly equal (the methodological possibility of making such calculations of growth, without data and comparable economic institutions and modes of calculation, seems somewhat doubtful). Global exponential growth of the economy and the unequal access and appropriation of natural resources occurred only in the last two centuries, privileging the industrialising countries of the Global North where most of the natural resources of the world are used. The last phase of this acceleration was the economic globalisation and deregulation of markets in the second half of the twentieth century, leading to a global "ecological divide" between those with access to fresh water and other uncontaminated resources and those without such access. The message of McNeill's book coincides with that of the "limits to growth" debate: exponential growth cannot continue indefinitely.

A classical discussion about property-related conflicts between collectively owned common land and private land during modernisation is termed the "tragedy of the commons" (see Chap. 4), complemented by the "tragedy of enclosures", or the privatisation of common land (Martinez-Alier 2001). The discussion of the commons is connected with broader questions of the consequences of human use of natural resources: what are the factors affecting sustainable management of natural resources in the long run? Beyond property rights and institutions of resource management, other topics of discussion include generalised assumptions about human nature (for example, as egoistic or altruistic), type or quality of the resource, scarcity or abundance, cultural values and worldviews, institutional and organisational forms of the economy or society, modes of production and socio-metabolic regimes, the size of resource user groups, the conditions of access to the resource, the ease or difficulty of obtaining individual benefits or profit form the resource, the difficulties of extracting and processing the resource, access to the resource and others. These same questions were asked in the earlier discussion about protecting the environment and maintaining the state of natural resources on the earth. An example of a detailed analysis of the degradation of nature due to human use of natural resources is given by Marsh (1965 [1864]; Box 2.9).

Box 2.9 Anthropogenic Transformation of the Earth: Marsh (1864)

Against the dominant interpretation offered by geographers in the nineteenth century, Marsh identified humans as agents in the transformation of nature and argued for the repair of the damage inflicted on nature and the prevention of further damage through human domination of the earth. He saw deforestation as a main form and cause of physical degradation of the earth. In his introduction to the book "Man and Nature" he describes the destructiveness of man as follows (Marsh 2002: 170–177):

> … a certain measure of transformation of terrestrial surface, of suppression of natural, and stimulation of artificially modified productivity become necessary. This measure man has unfortunately exceeded. He has felled the forests whose network of fibrous roots bound the mould to the rocky skeleton of the earth; but had he allowed here and there a belt of woodland to reproduce itself by spontaneous propagation, most of the mischiefs which his reckless destruction of the natural protection of the soil has occasioned would have been averted. (171)

(*continued*)

Box 2.9 (continued)

Comparatively short as is the period through which the colonization of foreign lands by European emigrants extends, great, and, it is to be feared, sometimes irreparable, injury has been already done in the various processes by which man seeks to subjugate the virgin earth; and many provinces, first trodden by the homo sapiens Europae within the last two centuries, begin to show signs of the melancholy dilapidation which is now driving so many of the peasantry of Europe from their native hearths. (174)

… the effects of human action on the forms of the earth's surface could not always be distinguished from those resulting from geological causes, and there is also much uncertainty in respect the precise influence of the clearing and cultivating of the ground, and of the rural operations, upon climate. It is disputed whether either the mean or the extremes of temperature, the periods of the seasons, or the amount of distribution of the precipitation and of the evaporation, in any country whose annals are known, have undergone any change during the historical period. It is, indeed, impossible to doubt that many of the operations of the pioneer settler tend to produce great modifications in atmospheric humidity, temperature, and electricity; but we are at present unable to determine how far one set of effects is neutralized by another, or compensated by unknown agencies. (175f)

Source: Marsh 1965 [1864], 2002; Clark and Foster 2002

The questions about problems, tragedies or dilemmas in the human use of natural resources are in some form connected to the sustainability discourse. With sustainable development the problems of collective resource use were raised in the form of connected problems and crises that require global solutions: they cannot be solved locally or nationally but only through global action and change.

Environmental and social problems, dilemmas, conflicts and crises gave rise to the search for solutions through transformation of resource use regimes. Three connected and interacting factors had a salient role in the political-economic debates on resource use regimes and their change: *population growth, availability of natural resources for subsistence, and the economic growth and accumulation regimes*. For a long time the Malthusian approach, reducing the problem to one of population growth exceeding the capacity of agricultural production, was influential, although it has long been criticised and the weak points of Malthusian thinking were already laid bare in classical political economy. The critique of Malthus focused mainly on its deficiency as an economic theory and explanation of the modern economic system, whereas the ecological aspects of the theory were neglected. An anti-Malthusian perspective was developed in the late analysis of Boserup (1965), in a model for historical agricultural societies where the dynamics of population growth and technological innovations to increase agricultural yields together allowed for a better understanding of the interaction between growth and innovation to solve resource crises within agricultural economies. A renewal of the Malthusian discourse occurred in parallel to the new social and environmental movements in Western countries in the second half of the twentieth century. The debate was opened again with the neo-Malthusian approach of the "limits to growth" report from 1972, broadening the perspective on natural resource use crises from one of population growth and subsistence to a global model of the resource use dynamics in the late industrial global economy. As a result, differentiated views of natural and social limits to growth (Hirsch 1976), and more systematic analyses of the dynamics and interaction of economic and ecological factors in complex social-ecological systems (human, social and political ecology), were developed.

2.4 The Building of the Modern World System and Its Transformation

The historical epoch of modernity and the building of the modern global capitalist system since the fifteenth century (Wallerstein 1974, 1980) introduced complex, differentiated and ramifying processes of societal development, making it difficult to identify unifying trends in the diversity and particularity of the many social processes and changes. This complexity is experienced in the contrasting global change processes in recent decades: economic globalisation under conditions of cultural and political diversity; global environmental governance and the fragmented "architecture of international governance"; and sustainable development under conditions of global climate change and biodiversity loss. In the sustainability process, the complexity of problems has brought environmental and resource use conflicts and difficulties, relapses and failures of sustainable development (see Chap. 7).

World system analysis is one of the few interdisciplinary approaches that integrate knowledge from different social-scientific theories and disciplines, including ecological knowledge. Within this framework the historical differences between countries and cultures and the particularity of social and ecological processes are clear: they cannot be explained by one universal and general theory. In the long historical process of human societies, many culturally and socially different forms of development and change have coexisted, and societies have followed different development pathways, at different times, that cannot be fully integrated in theoretical analyses. In previously existing world systems, such as the territorially limited empires of the Chinese, Egyptians, Romans and Amerindians, cultures and societies remained isolated in limited parts of the world that had no or limited contacts with each other. The isolated cultures across the world first established continuous contact exchange of economic goods, people and information in early modernity through the process called in environmental history "Columbian exchange" (Crosby 1986): the global flow of natural resources, people, knowledge and capital initiated by Europe through the colonial conquest of the rest of the world. The development of a global economy and global exchange constituted modern society and led, in late modern society, to global environmental problems such as climate change and limits on the natural resource base of the earth which undermine further expansionism and economic growth.

The global societal system, which exists today as a multi-scale system, is functionally differentiated and spatially divided and fragmented. Within the discourse of world system theory, Wolf (1982) showed the limits of theoretically based reasoning, which can only explain part of the processes and the changes in the coupled systems of global society. The macroscopic perspective of world system theory is complemented by Wolf with a microscopic view of the local differences in societal development that can only be described in their temporal, spatial, social and ecological specificities, elements that fall outside "grand theory" and "big history". The patchwork of modernisation includes local differences, historical asynchrony and limited integration of locally and culturally specific economies and modes of production in the global dimension of the global economic system of modern capitalism. These "imperfections" of economic rationalisation and systemic integration of economic processes continue also in the global economic and ecological change processes in late modernity.

The epoch of modernity can be seen to involve the global spread of the specific capitalist mode of production through the development of a global system, first as an economic system, and later developing as a global society encompassing all politically and culturally different systems. In the course of modernisation the economic relations between countries, cultures and societies that were isolated in earlier phases of human history as historical world systems were changed. Now the countries came together for the first time in a global economy, in violent and non-violent processes of conquest, colonisation, imperial wars, economic trade, and cultural and social exchange and mobility. During the eighteenth century the European, Chinese and Indian economies, the largest parts of the global economy, were at the same level of development in terms of gross domestic production: in 1700, each of them accounted for around 23% of the global economy. During the nineteenth century, with the industrialisation in European countries, this changed dramatically. In 1900, the share of global production was 2% for India, 7% for China, 60% for Europe and 20% for the United States (Marks 2007: 123). The economic, social and environmental changes that occurred during modernity are described by environmental history and interdisciplinary environmental research (in environmental history: Crosby 1986; MacKenzie 1990; McNeill 2000; Radkau 2002; Davis 2001; Marks 2007; Ponting 2007; Radkau 2012). More difficult is the theoretical analysis and explanation of the multifaceted change processes.

The social, cultural and economic reality of the modern global economic system built in the past five centuries, starting with the global colonisation by European countries, is described in empirical studies in cultural anthropology and sociology as a system with contradictory forms of unity and diversity. The world system is incoherent, with asynchronous development, continuous social, economic, cultural and political divides, and continuous conflicts over access to, use and sharing of natural resources.

The world ecology that emerged amid the processes of global social and environmental change (Moore 2000, 2015; Moore and Patel 2017) is as contradictory and incoherent as the modern global economic system that created it. Moore and Patel describe a form of ecological change under conditions of unequal development and appropriation of natural resources, externalisation of environmental effects, economic growth and global divides deepening with the digital economy. Global sustainability as social-ecological transformation of the modern world system requires the solution of problems that have been neglected in the early sustainability process: unequal exchange and the divide between the Global South and the Global North.

2.5 Discussion and Conclusions

The current sustainability discourse is a response to the unintended and unwanted consequences of the development of modern society with the development of a global economic system on the basis of European colonialism and the uneven development which has accelerated since industrialisation. The history of modern society is not a central theme in the discourse that followed the Brundtland Report, although it is necessary to understand and change the destructive course of exponential growth and to solve the accumulated environmental problems created in the era of modernity. Environmental change reached global dimensions as industrial society developed. The decoupling of natural resource use and environmental pollution from local and national spaces, a consequence of the building of the modern global economic system, made the "globalisation of environmental destruction" possible. The connection between cultural, political, economic and ecological processes created the global nexus of human activities: the heliocentric scientific worldview and the dominant Christian worldview, and the development of modern natural sciences (starting with physics). The creation of new technologies supported an imperial culture of conquest and colonisation, enabled global trade and exchange, made industrialisation possible and kept the world divided between the wealthy nations of the Global North and the poor countries of the Global South. Eurocentrism is not only a cultural construction and worldview; it became a global practice in the social spheres of science, politics and economy.

The historical development of the sustainability discourse in modern society is reflected in various economic analyses and discussions. Economics influenced the sustainability discussion in analyses of the "tragedy of the commons" and of forms of property rights for the use and management of natural resources. In the economic mainstream, under the influence of neoclassical economics, the dominant model is one of market processes based on private property relations. The sustainability discourse has opened up the perspective to allow for public economies and forms of collective economic action that are dealt with in critical economics, institutional economics and ecological economics (Ostrom 2007; Jacobs 1994). Interdisciplinary economic research incorporates other social-scientific and ecological knowledge in analyses of complex, multi-scale forms of collective and institutional action (Ostrom 2009). This interdisciplinary economic research produces relevant knowledge for the sustainability discourse and process: its focus is not on individual consumers, actors and their economic motives, but on institutionalised structures and processes that influence economic and environmental action.

Putting environmental and economic thinking in science and practice in a historical perspective has two purposes. The first is to identify the social and cultural limits of knowledge and its particularity. Assumptions about humans and their behaviour as universally valid or valid throughout history are dubious, but they can be found in some forms of psychological, anthropological and evolutionary reasoning and in economic research, for example the assumption of individual rationality. The second purpose of historical analysis is to generate a critique of the development of the modern economy and society. It is not a continuous process involving linear development and improvements in terms of knowledge and the creation of welfare. The possibilities of changing human behaviour accelerated with the environmental discourse. In contrast to the popular thinking that history is a source of learning for current problems, it becomes more important to think of the future in terms of possible changes in resource use behaviour and system changes that are relevant for the transformation to sustainability. The modern market economy, and the commercialisation and capitalisation of everything that occurred with economic globalisation, are not without alternatives, even if they are not yet visible.

Summary of Chap. 2: History of Sustainability Thinking

1. The sustainability discourse arose in early modern European society in the seventeenth century, with the paradigmatic examples of analyses of the forest crises and their solution through reforestation (Evelyn in England, Carlowitz in Germany). At that time the idea of maintaining the natural resource base for the continuous development of society and economy was based on principles of economic production and consumption.
2. Beginning in the second half of the eighteenth century, with industrialisation and the rise of classical political economy, a critical discourse emerged about economic growth and its consequences for the environment (Malthus, Ricardo, Mill, Marx), which became a core theme in the current debates about limits to growth and sustainable development.
3. The reasoning behind the critique of modern industrial society varied in the philosophical, religious, political, scientific and cultural discourses and social movements accompanying the development of modern society. It took until the end of the twentieth century before the fragmented knowledge was synthesised in the sustainability discourse, in an ecological critique of exponential growth.

Appendices: Further Information and Material

1. Questions and Individual Exercises

Learning Exercise 1: "Industrial Society and Ways Beyond"

Organise a discussion about the ways out of the industrial society in the form of a discussion between two groups: one defending and one rejecting the possibility of phasing out the industrial society.

The group of defenders should prepare their arguments based on the "transformation hypothesis" (using arguments from the debate about post-industrial and post-capitalist societies). The socio-economic transformation towards a post-industrial society has been under way since the second half of the twentieth century in advanced industrial societies and is visible in a series of changes: the continuous reduction of the second or industrial sector in the national economies, the growth of a third or service sector, the economic change of accumulation regimes (Post-fordism) and technological change through the internet-based knowledge economy and society.

The opposing group should prepare t arguments for the permanence of industrial society with other interpretations of the socio-economic changes observed: the technological innovation of the knowledge society as a new form of industrialisation (hypothesis of the "fourth industrial revolution": Klaus Schwab); the late industrialisation of the BRICS countries in the Global South that export large parts of their industrial production to the early industrialising countries in the North—industrial production is not being phased out but is being shifting geographically and differentiated into "high tech industries" in the Global North and old forms of polluting industrialisation in the BRICS countries.

In the discussion (30–45 minutes) each group also needs to deal with the responses and arguments of the other group, not just defend their own positions. One or two participants will guide the discussion and document the main points (in a flipchart); another participant will write a protocol. In a follow-up discussion all participants will discuss their learning process (this should also be documented).

Introductory statement for the discussion:

The discussion of alternatives to industrialisation during the building of industrial economies in the Western world in the nineteenth and twentieth centuries did not advance. Ideas about the agricultural welfare state and discussions of the peasant economy and simple commodity production (Chayanov) as another mode of production in the modern capitalist economy did not turn out to be alternatives. The agricultural economies became part of asynchronous process of industrialisation, where older forms of economic production were integrated in the industrial division of labour (shown in the anthropological analysis of modern capitalism by Eric Wolf 1982). Only with the current environmental and sustainability discourses did the trend towards post-industrial societies accelerate in the early industrialising countries of the West. This has occurred in contradictory forms of "super-industrialisation" (high-tech and knowledge economy) and the ideas of post-growth societies and post-capitalist societies.

Additional reading: two classical books on the post-industrial society—Bell (1973), Touraine (1971).

Learning Exercise 2: "Agenda 2030" for Sustainable Development

Suggested tasks and questions for further reflection and discussion (individually or in a working group): Download "Agenda 2030" from the internet and discuss the development goals and the process to achieve them; do a SWOT analysis (descriptions for the method available on the internet) of the strengths and weaknesses of the Agenda as a political document: (How) can the achievement of goals be measured (indicators, criteria)? Which goals can be realistically achieved by 2030? What would be required to make the process effective and successful (in terms of: policy instruments, financing, institutional change, resource sharing, coordination, monitoring, assessment, cooperation of governments, empowerment of actors, responsibility of actors and institutions, removing hindrances, institutional change)? Who are the key actors in the process that need to take the initiative and responsibility?

The UN's "Agenda 2030", "Transforming our world", is currently the guiding document for global sustainability policy. It does not have the status of a legally binding document; it is a shared global vision adopted by the governments at a global summit in 2015 as a guideline for all countries. At the core of the agenda are seventeen Sustainable Development Goals as decided at the "Rio+20" conference in 2012. A summary:

1. *The global Sustainable Development Goals*: the goals are specified for the three dimensions of social, economic and environmental sustainability.
2. *The guiding principles and messages*: "5Ps"—People—human dignity; Planet—protecting; Prosperity—for all; Peace—securing; Partnership—global cooperation and partnership.
3. *How to achieve the goals*: no specific regulations are formulated, only goals and principles; all countries/governments are asked to report on their efforts and progress in achieving the goals; the main actors cooperating in the process should include individuals, civil society, local communities, private economy, governments and non-governmental organisations.

If you are interested in a deeper discussion of the problems, you should work with the Global Sustainable Development Report (UN 2019): try to find out more about the transformative changes of the global economic system to achieve sustainability—markets and their limitations regarding sustainable development, possibilities of global redistribution and sharing of resources, of new forms of sustainable growth and of a sharing economy.

2. Further Reading Suggested: Deepening, Thematically Specialised

Books—environmental policies, governance and natural resource management:

Caradonna, J. L. ed. (2018). *Routledge Handbook of the History of Sustainability*. London: Routledge.

Isenberg, A. C., ed. (2014). *The Oxford Handbook of Environmental History*. New York et al.: Oxford University Press.

Worster, D. (1994 [1970]). *Nature's Economy: A History of Ecological Ideas*. Cambridge: Cambridge University Press.

Scientific journals (international, peer reviewed) for the themes of this chapter, showing the advancing historical debate and research: "Environmental History", "Environment and History" (interdisciplinary), "*Global Environmental Change*", "The Journal of Economic History", "The Journal of Peasant Studies", "Sustainability Science"

References

Balée, W. (Ed.). (1998). *Advances in Historical Ecology*. New York: Columbia University Press.

Balée, W. (2006). The Research Program of Historical Ecology. *Annual Review of Anthropology, 35*, 75–98.

Beck, U. (1996). World Risk Society or Cosmopolitan Society? Ecological Questions in a Framework of Manufactured Uncertainties. *Theory, Culture & Society, 13*, 1–32.

Beck, U. (2009). Critical Theory of World Risk Society: A Cosmopolitan Vision. *Constellations, 16*(1), 3–22.

Bell, D. (1973). *The Coming of Post-Industrial Society*. New York: Basic Books.

Bennett, J. W. (1976). *The Ecological Transition: Cultural Anthropology and Human Adaptation*. New York: Pergamon Press.

Bercé, Y.-M. (1987). *Revolt and Revolution in Early Modern Europe: An Essay on the History of Political Violence*. Manchester: Manchester University Press.

Bertram, C. (2018). Jean Jacques Rousseau. In E. N. Zalta (Ed.), *The Stanford Encyclopaedia of Philosophy* (2018th ed.) Retrieved from https://plato.stanford.edu/archives/fall2018/entries/rousseau/.

Bess, M. (2005). Artificialization and Its Discontents. *Environmental History, 10*(1), 31–33.

Blickle, P. (1981). *The Revolution of 1525: The German Peasants War from a New Perspective*. Baltimore: Johns Hopkins University Press.

Boserup, E. (1965). *The Conditions of Agricultural Growth*. Boston, MA: Allen and Unwyn.

Braudel, F. (1949). *La Méditérranée et le monde méditérranean à l'époque de Philippe II*. Paris: Colin.

Bruckmeier, K. (2019). *Global Environmental Governance: Social-Ecological Perspectives*. Cham, Switzerland: Springer Nature and Palgrave Macmillan.

Burns, T. J., & Schaefer Caniglia, B. (2016). *Environmental Sociology: The Ecology of Late Modernity*. Norman, OK: Mercury Academic.

Caradonna, J. L. (Ed.). (2018). *Routledge Handbook of the History of Sustainability*. London: Routledge.

Carlowitz, H. C. v. (1713). *Sylvicultura Oeconomica*. Leipzig: Braun.

Catton, W. J. R., & Dunlap, R. (1978). Theories, Paradigms and the Primacy of the HEP-NEP Distinction. *The American Sociologist, 13*, 256–259.

Chakravarty, S., Ghosh, S. K., Suresh, C. P., Dey, A. N., & Shukla, G. (2012). Deforestation: Causes, Effects and Control Strategies. In: Akais, O. C. (Ed.), Global Perspectives on Sustainable Forest Management: 3-28. Rijeka: InTech.

Clark, B., & Foster, J. B. (2002). George Perkins Marsh and the Transformation of Earth. *Organization & Environment, 15*(2), 164–169.

Clark, B., & York, R. (2007). The Restoration of Nature and Biogeography: An Introduction to Alfred Russell Wallace's "Epping Forest" in 1873. *Organization & Environment, 20*, 213–234.

Cooper, L. (1999). *Rousseau, Nature and the Problem of Good Life*. Pennsylvania: Pennsylvania State University Press.

Costanza, R., Graumlich, L., Steffen, W., Crumley, C., Dearing, J., Hibbard, K., et al. (2007). Sustainability and Collapse: What Can We Learn from Integrating the History of Humans and the Rest of Nature? *Ambio, 36*(7), 522–526.

Crosby, J. (1986). *Ecological Imperialism: The Biological Expansion of Europe, 900–1900*. Cambridge: Cambridge University Press.

Daly, H. E. (1990). Toward Some Operational Principles for Sustainable Development. *Ecological Economics, 2*(1), 1–6.

Dauvergne, P. (2008). *The Shadows of Consumption: Consequences for the Global Environment*. Cambridge, MA and London, UK: The MIT Press.

Davis, M. (2001). *Late Victorian Holocausts*. London: Verso.

Debeir, J.-C., Deléage, J.-P., & Hémery, D. (1991 [1986]). *In the Servitude of Power*. London and Atlantic Highlands, NJ: Zed Books.

Derham, W. (2009 [1713]). *Physico-Theology, or, Demonstration of the Being and Attributes of God from His Works of Creation*. Newcastle upon Tyne, UK: Cambridge Scholars Publishing (Reprint).

DesJardins, J. R. (2013). *Environmental Ethics: An Introduction to Environmental Philosophy*. (5. Edition). Boston et al.: Wadsworth: Cengage Learning.

DesJardins, J. R. (2006). *Environmental Ethics. An Introduction to Environmental Philosophy*. Belmont, CA, et al.: Thomson Wadsworth.

Diamond, J. M. (1997). *Guns, Germs, and Steel: The Fate of Human Societies*. London: Random House.

Diamond, J. M. (2006 [2005]). *Collapse: How Societies Choose to Fail or Succeed*. New York et al.: Penguin Books.

Diergarten, Y., & Krieger, T. (2015). Large-Scale Land Acquisitions, Commitment Problems and International Law. *Law and Development Review, 8*, 217–233.

Evelyn, J. (1972 [1664]). *Sylva, or A Discourse of Forest-Trees and the Propagation of Timber in His Majesty's Dominions*. Scolar Press (Reprint of the Original).

Evelyn, J. (2018 [1661]). *Fumifugium; Or, the Inconvenience of the Aer and Smoake of London*. Franklin Classics.

Fischer-Kowalski, M. (1997). Society's Metabolism: On the Childhood and Adolescence of a Rising Conceptual Star. In M. Redclift & G. Woodgate (Eds.), *The International Handbook of Environmental Sociology* (pp. 119–136). Cheltenham, UK and Northampton, MA: Edward Elgar.

Fischer-Kowalski, M. (2009). Den Zusammenbruch von Zivilisationen konzeptualisieren: Plädoyer für eine soczalökologische Gesellschaftstheorie. *GAIA, 18*(1), 13–14.

Fischer-Kowalski, M., & Haberl, H. (1998). Sustainable Development: Socio-Economic Metabolism and Colonization of Nature. *International Social Science Journal, 158*, 573–587.

Freedman, B. (Ed.). (2014). *Global Environmental Change*. Switzerland: Springer Nature.

Glotfelty, C., & Fromm, H. (Eds.). (1996). *The Ecocriticism Reader: Landmarks in Literary Ecology*. Athens, GA and London: University of Georgia.

Grober, U. (2012). *Sustainability: A Cultural History*. Totnes, Devon, UK: Green Books.

Guha, R., & Martinez-Alier, J. (1998). *Varieties of Environmentalism: Essays North and South*. Delhi et al.: Oxford University Press.

Gura, P. F. (2007). *American Transcendentalism: A History*. New York: Hill and Wang.

Hagedorn, G., Loew, T., Seneviratne, D. I., Lucht, W., Beck, M.-L., Hesse, J., et al. (2019). The Concerns of the Young Protesters Are Justified. *GAIA, 28*(2), 79–87.

Hirsch, F. (1976). *The Social Limits to Growth*. Cambridge, MA: Harvard University Press.

Hobley, M. (2007). *Forests – The Poor Man's Overcoat: Foresters as Agents of Change?* 6th Jack Westoby Lecture. The Fenner School of Environment and Society, The Australian National University. Retrieved from https://fennerschool-associated.anu.au/fenner-publications/occassional/westoby_lectures/6th_westoby_lecture07.pdf.

Hornborg, A., McNeill, J. R., & Martinez-Alier, J. (Eds.). (2007). *Rethinking Environmental History – World-System History and Global Environmental Change*. Lanham, MD et al.: Altamira Press.

Howarth, R. B. (2007). Towards an Operational Sustainability Criterion. *Ecological Economics, 63*, 656–663.

Hughes, J. D. (2001). *An Environmental History of the World: Humankind's Changing Role in the Community of Life*. Oxford: Routledge.

Hughes, J. D. (2005). Global Environmental History: The Long View. *Globalizations, 2*(3), 293–308.

Inglehart, R. (1990). *Culture Shift in Advanced Industrial Society*. Princeton, NJ: Princeton University Press.

Inglehart, R. (1977). *The Silent Revolution*. Princeton, NJ: Princeton University Press.

Israel, J. (2014). *Revolutionary Ideas: An Intellectual History of the French Revolution from the Rights of Man to Robespierre*. Oxford and Princeton: Princeton University Press.

Jacobs, M. (1994). The Limits to Neoclassicism: Towards an Institutional Environmental Economics. In M. Redclift & T. Benton (Eds.), *Social Theory and the Global Environment* (pp. 67–91). London and New York: Routledge.

Jo, T.-H., Chester, L., & D'Ippoliti, C. (Eds.). (2019 [2018]). *The Routledge Handbook of Heterodox Economics*. Abingdon, UK and New York, NY: Routledge.

Kapp, K. W. (1976). *Staatliche Förderung "umweltfreundlicher" Technologien*. Göttingen: Schwartz.

Kennedy, G. (2008). *Diggers, Levellers and Agrarian Capitalism: Radical Political Thought in Seventeenth Century England*. USA: Lexington Books.

King, F. H. (1911). *Farmers of Forty Centuries or Permanent Agriculture in China, Korea and Japan*. Emmaus, PA: Organic Gardening Press.

Krech, S., III, McNeill, J. R., & Merchant, C. (Eds.). (2004). *Encyclopedia of World Environmental History*. 3 vols. New York: Routledge.

Lee, K., Holland, A., & McNeill, D. (Eds.). (2000). *Global Sustainable Development in the Twenty-First Century*. Edinburgh: Edinburgh University Press.

Liebig, J. (1840). *Die organische Chemie und ihre Anwendung auf Physiologie und Pathologie*. Braunschweig: Vieweg.

Liebig, J. (1856). *Über Theorie und Praxis der Landwirtschaft*. Braunschweig: Vieweg.

Linné, C. v. (2010 [1749]). *Oeconomia naturae*. Whitefish, Montana: Kessinger Publishing (Kessinger Legacy Reprints).

MacKenzie, J. M. (1990). *Imperialism and the Natural World*. Manchester: Manchester University Press.

Maddison, A. (1995). *Monitoring the World Economy 1820–1992*. Paris: OECD.

Marks, R. B. (2007). *The Origins of the Modern World: Fate and Fortune in the Rise of the West*. Lanham et al.: Rowman & Littlefield.

Marsh, G. P. (1965 [1864]). *Man and Nature: Or, Physical Geography as Modified by Human Action* (ed. D. Lowenthal). Cambridge, MA: Belknap Press of Harvard University Press.

Marsh, G. P. (2002). Man and Nature (Introduction to: Marsh 1864). *Organization & Environment, 15*(2), 170–177.

Martinez-Alier, J. (2001). Ecological Conflicts and Valuation: Mangroves versus Shrimps in the Late 1990s. *Environment and Planning C: Government and Policy, 19*(5), 713–728.

Martinez-Alier, J. (2002). *The Environmentalism of the Poor: A Study of Ecological Conflicts and Valuation*. Cheltenham, UK and Northampton, MA: Edward Elgar.

McNeill, J. R. (2000). *Something New Under the Sun: An Environmental History of the Twentieth Century World*. New York: W.W. Norton and Company. San Francisco: Harper & Row.

Merchant, C. (1980). *The Death of Nature: Women, Ecology, and the Scientific Revolution*. San Francisco: Harper & Row.

Merchant, C. (2005). *Radical Ecology: The Search for a Livable World*. New York: Routledge.

Mol, A., Sonnenfeld, D., & Spaargaren, G. (Eds.). (2009). *The Ecological Modernization Reader: Environmental Reform in Theory and Practice*. London and New York: Routledge.

Moore, J. W. (2000). Environmental Crisis and the Metabolic Rift in World Historical Perspective. *Organization and Environment, 12*, 123–156.

Moore, J. W. (2015). *Capitalism in the Web of Life. Ecology and the Accumulation of Capital*. London: Verso Books.

Moore, J. W., & Patel, R. (2017). A History of the World in Seven Cheap Things. In *A Guide to Capitalism, Nature, and the Future of the Planet*. Berkeley: University of California Press.

Moscovici, S. (1968). *Essai sur l'histoire humaine de la nature*. Paris: Flammarion.

Mulvihill, P. R. (2009). Endless Paradox: Environmentalism in Transition. *Futures, 41*, 502–506.

Mumford, L. (1934). Technics and Civilization. New York: Harcourt, Brace & Company.

Needham, J., & Bray, F. (1984). *Science and Civilization in China, Vol 6, Part 2: Biology and Biological Technology: Agriculture*. Cambridge: Cambridge University Press.
Ostrom, E. (2007). Challenges and Growth: The Development of the Interdisciplinary Field of Institutional Analysis. *Journal of Institutional Economics, 3*(3), 239–264.
Ostrom, E. (2009). A General Framework for Analyzing Sustainability of Social-Ecological Systems. *Science, 325*, 419–422.
Polanyi, K. (1944). *The Great Transformation*. New York and Toronto: Farrar & Rinehart.
Polanyi, K., Arensberg, K., & Pearson, H. (1957). *Trade and Market in the Early Empires*. New York: The Free Press.
Ponting, C. (2007). *Anew Green History of the World: The Environment and the Collapse of Great Civilizations*. London: Penguin.
Radkau, J. (2002 [2000]). *Natur und Macht. Eine Weltgeschichte der Umwelt*. München: Beck.
Radkau, J. (2012). *Die Ära der Ökologie. Eine Weltgeschichte*. München: Beck.
Radkau, J. (2018 [2007]). *Holz. Wie ein Naturstoff Geschichte schreibt*. München: Oekom.
Radkau, J., & Schäfer, I. (1987). *Holz. Ein Naturstoff in der Technikgeschichte*. Reinbek: Rowohlt.
Rice, J. (2007). Ecological Unequal Exchange: Consumption, Equity, and Unsustainable Structural Relationships Within the Global Economy. *International Journal of Comparative Sociology, 48*(1), 43–72.
Richards, J. (2003). *The Unending Frontier. An Environmental History of the Early Modern World*. Berkeley: University of California Press.
Rudel, T. K. (2005). *Tropical Forests: Regional Paths of Destruction and Regeneration in the Late Twentieth Century*. New York: Columbia University Press.
Runyan, C. W., & D'Odorico, P. (2016). *Global Deforestation*. New York: Cambridge University Press.
Sahlins, M. (1972). *Stone Age Economics*. New York: De Gruyter.
Sieferle, R. P. (2008). Zusammenbruch von Zivilisationen: Eine konzeptionelle Analyse am Beispiel des Imperium Romanum. *GAIA, 17*(2), 213–223.
Sieferle, R. P., Krausmann, H., Schandl, V., & Winiwarter, V. (2006). *Das Ende der Fläche: Zum gesellschaftlichen Stoffwechsel der Industrialisierung*. Wien: Böhlau.
Smith, R. (2005). Capitalism and Collapse: Contradictions of Jared Diamond's Market Meliorist Strategy to save the Humans. *Ecological Economics, 55*, 294–306.
Solow, R. M. (1993). Sustainability: An Economist's Perspective. In R. Dorfman & N. S. Dorfman (Eds.), *Economics of the Environment*. New York: Norton.
Stavrianos, L. S. (1997). *Lifelines from Our Past: A New World History*. Armonk, NY and London, UK: Sharpe.
Tainter, J. A. (1988). *The Collapse of Complex Societies*. Cambridge, UK: Cambridge University Press.
Tainter, J. A. (2006). Archeology of Overshoot and Collapse. *Annual Review of Anthropology, 35*, 59–74.
Touraine, A. (1971). *The Post-Industrial Society*. New York: Random House.
Trepl, L. (1994 [1987]). *Geschichte der Ökologie vom 17. Jahrhundert bis zur Gegenwart*. Weinheim: Beltz Athenäum.
Wakefield, A. (2009). *The Disordered Police State: German Cameralism as Science and Practice*. Chicago: Chicago University Press.
Wallace, A. R. (1878). Epping Forest. *The Fortnightly Review, 24*, 628–645.

Wallerstein, I. (1974). *The Modern World System: Capitalist Agriculture and the Origins of the European World-Economy in the Sixteenth Century*. New York et al.: Academic Press.
Wallerstein, I. (1980). *The Modern World System II. Mercantilism and the Consolidation of the European World-Economy, 1600–1750*. New York et al.: Academic Press.
Weber, M. (2019 [1921/1922]). *Economy and Society*. Cambridge, MA: Harvard University Press.
Williams, M. (2006). *Deforesting the Earth: From Prehistory to Global Crisis. An Abridgement*. Chicago: University of Chicago Press.
Wittfogel, K. A. (1957). *Oriental Despotism. A Comparative Study of Total Power*. New Haven: Yale University Press.
Wolf, E. (1982). *Europe and the People Without History*. Berkeley, CA: University of California Press.
Zapf, H. (Ed.). (2008). *Kulturökologie und Literatur: Beiträge zu einem transdisziplinären Paradigma der Literaturwissenschaft*. Heidelberg: Winter.
Zinn, K. G. (1980). *Die Selbstzerstörung der Wachstumsgesellschaft: Politisches Handeln im ökonomischen System*. Reinbek: Rowohlt.

The Knowledge Context of the Sustainability Discourse

3

The chapter describes problems and practices of inter- and transdisciplinary knowledge integration in sustainability research and in the governance process. Knowledge problems in the sustainability process include: (a) analysis of complex systems and processes for which interdisciplinary knowledge generation and integration is required; (b) epistemological and methodological problems of knowledge integration, including concepts and theory construction; (c) problems in the transfer and application of knowledge in the sustainability process; and (d) controversial themes in economic research about natural resource use and sustainable development.

The web of concepts that structure the interdisciplinary sustainability discourse can be seen in Fig. 3.1.

Sustainable development as a process of social, economic and ecological change cannot be managed without knowledge from many sources and actors. The problems that must be solved in sustainable development can be understood as unforeseen consequences of changes in the relations between society and nature. The social-scientific terminology for these problems includes unanticipated and unintended consequences of action, side effects, indirect effects, long-term effects, externalities, risks and dangers, dysfunctionality of systems or disturbance, and maladaptive change. These concepts refer to effects in the near or distant future—to potential catastrophes or the collapse of societal and economic systems that could be avoided through sustainability transformation. The solutions, in the form of resolving conflicts, managing crises and preventing collapse, require capacities of governance and regulation that currently do not exist and thus need to be created in the sustainability process. The forms of collective action and agency needed for the transformation differ from the social action that currently occurs in the routines of social life, in policy and the economy, in production and consumption, in short

K. Bruckmeier, *Economics and Sustainability*,
https://doi.org/10.1007/978-3-030-56627-2_3

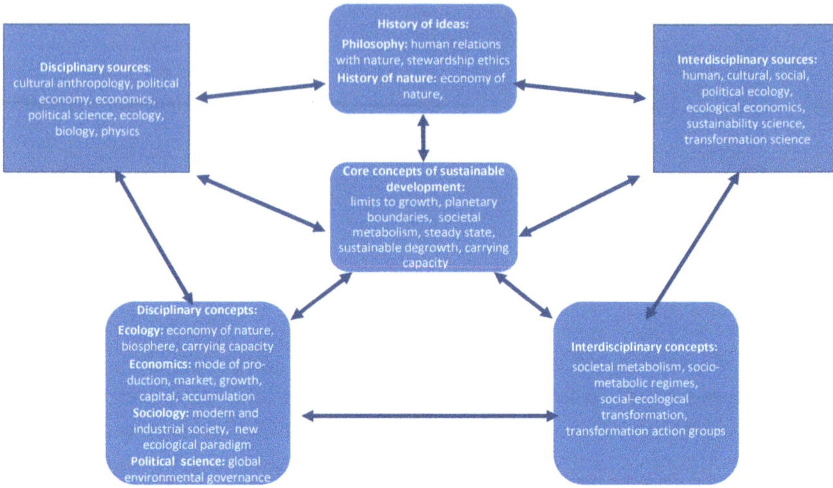

Fig. 3.1 Conceptual web of the sustainability discourse

temporal perspectives and in specialised forms of planning. The potential risks and consequences that must be dealt with in the sustainability process extend beyond the typical time horizons of foresight and planning. Climate change and biodiversity reduction are paradigmatic forms of unintended effects of human action occurring with time lags, unforeseen and unperceived by humans, and documented only through research in recent decades. This knowledge is still insufficiently dealt with in the public policies and in the social practices of resource use. Much of this knowledge was not yet available when the global sustainability discourse began in the 1980s: although the processes of environmental change have long been recognised by science, they were widely ignored by scientists, the public and politicians.

3.1 Deficient Knowledge Processes in Sustainable Development and Their Improvement

The discourse and the practice of sustainability governance are confronted with interconnected environmental and social problems that built up during the twentieth century, from local or regional to global problems. The knowledge for sustainable development includes that about new themes of global social and environmental change: economic globalisation, uneven development, urbanisation, climate change, loss of biodiversity or species reduction, land use change and deforestation. Social-scientific knowledge relevant for sustainability governance includes the policy-relevant knowledge about knowledge transfer and policy instruments in economic research. Furthermore, economic and other social research on the sustainability transition, the consequences of economic growth and the possibility of degrowth is required (see Chaps. 5 and 6). The knowledge from multi- and interdisciplinary research on transition or transformation (summarised by Nesshöver et al. 2017; Köhler et al. 2019) and reconnecting society with nature (summarised by Ives et al. 2018; Zylstra et al. 2019) still needs to be applied in the sustainability process.

The participants in the sustainability discourse and process, most of whom are not practitioners of inter-and transdisciplinary knowledge production and application, are confronted with an inextricable mass of scientific and other knowledge from many disciplines and sources that requires criteria, heuristics and methodological tools to identify the relevant knowledge to be synthesised and integrated in the processes of sustainability governance. The processes of creating, maintaining and improving the knowledge for policy and governance, of formulating and assessing strategies and programmes of action in the long transition to global sustainability, require new capacities and knowledge practices in science and policy—that is, continued collective learning of all actors in the sustainability process. Creating and using scientific knowledge for sustainable development can no longer be done via the traditional division of labour, according to which scientists produce valid knowledge, and decision-makers and practitioners apply the knowledge for solving social and environmental problems. The relations between science and practice changed significantly with the environmental and sustainability movements and policies, which is insufficiently reflected in science and in the policy process (for a detailed discussion see Chap. 8). The practices of searching for, integrating and applying knowledge happen to a large degree without epistemological or methodological guidance. A precondition of sustainable development is the building of an interdisciplinary knowledge base through knowledge integration and synthesis. For the international climate policy this is done systematically: the knowledge is reviewed and continually updated and communicated to the policy process, the public and the civil society, a process that is managed by the Intergovernmental Panel on Climate Change (IPCC). This knowledge is important for sustainable development, but much more knowledge is required that is widely scattered, fragmented, difficult to integrate and contested and uncertain. Furthermore, the dominance of powerful political institutions and global players with vested interests creates additional problems in knowledge transfer and application.

The established routines in public policy processes, knowledge transfer, and monitoring and review of policy programmes are not sufficient for building a knowledge stock for the complex and long processes of sustainability transition. Too much knowledge is excluded, and too many questions remain unanswered in the transformation process: how to deal with knowledge selection in complex processes, in situations of competing knowledge and continuous knowledge controversies? Who manages and controls the knowledge flows between science and governance, and with what kinds of legitimation, expertise, experience, value commitments, loyalties, interests and vested interests? The many research institutions, think tanks, and advisory and lobby organisations mediating and brokering knowledge in the sustainability process need to be assessed according to their legitimacy, responsibility and practices of action. The "policy business" of knowledge brokering, of buying and selling advice and information services, has become part of the knowledge-based digital economy, and the sustainability process is deformed by this commercialised research and knowledge production. Another complication of the knowledge process is that far more information is accessed and exchanged on the internet than is communicated through scientific journals, publications and communication media, meaning it is often unclear and has not undergone review or quality control processes.

3.1.1 Inter- and Transdisciplinary Knowledge Integration in the Sustainability Process

Inter- and transdisciplinary knowledge generation and integration is a response to the decoupling of scientific knowledge production in disciplinary and specialised research from social reality and the real problems in modern society. The knowledge that must be combined and integrated for sustainability spans a series of social- and natural-scientific disciplines, specialised research areas, interdisciplinary fields of research and non-scientific, managerial and practical knowledge. The process of knowledge integration is influenced by contrasting and competing views and interpretations of sustainability. The dominant view currently differentiates sustainability into three forms: social, economic and ecological sustainability. Other views include weak and strong sustainability in the early phase of the discourse, maintenance of the natural capital, integration of the economy and ecology, sustainable livelihoods, system transformation and social-ecological transformation, behavioural or value changes, redistribution and sharing of resources, and the process of sustainable development as one of connected changes, with different views of their relative importance: ecological, technological, economic, political, social or cultural (further discussion: Bruckmeier 2016: 125ff). These views of sustainability are embedded in general views about nature and society (Box 3.1).

In the sustainability discourse the "relational themes" of dichotomies, distinctions, contrasts, the disintegration of nature and society, the coupling and decoupling of ecosystems and social systems that accompany the history of modern so-

Box 3.1 Nature–Society Relations in Modern Society

The sustainability discourse is influenced by the broader discourse of nature–society relations in modern society, which is found across in different fields of research (summarised in Bruckmeier 2016: 15ff, 69ff).

Philosophical discourses about the relations between nature and humans, society or culture were transformed with the development of modern society into questions of belief and worldviews, natural-scientific research and positive knowledge. The philosophical and the natural-scientific discourse came together again in the new ecological discourse initiated by the new environmental movements. Philosophical themes relevant for ecological science (Haila and Levins 1992; Haila and Heininen 1995; Colyvan et al. 2009) include that of the integration or separation of humans and nature, or of dualisms and dichotomies in Western thinking about nature and humans. Dichotomies of nature and man or society, of structure and agency, of masculinity and femininity, of reason and emotion, are critically reflected in the feminist, the postmodernist and the ecological discourses, where the dichotomies are often seen as value- and power-based social constructions, not as natural, biologically or ecologically given (Warren and Erkal 1997; Cudworth 2005). This can be seen as a renewal of the older discourse about humankind's alienation from nature in modern society. In a classical formulation of the ecofeminist framework (Gaard and Gruen 1993: 236ff, 248ff), the separation between nature and culture is seen as preventing global justice and planetary health. Four primary, mutually reinforcing forces of oppression are responsible for the separation of culture from nature:

1. the modern scientific revolution, resulting in the mechanical model of nature (Carolyn Merchant);
2. the rise of patriarchal religions that brought gender hierarchies into society;
3. the rise of ideological conceptions of nature and culture based on dualisms such as self and other, man and woman, human and non-human, with inbuilt valuations and relations of dominance by privileged males;
4. the rise of capitalism as exploiting and suppressing people, animals and the earth for the purpose of creating profit and wealth.

(*continued*)

Box 3.1 (continued)

This framework takes up arguments from many movements and theories; the ideas and concepts should not be integrated in the form of a unifying and dominant theory, but as "unity in diversity", respecting diverse voices, experiences and ways of life.

Controversial questions addressed in critical debates include how power-based dichotomies should be reinterpreted, how epistemologically privileged social perspectives should be dissolved and whether all dichotomous constructions can or should be dissolved (on the controversy between essentialism and anti-essentialism see Gaard 2011). The questions cannot be answered only empirically; they require theoretical reflection and awareness of the values in the reasoning.

Sources: mentioned in the text

ciety reappear. Three problems resulting from the nature–society relations in modern society show the difficulties of sustainable development and sustainability transformation:

1. *The modern economy and society are not integrated in natural processes and ecosystems.* They are organised as specific forms of social systems, based on the social division of labour and structured through social, power and property relations, whereas the relations of society and humans with nature are neglected in the organisation of society. The relevant interactions between nature and society are mediated through these historically specific economic systems of society and economy and institutions for collective behaviour. The subjective and individual awareness of environmental destruction is only a weak condition for the solution of the problems of overuse of resources and pollution. Environmental pollution and resource use can only be reduced when the systemic structures of the economy and the ways of life of social groups that are overconsuming are changed. The decoupling of the markets and the economy from ecosystems happened with the institutionalised forms of social and ecological dysfunctionality or maladaptation described in critical economics as the great transformation and disembedded economy (Polanyi 1944) or as the great transition in modernity (Boulding 1965).
2. *Sustainable development is an inexact term for repairing the disturbed interactions between nature and society; sustainability can only be achieved through a social-ecological transformation.* A critical system analysis of the ecologically

dysfunctional global economy is necessary to identify the forms and paths of transformation. This cannot be done by describing the overuse of resources and environmental damages quantitatively and qualitatively. Strategies of sustainable development require ecological, economic and social indicators, and conditions and threshold values, in order for the social-ecological systems to change. This is easier for the ecological than for the social and economic conditions, which are more controversial. The ecological criteria are described by Bennett (1976: 298ff); the main social condition for achieving sustainable resource use is the redistribution and fair sharing of resources between the Global North and the Global South—the rich and the poor economies. How to achieve this needs to be determined through critical analyses of the system of unequal development (in inexact form in the Brundtland Report and other reports in the global policy process; in more theoretical and detailed form in Altvater [1992: 222ff], and in the research in heterodox and critical economics: see Chap. 6). Actors from the Global North, who are predominant in the sustainability process, block the transfer of resources to the South. Their vested interests as decision-makers and members of affluent and industrialised countries and economies (Ross 1998: 214ff) do not allow for seeking alternatives to a growth-based economy. They defend growth, for example, in the form of the "green economy" (UNEP 2011) with ideas that serve the interests of the economic elites, including ecological modernisation, strengthening the markets and technological innovations (Brockington 2012: 414).

3. *In the sustainability process, the key changes required for successful transformation involve redistribution, sharing of resources and reducing uneven development in the modern world system.* These changes are not yet on the sustainability agenda. The abstract value of solidarity, or changing worldviews in the sense of moving from the human exceptionalism paradigm towards the new ecological paradigm, or individual changes in ways of living, cannot substitute for the system transformation. The redistribution debate that has been occurring since the Brundtland Report has been insufficient, mainly focused on the concept of basic needs or the human development index, which do not reflect the necessary forms of system change. They do not measure the transformative effects of sustainable development policies and programmes, only annual changes in the conventional process of development—economic growth and its indicator, gross domestic product (GDP), supplemented by some indicators of well-being such as health and education. With the renewal of sustainability as the social-ecological transformation of the global economy, it will become necessary to find effective forms of redistribution and resource sharing.

To deal with these difficulties of sustainable development, new forms of knowledge production and a rethinking of the relationship between science, politics and society are important steps. Interdisciplinary knowledge practices for sustainability transformation include open, plural and flexible forms of integration of knowledge from different approaches and disciplines, across the borders between the social and natural sciences. The interaction of complex social and ecological systems becomes a core theme in environmental and sustainability research.

3.1.2 Complex Systems and Processes

Complexity is traditionally studied in research on non-linear dynamic systems and complex adaptive systems (Funtowicz and Ravetz 1994; Lansing 2003). The system properties of complex systems include large numbers of elements, feedback and interaction, information exchange between elements, non-linear interaction and unpredictability of the behaviour of the system as a whole. Research on complex adaptive systems describes the specific behaviour and properties of such systems with metaphorical terms ("adaptive landscapes", "basins of attraction", "edge of chaos"). The properties of complex adaptive systems, for example the social and ecological systems that interact in the sustainability process, limit the possibilities for regulation and sustainability governance. This is presented in soft form as "wicked problems and clumsy solutions" (Hartmann 2012) and in hard form as the impossibility of regulation and the abandonment of the concept of sustainability in scientific research (Benson and Craig 2014). Given the complexity, uncertainty and unpredictability of environmental change, sustainability is seen as unrealistic; it should be replaced by aims that can be achieved more easily, for example resilience, with its focus on disaster mitigation. The ecological critique of the formulation of planetary boundaries devolves into a similar line of reasoning, reducing the aspirations and the scope of environmental and sustainability policies. Furthermore, since the Johannesburg Summit in 2002 sceptical views such as "it is too late and impossible to achieve sustainability" are spreading, based on the idea that environmental change in the Anthropocene has caused irreversible environmental damage and change.

Many processes in nature and in society are too complex to be regulated and controlled by humans. But the overuse of natural resources and the pollution of the environment are caused by humans, and based on the indicators and measurements that exist for the sustainability process, potential solutions can be imagined and implemented through governance programmes. Even under deteriorating conditions and ongoing global environmental change, sustainability is possible, although it is more difficult and is made more so by environmental disasters. Sustainability can be broken down into a set of problems to regulate, and the limits of regulation can be specified. An abstract argument about its inscrutable complexity translates easily into defensive reasoning for refusing to take action or make changes to one's lifestyle. The complexity of the systems and processes that must be dealt with in sustainable development requires continuous improvement of knowledge through research, inter- and transdisciplinary knowledge integration and joint learning from failures of the past. The science of complexity, which Ostrom (2009) argues is needed to achieve sustainable development, is a science of learning to deal with complexity.

The planning philosophy of "wicked problems and clumsy solutions" (Hartmann 2012) seems to deal with complexity from a reductionist and simplified perspective, reducing it to planning and its short time horizons, to small and piecemeal solutions, and thus reducing sustainability to piecemeal engineering and muddling through. The problems of ignorance and the unknown future cannot be sufficiently dealt with in such forms. The future needs to be deconstructed in the transformative action for sustainability into phases and sequences of action that extend over a long period of time, beyond the time horizons of environmental, landscape, community or marine planning. The analyses of the historical development of complex resource use regimes (see Chap. 2) can help deal with the complexity of sustainability transformation, as was done, for example, by Bennett (1976) in cultural anthropology before the political sustainability discourse even began. He set sustainability-related resource use problems in different societies in their historical and social context when he reconstructed *ecological transitions* in human societies and the increasing complexity of natural resource use regimes (Box 3.2).

Dealing with the complexity of global processes and the integration of ecological, economic and other social-scientific knowledge for the sustainability process

Box 3.2 Changing Relation of Nature and Culture: Bennett

1. *The historical trend of "absorption" of nature by culture*: In the coupled social-ecological systems (or "socionatural systems" as Bennett calls them) the exploitation of natural resources and their economic processing have become more complex and risky processes, with ramifications in many institutions in society and many ecosystems in the natural world. Large-scale environmental destruction and economic catastrophes, and environmental and social dislocation, make solutions more difficult (Bennett 1976: 22).

(*continued*)

Box 3.2 (continued)

2. *Unravelling the many feedbacks between the coupled systems to generate solutions through reduction of resource use*: The advanced processes of adaptive and maladaptive social-ecological change in the course of human history and the ecological transition at global scale made the ecological problems too big to synthesise solutions in a general theory or in one subdiscipline. In anthropology this complexity was first discussed in the 1940s in the cultural anthropology of Steward, and thereafter by Geertz in the 1960s regarding ecosystems as open systems that interact with social systems. The ecological transition is interpreted by Bennett as one from adaptation to maladaptation (in Bennett's formulation, abusing the environment), from equilibrium to disequilibrium of social and ecosystems (Bennett 1976: 2, 9f, 22, 123ff).

Sustainable development and the sustainability discourse appear in the anthropological perspective of Bennett as the advanced phase of the complex and conflicting processes of ecological adaptation and maladaptation. The ecological transition as described by Bennett is a multifaceted transition in various forms: material and energetic transitions, demographic transition, and technological and economic change and innovation (Bennett 1976: 123ff). This transition resulted in its late phase in the twentieth century in a globally disturbed nature, with exponential growth of the population, resource use, economy and the incidence of disasters—forms of maladaptive change that are to be corrected through sustainable development.

Source: Bennett 1976

requires tools and methods to analyse and assess the changes in nature–society interactions in modern society at different levels in space and time, including the coupling and the possibilities of co-evolutionary development of social and ecosystems; the dynamics of short and long, fast and slow changes; the consequences of economic growth and its transformation; the combination of economic and ecological valuation of natural resources in strategies for regulating social and environmental change (see Chap. 4); and potential changes in the distant future (see Chap. 9).

The new interdisciplinary subjects established in recent decades, with rapidly growing research adding up to the disciplinary knowledge production, include human, cultural, social and political ecology and sustainability science. Methods of assessing the deteriorating or improving interaction between nature and society were developed along with the interdisciplinary knowledge production (for example ecological indicators, see Table 3.1). Other methods relevant for sustainability transformation include, for example, participatory research developed in citizen science, transdisciplinary and "mode two" research in the context of knowledge application, post-normal science, and research connected with social movements such as feminism and ecofeminism. Although these interdisciplinary subjects and approaches are not yet dominant in environmental and sustainability research, they generate more and more of the knowledge for sustainable development, especially on themes and problems neglected by disciplinary research.

The conditions in which social-ecological systems function and the conceptualisation of future systems change are key themes of interdisciplinary sustainability research:

1. *In order to understand the functioning of complex systems*, cognitive limits, conditions of application of knowledge in sustainability policy and governance, and normative criteria and principles of knowledge selection and use need to be dealt with. The changes to social, economic and ecological systems in sustainable development require more than principles, heuristics, policy instruments and empirical knowledge from policy programmes and documents for sustainability governance. The "invisible knowledge", reflection and discussion of the forms and methods of the search for knowledge, and the identification of "existential needs" and "objective exigencies" of society, is insufficiently discussed in the specialised and theoretical literature on functionalism (Becker 1988: 873ff, 880). The criteria and conditions for generating solutions to disturbed functions, dysfunctions and maladaptation tend to be simplified in environmental and sustainability research. In environmental sociology, Dunlap and Catton (2002) identified three functions of ecosystems for human societies, which they describe as supply depot, waste repository and living space. In the research about complex adaptive systems, more functional relations and their disturbance need to be taken into account. In the ecological and biological research about living or biological systems (Wagner 1999, 2005, 2011) and ecosystems in the larger context of the earth system (for example, Malhi et al. 2020), the capacities for systems transformations are described as robustness of systems, ability to withstand disturbance from mutations or environmental change, and capacity to innovate and develop through trial-and-error processes that become relevant for sustainability research and governance. This knowledge can be

applied in regulatory practices, in strategies of adaptive management and co-management (see Chap. 4).

2. In order to conceptualise future possibilities of adaptation and transformation of social-ecological systems, it is necessary to reflect the epistemological conditions of producing knowledge about the future and the relations between knowledge and ignorance (Gross and McGoey 2015). With the long temporal perspective adopted in sustainability research and governance, beyond the short time horizons of planning, ignorance becomes a research theme. Ignorance about future events is only one form of ignorance that must be dealt with in sustainability research; other forms, resulting from intended or unintended effects of research and knowledge production, affect the sustainability discourse and process as well. In the epistemology of ignorance (Proctor and Schiebinger 2008; Fernández Pinto 2019) different forms of ignorance are discussed; these include socially constructed ignorance and ignorance generated by science itself (Ravetz 1987; Sarewitz 2004). Both forms result from selective choice of literature and misleading assumptions in studying complex processes. For the sustainability discourse the neglected theme of ignorance becomes relevant for the development of transition and transformation research in long temporal perspectives where the future is dissolved in wanted and unwanted futures as in scenarios of potential, foreseeable or unforeseeable events and changes (see Chap. 9).

Procedures of time management and conflict mitigation need to be developed for the building and implementation of policies and policy regimes in sustainability governance.

Time management: The strategies and political programmes for the implementation of sustainable development do not adopt long-term perspectives in which the transformation process is constructed in phases, cycles and sequences of activities and with consideration of the transformative capacities required. Policy programmes are formulated for short-term action in simple forms and define the goals to be achieved. Processes that are decisive for the success of sustainability transformation, such as practices of knowledge integration and application, cooperation of actors, joint learning and knowledge sharing, international coordination and networking of activities, are not systematically researched. The policy processes advance from one preliminary and time-limited activity to another, with limited perspective. Global scenarios are the main instrument for unfolding long-term perspectives that imply different phases and changing conditions of transformation; they can replace the simple concepts of goals and visions to describe the future that are found in policy programmes and strategies. Improved time management in the policy and governance processes requires the formulation of concrete and action-specific process models where the phases of action are designed with specific criteria for the progress of transition by conceptualising long-term processes of social-ecological transformation as they are known from the history of human societies and imagining time models for a future sustainable society (Chap. 9).

Conflict mitigation: In political and civil society action for sustainable development, conflicts in the process resulting from the conflicting needs and interests of different social groups, resource users and political and economic actors need to be dealt with continually. Conflict mitigation and mediation is not an exceptional requirement in natural resource management and sustainable development; it is part of action strategies that aim to break through the roadblocks to social and economic transformation. The conflicts that need to be mitigated include violent and non-violent conflicts, manifest and latent conflicts created through asymmetrical power relations in the local, national and global sustainability processes, and conflicts resulting from property rights, privatisation and enclosures. Dieng and Pesqueux (2017) discuss "green governance" as a conceptual framework for the integration of normative and positive knowledge, bridging knowledge production and application in policy processes and in conflict mitigation (see Chap. 7).

In order to learn from the failure to build a new economic order under the guiding idea of sustainable development, it is not enough to incorporate the neglected concept of transformation into strategies of sustainable development, as frequently happens, even in UN documents. To provide knowledge for the future transformation processes, it will be necessary to analyse the inherent conflicts and maladaptive changes of the modes of production, the accumulation regimes and the socio-metabolic regimes in the global economic system. These are core themes of interdisciplinary research in social ecology about the connections of social and ecological systems, but they have not yet been sufficiently incorporated in sustainability policy and governance. It seems impossible to transform economic and societal systems without knowledge of their complex systemic constitution, their functions and interaction with other systems, the dysfunctionality of social and ecological systems and processes, and their side effects and unanticipated consequences. *Sustainable economic production is not only a better combination of production factors explained with conventional economic theory; it is part of larger processes in social and ecosystems where the interaction and resource flows between society and nature are structured in modes of production, socio-metabolic regimes and accumulation regimes.*

3.2 Knowledge Integration: Epistemological and Methodological Problems

The epistemological and methodological problems of knowledge integration in the sustainability discourse and in the governance process are manifold (overview in Bruckmeier 2019: 153ff). Knowledge practices that include inter- and transdisciplinary knowledge integration to formulate effective transformation strategies are required not only in the scientific sustainability discourse, also in the policy and governance process. Interdisciplinary knowledge practices vary, from the simple forms of adopting new knowledge from continuing empirical research, evaluation and assessment of policy programmes through new forms of knowledge integration and synthesis for sustainability transformation with a variety of methodologies and procedures.

In interdisciplinary knowledge practices, economic knowledge does not simply compete with knowledge from other disciplines in the sustainability discourse and governance process, in science and in politics. The processes influencing the transformation to sustainability are interconnected and interactive, the knowledge processes are complex, and the knowledge transfer cannot be reduced to the "green economy" strategies of the powerful global players. It is necessary to pass through controversies about growth and alternative forms of development in the course of transformation. Furthermore, it is necessary to connect and apply knowledge that cannot be synthesised in one coherent perspective, theory or approach: open, discursive and pluralistic knowledge practices need to be developed for the sustainability process. The relations between economic (or other social-scientific) and ecological (or other natural-scientific) knowledge can vary along a continuum from overlapping and complementary to excluding and incompatible forms, but this remains a one-dimensional knowledge structure. What will be required in the future sustainability process are pluralistic strategies and knowledge practices that allow for the use of different and competing forms of knowledge. This pluralisation begins with the choice of guiding scientific concepts for the sustainability discourse, such as the concepts used to describe modern society (Box 3.3).

Interdisciplinary knowledge integration does not imply levelling down contrasts, tensions and contradictions in the knowledge process. It seeks ways to work with open epistemologies and methodologies that use knowledge from different disciplines and sources which is not fully compatible and coherent. Instead of negotiated compromises as in the policy process, knowledge integration and synthesis require epistemic anchoring of knowledge and methodological rules for differ-

Box 3.3 Concepts for the Analysis of Modern Society

There is no consensus in sociology about how to describe modern society in theoretical terms; the concept remains inexact in the social sciences and in its use in policy and in everyday life. This inexact quality remains when using modern society as an overarching concept for other notions of the global society. Although it is intuitively clear that the global society is one societal and economic system, the historical reality is and will remain more difficult to understand. Currently large parts of the global population still live in pre-industrial societies and economies, but the industrial society and economy are global systems established long ago, and the post-industrial society is a vague notion of a future society that will develop in the transformation to sustainability. The political and cultural fragmentation of nation states and of different cultures remain and will also exist in the future, in a fragmented global society with different forms and spatial scales of organisations; whereas the economy and the markets are globally connected, the states, the cultures and religions remain particularistic and fragmented, their global ties are weak. The deregulation of the economy and markets during globalisation and the loss of functions of the nation state may be seen not only as a weakening of political systems (described as a "post-political situation") but also as a transition to post-national and cosmopolitan forms of governance developing beyond the national state.

The complicated state of societal development in modernity can be explained by several concepts in combination. The first set of concepts, including modern society, capitalist society, industrial society, world society, risk society, and knowledge and information society, are intuitively clear, self-explanatory, and more descriptive than theoretical. Furthermore, concepts from theoretical analyses of modern society, where the concept of society is not the theoretical marker, are necessary to describe societal dynamics, such as modern world system, world ecology, globalisation and global social change; or to describe the theoretical concepts in human and social ecology, such as societal and human relations with nature, societal metabolism and interacting social-ecological systems. The relations with nature and the environmental problems are neglected in most theoretical concepts of modern society—as if they would not influence the functioning of society.

The example of world system analysis illustrates the complexity of analysing modern society. Wallerstein (1991, 1995) argued for interdisciplinary concepts and knowledge production, doubting that the differentiation between the social and natural sciences can be maintained in the future.

(continued)

Box 3.3 (continued)

This argument was motivated by the search for new concepts and a new language to show the interactions between systems—dispelling the illusion that there are three distinct spheres of society, economy and politics in the modern world system. The three spheres are better described in transitory processes and continuous change, interconnecting, overlapping and interacting. World system analysis requires more than a formal concept of society; it also requires a series of additional concepts such as division of labour; countries in the core, periphery and semi-periphery; unequal development and exchange; and transformation. In social ecology (Fischer-Kowalski and Haberl 2007) the term society is less important as a theoretical marker than the concepts of societal metabolism, societal relations with nature and colonisation of nature.

The concepts of society mentioned above are not always used in their theoretical context and as part of a theory; they are often used as descriptive terms, as intuitively clear and self-explanatory concepts, in interdisciplinary discourses. The global cleavage between the industrialised North and the developing South is not only a line of conflict separating the rich and the poor countries; it makes clear that concepts of society such as modern, industrial, post-industrial and information society are not universal concepts but are oriented toward European historical time. As European countries are deindustrialising and on a path to creating post-industrial societies, most countries in the Global South and the larger part of the global population are not yet or just entering the industrial society. The time lags and the asynchronous occurrence of most social change processes is relevant for sustainable development, which must deal with the conflicts between countries in order to connect societies at different stages of modernisation to a global policy process.

Sources: mentioned in the text

ent forms and degrees of integration: epistemological, theoretical, conceptual and empirical integration, integration in loose forms of connecting knowledge generated with different epistemologies (for example causal analyses and subjective interpretations) and methodologies (for example quantitative and qualitative research).

The inter- and transdisciplinary knowledge practices for the sustainability process in science and governance are developing through the continuous updating and improvement of knowledge. Joint reflection and learning by the actors involved is a core component of the transformation. Methods to connect knowledge from different disciplines and areas of research, including intuitive methods and heuristics, have not yet been scrutinised and refined. The synthesis processes in the global synthesis and assessment reports about sustainability, ecosystem assessment (Millennium ecosystem assessment 2005), agriculture, climate change adaptation and mitigation provide examples of pragmatic forms of synthesis done by expert groups. These have been carried out via different modes of communication—oral, written and internet-based— and in different scientific and political institutions, including research institutions, think tanks, thematically specialised organisations for knowledge synthesis, and knowledge networks, using databases and internet-based knowledge platforms. The methodological components of interdisciplinary knowledge syntheses include process methods of synthesis (synthesis workshops, oral synthesis, synthesis reports), conceptual methods (models and schemes, frameworks, classifications and typologies, databases), inductive methods (systematisation guided by principles, codification, generalisation, abstraction) and, for synthesis of knowledge about complex interactive systems, methods such as meta-synthesis of qualitative data (described in: Nakamori 2016: 55ff; Leary and Walker 2018). The results of knowledge syntheses are documented in the form of periodical reports (synthesis reports, assessment and evaluation reports). The *co-existence of multiple, conflicting and contradicting ideas, interests and knowledge forms in the global sustainability discourse cannot be levelled down and harmonised in the synthesis process.* They need to be made applicable by way of transforming the "brute force" of competing truths, clashes of interests and power struggles into practices of plural knowledge use that reflect the complexity of the problems, systems and processes that are being studied and regulated. To improve the use of knowledge and the strategies for sustainable development in situations where no consensus is achieved about future governance practices, it is possible to experiment with different strategies in the sense of adaptive governance in order to find ways to improve the sustainability transformation. In the "laboratories for green growth" in the Organisation for Economic Co-operation and Development (OECD) countries, experimenting with a variety of models and strategies that include variants of growth and degrowth can create more and better options for future transformation strategies.

For the integration of economic with other knowledge in the sustainability process, five epistemological and methodological difficulties must be overcome in the interdisciplinary knowledge practices:

1. *Context-related knowledge integration (which social, ecological and knowledge contexts influence sustainable development)*: In complex interacting systems such as the sustainability process, the core requirements are to specify the primary sources of knowledge for reducing the over-abundance of knowledge from disciplines and developing local strategies of action. The thematic scope of the interdisciplinary sustainability discourse can be specified through the description of problems to be dealt with and goals to be achieved. The differentiation of social, economic and environmental sustainability and the seventeen Sustainable Development Goals of the UN show the thematic contours of relevant knowledge fields. Sustainability-relevant knowledge can be described with concepts from interdisciplinary ecology, including interdisciplinary research about the interaction of modern society and nature, human and social relations with nature, coupling of social and ecological systems, natural resource management, the combination and integration of economic and ecological knowledge, governance-related and normative knowledge about regulation and the formulation of international environmental regimes. The social and ecological contexts of sustainable development can be specified for the relevant subjects and disciplines in sustainability research with regard to social- and natural-scientific and interdisciplinary research. The general problems of finding a common language and terminologies for interdisciplinary communication and cooperation cannot be solved with one single method or practice. A variety of communication processes are required that are part of the more specific procedures and methods of knowledge integration. These comprise the forms already practised (only a minor part of synthesis involves quantitative synthesis of statistical data; the larger part involves work with qualitative data and methods such as oral synthesis, synthesis workshops, concept transfer, conceptual frameworks, interdisciplinary assessments, policy evaluation and reviews). The institutional forms of knowledge integration include think tanks, advisory councils, and participatory methods and methods for future research. The organisation of the sustainability process at different levels, from local to global processes and action, suggests that knowledge integration needs to be institutionalised at every level of research and governance.
2. *Complexity-related knowledge integration (how complexity is created and reduced)*: The contexts of the sustainability process enlarge the complexity of

interacting social and ecological systems and processes. In which parts of the sustainability process, at which levels and for which purposes is social-scientific and ecological knowledge required? How can knowledge from different disciplines be integrated to build the interdisciplinary knowledge base that the sustainability process requires? Dealing epistemologically and methodologically with the social and ecological complexity requires knowledge integration in terms of formal methods of complexity analysis and in terms of integration of knowledge from different thematic areas. Identification of relevant knowledge implies value-based selection processes, a controversial part of knowledge integration through negotiation. Regarding economic knowledge and its integration with other knowledge, a decisive step forward has been made in the economic discourse itself, implying that environmental problems are a specific component, with the concept of "embeddedness" and the embedding of economic processes in the societal context of the economy. This made visible the complex processes and interactions between systems that are required for the functioning of economic systems and for regulating, socially, culturally or politically, the economic and resource use processes. The discussion of embeddedness occurred in theoretical forms in institutional economics (Polanyi 1944; Polanyi et al. 1957) and in economic sociology (Granovetter 1985; Granovetter et al. 2004; Beckert 2007). Beckert (2007: 21) reviewed the limits of the embeddedness concept in economic sociology and re-established it with more inclusive concepts that refer to networks of social relations, to coordination problems in economic action, and to the connected micro-, meso- and macro-social levels of change. Furthermore, he discussed the normative and social-reformist concerns that shaped Polanyi's concept of the great transformation.

3. *Dealing with the conceptual and theoretical requirements of framing interdisciplinary knowledge integration*: The forms of inter- and transdisciplinary knowledge production and integration for the sustainability process have not yet been defined and structured in more concrete forms of designing knowledge practices and knowledge regimes for sustainability transformation. These cannot be reduced to repertoires of methods for knowledge production, integration and application, but need to be specified for the different parts and phases of the sustainability discourse and process. A widely neglected component of structuring the knowledge processes is their theoretical conceptualisation and framing; this requires identification and application of knowledge from theories of interaction of society and nature, of societal and economic transformations, and of transformative activities and action groups. Given the complexity, lengthy time frame and multi-scale process of sustainability transformation, the framing of knowledge practices cannot be done in conventional forms with singular or uni-

tary concepts as in systems theory; the conceptual framing requires combinations of concepts, inter-theoretical models and pluralistic strategies.

4. *Dealing with multiple criteria and plurality—redundancy, overlapping research, competing and contradictory, mutually excluding approaches and knowledge*: Redundancy refers to the functioning of complex systems for which critical components, structures, functions or processes are duplicated to increase the safety, reliability or resilience of a system. The term has a variety of meanings in different disciplines; it is used in this sense in engineering for technical and computer systems (Koren and Krishna 2007) and can be applied to the sustainability process to deal with problems of regulation of interacting social and ecological systems. Perrow (1984) analysed complex technical systems in their social context and in a risk perspective, concluding that complex systems and tightly coupled systems with redundancies are prone to "normal accidents"—accidents that are unavoidable in extremely complex systems. Redundancy may produce the unwanted effects of reducing safety and reliability because it increases the complexity of systems and thus the possibility of errors and accidents (Sagan 2004). Furthermore, the broadening of the knowledge contexts, interdisciplinary communication and knowledge integration make it necessary to deal with a greater variety of concepts that are similar to or compete with concepts in the environmental discourse. Examples of such overlapping or competing concepts are different definitions and interpretations specifying the idea of sustainability: weak and strong sustainability; shallow and deep ecology; the concepts of development, transition and transformation; the forms of ecological and social-ecological resilience; the economic concepts and models of the dynamics of economic change; and abstract interdisciplinary concepts such as resources, complex adaptive systems and co-evolution. With the operationalisation of such concepts come different and competing interpretations and controversies about different norms and values guiding the transformative action, which need to be mediated. Epistemological pluralism in interdisciplinary research can tackle the conflicting processes of competing, contradictory or mutually exclusive approaches and knowledge practices with epistemological and methodological rules and heuristics.
5. *Dealing with the inherent normativity of the sustainability discourse and the clash of heterogeneous norms and values in knowledge production*: Cultural values (Ginsburg and Throsby 2006) and social norms (Bicchieri 2006) influence scientific and other knowledge production, for example through worldviews and visions that guide the knowledge practices of participants in scientific and political discourses. How cultural values structure social action is the theme of the sociology of Parsons (Loubser et al. 1976). The influence of values and

norms on scientific research has been a topic of intense discussion since Kuhn's (1962) analysis of scientific revolutions, especially in science and technology studies (Felt et al. 2017). Values are not always consciously reflected in social action or in scientific research. In the sustainability process it becomes necessary to discuss values, pre-analytical visions and value-loaded concepts explicitly and to reflect on their influence on research and political and social action. Values and norms that are rooted in different ethics are omnipresent in the sustainability process and clash in the controversial discussions about the concepts, strategies and forms of joint action. The ethical and normative aspects have to be discussed in all forms and phases of the process, including inter- and transdisciplinary knowledge integration (for example in synthesis workshops). The discourse ethics of Habermas (1991) is a methodologically structured approach for communicating ethical norms that can also be practised in the environmental and sustainability discourses. It specifies the rules for arguing about ideas, values and norms and justifying the validity of norms in public conversations. In the future sustainability process, the rules and procedures for a discourse-specific ethical communication of sustainable development need to be specified for the many forums and platforms of discussion in this global process (see Chaps. 8 and 9).

3.3 Application of Economic Knowledge in the Sustainability Process

Many questions arise in the policy process, such as which economic institutions are functional for sustainable development and which ones need to be changed, developed or newly built? How should market mechanisms to be integrated with other economic and political mechanisms? The knowledge from political science is relevant for the policy and governance processes of which sustainability governance is a part, following the organisation of the policy processes, and in this regard it is similar to economics: both disciplines produce knowledge about the functionally differentiated societal subsystem of policy and economy in modern society. This knowledge needs to be complemented by ecological knowledge about the effects of resource use on ecosystems and ecosystem processes. Sociology and cultural anthropology create knowledge about many social institutions and processes at different levels of society and social organisation, including political and economic processes, in research that overlaps with political scientific and economic research. A dilemma facing knowledge application for the sustainability process is that there is too little applicable knowledge about the complexity of systems, and too much knowledge is created that is not applied in the sustainability process (a consequence of overlapping and competing research in many disciplines).

The knowledge about natural resource use and management and the modern economy does not come exclusively from the academic discipline of economics, but from other disciplines as well. Economics is a discipline with inexact boundaries, overlapping to some degree with other social-scientific disciplines such as political science, sociology and cultural anthropology, and interdisciplinary subjects such as political ecology, as well as research on decision-making and problem-solving. Moreover, economics, like other disciplines, does not create an internally coherent knowledge system; it is breaking into ever more specialised subdisciplines that do not generate coherent knowledge. Throughout the history of economics there have been competing and contradictory forms of knowledge, concepts, theories and approaches, making knowledge transfer and application difficult. In the sustainability process, the fact that economics does not have a monopoly on exclusive knowledge production about the modern economy leads to the methodological difficulties discussed above. In the sustainability process, knowledge about economic production and decision-making often clashes with knowledge about natural resource use and environmental problems from other social-scientific and natural-scientific disciplines, especially from ecological and environmental research.

In the sustainability debate since the Brundtland Report, it has become evident that economic processes interact strongly with ecological processes. Economic knowledge alone will not be sufficient to achieve sustainability, nor should economic sustainability be restricted to a specialised part of sustainability research under the guidance of economics. If sustainability is separated into social, economic and environmental sustainability and managed by specialised knowledge and expertise, the integration problems and processes will tend to be ignored or externalised to the policy and governance processes. Economic and ecological processes are often conflicting. Overuse of resources and environmental pollution cannot be managed through economic knowledge only, require the integration with ecological knowledge to observe resource limits. To regulate economic resources, economic knowledge needs to be combined and integrated with other specialised knowledge in the sustainability process, which is mainly done in interdisciplinary research and knowledge practices.

1. *Ecological knowledge in the sustainability process*: Sustainable development is a process that requires the regulation of functions and services of ecosystems. Clearly these ecosystem functions and processes cannot be fully understood based on economic knowledge: this is due to the nature of ecosystems, of which

humans are a part, but only as part of ecological communities with other species that cannot be subjected to the same rules of economic systems in society as humans. Ecological research for sustainable development is, therefore, more important than economic valuation of ecosystem functions and processes. Specialised economic research can explain only fragments of the social-ecological processes that constitute sustainable development. The limits of economic knowledge are evident in the discussion of limits to growth.

The problem in analysing global resource flows is not one of lack of data and information. The production of data and statistical information within the routines of economic processes and the increased data production and communication through the internet is sufficient to assess resource flows at every level: for a specific household, enterprise or organisation; for an economic branch or sector; for a community, region or country; and for the global flows in the world economy. These data can be combined with data about emissions and pollution of the environment.

2. *Social-scientific knowledge in the sustainability process—sustainable development as a process framed by social institutions*. The economic processes of production, reproduction, distribution and consumption are part of other social systems in the social lifeworld and in the functionally differentiated organisation of modern society. Moreover, in the globalising and disembedded market economy the different functions of societal systems cannot be ignored. Attempts to subject all social and ecological processes and resources to a generalised economic valuation and to commodification for the sake of economic growth are increasingly criticised in the sustainability discourse. The economy is not only a profit-based market mechanism; it is a social system with multiple functions and institutions beyond markets, ramifying throughout modern society and connecting with functions of other systems that need to be maintained to prevent social and ecological destruction. Economic actors act in multiple economic and non-economic roles. Political regulation and coordination of economic processes, the territorial divisions of labour in economic processes and their multi-scale operation cannot be managed without further political and social institutions and the knowledge necessary for their functioning. Buchanan's "constitutional economics" (Buchanan and Tullock 1962; Buchanan and Brennan 1985) analyses the regulation of the economy, the legal rules, norms and constraints of economic functions and processes, but only in an economic logic that regards them as functional requirements of the maintenance of economic systems, not of complex interacting systems. The key roles of individual and collective economic producers and consumers are connected with other roles that need to be maintained for the functioning of society. "Consumer sov-

ereignty" as a normative economic construction of consumers in the modern market economy ignores the many forms of influencing and manipulating consumption, and the negative consequences of the overuse of natural resources. The human needs of resource use demanded in the principles and goals of sustainable development of the UN cannot be reduced to "needs that can pay", as in economics. The sustainability process creates new roles and responsibilities for consumers and producers, ecological responsibilities in the form of ethically responsible resource use, resource saving, stewardship and responsibility for future generations.

3. *Economic knowledge in the sustainability process*: Having discussed non-economic knowledge from other social- and natural-scientific research that is relevant for the sustainability process, it seems easier to identify the components of economic knowledge and determine whether they clash with other knowledge. The economic processes to be assessed for sustainable development happen within enterprises and between enterprises in a market economy, including the forms of production, exchange and trade; the monetary flows in and between economic systems; and the flows of goods, capital, people and information. All these flows are connected with flows of material and energy resources in the global economy that can be defined in monetary terms as well as in physical terms. The functions of economic knowledge and the forms of economic valuation of natural resources need to be assessed to determine whether they are compatible with the functional maintenance of social and ecological systems.

Economic knowledge is mainly required in seeking solutions for the environmental and resource use problems that are institutionalised as economic processes in the national economies with their sector-specific organisation as agriculture, fishery, forestry (primary production), industry (processing or secondary production) and services. This is the domain of economic knowledge about macroeconomic systems and processes. At the microeconomic levels the producing firms and organisations in high-polluting sectors become the topic of the sustainability discourse, for example in ecological modernisation strategies. The (private) households and the consumers become a topic of the sustainability discourse in terms of changing resource-intensive ways of life that are part of the modern economy.

3.4 Controversial Themes in Economic Research About Natural Resource Use and Sustainable Development

In the theoretical and practical discussion of the economic valuation of nature, the main controversy can be described as that about "economising nature". This can be described as the monetary valuation and pricing of natural resources so that they can become goods to be exchanged and traded in markets and subjected to the functions of markets and the rules of the financial economy. One problem with this process is that the privatisation and commodification of natural resources is applied to heterogeneous types of resources that cannot easily be subjected to economic rules. Some natural resources, for example land, can more easily be economically valued and priced; land has long been a dominant form of traded and commercialised good. Resources such as water and air, referred to in economics as public goods or commons, consequently cannot be privatised. With the valuation of complex natural resources such as ecosystems or their functions and services, or the whole "natural capital" of the earth system, economic valuation becomes more difficult or arbitrary because of the complexity of ecosystems and their processes. It is accepted in heterodox and critical forms of economics that nature and ecosystems should not be reduced in their complexity through valuation or subjected to the economic logic of markets, but rather economic and market processes should be (re-)embedded in nature. This implies observing ecological functions and limiting growth and privatisation through ecological regulation of resource use—setting limits to markets and dealing with scarcity, overuse and pollution in other forms.

3.4.1 Economic Value of Nature, Natural Resources and Ecosystems

Nature and the value of nature have long been discussed in economics, which has struggled with many concepts of value that have come into use in economics since the neoclassical conceptualisation of the value concept in subjective forms. The more recent discussion in neoclassical theory is summarised by Brown (1984) with regard to resource allocation. A controversial debate about the nature of economic value re-emerged in the environmental discourse regarding an ecological value analysis in the theory of Marx (Burkett 1999), and in the debates on the value of nature in the new environmental discourse. This discourse is manoeuvring between ecological and economic criteria, taking up the older traditions of objective (labour-based) value in political economy and the critique of political economy by Marx with regard to the value of nature (Vlachou 2002; Park 2003; and the subsequent debate in "Science and Society", vol. 67, 2003–2004). The links to the value theory of Marx, its use for sustainable economic welfare and its relation to the value of nature have also been discussed in ecological economics (Douai 2009). These theoretical debates are connected to the broader critical debate in the sustainability process about the monetarisation and commodification of nature.

The attribution of monetary value to natural resources is a controversial theme in economics and ecology. When value is attributed to natural resources for the purposes of resource allocation (in production processes in a market economy), it is done in terms of prices. The price mechanism does not help to determine which parts of nature, which functions of ecosystems and which natural resources can be priced without negative consequences as overuse or functional disturbance? A presupposition of economic valuation and commodification is the attribution of private property rights to resources. This is difficult or impossible for certain types of natural resources, for example global commons such as the air or the oceans. In theoretical terms the attribution of value to natural resources and the valorisation of nature are more complicated than setting prices. A specific form of valorisation is politically influencing prices through governmental decisions in favour of producers and regulating market prices (practised, for example, in agricultural policies). This meaning of valorisation is not relevant for valorisation as a complex process of attributing economic value to nature. The whole economic system, its institutional structures and its functions in modern society, need to be analysed to identify the causes of what are called in economics "externalities". Environmental pollution and destruction are examples of externalities that are not paid for with the market price and need to be compensated for as social costs by the government or through public expenditure, if they are compensated for or restored at all. Externalities result from the interplay of the economic institutions, not only from a dysfunctional price mechanism. Economic growth as institutionaised process is closely connected with the regulation of property rights in the market economy, the systemic functions of the economy and its interaction with other systems in society and in nature. Assigning value through individual economic actors and social groups (Brown 1984: 244f) cannot solve the problems of externalities and of attributing value to nature, for which different value concepts are used in economics. For the valuation of complex resources and environmental problems such as climate change, it is necessary to apply ecological criteria and norms that are incompatible with the short-term economic processes for which costs and benefits can be specified. Managerial decision-making about nature requires a long-term perspective (Barkin 2006: 56). The debate continues and remains controversial when it becomes part of the broader sustainability discourse, where the integration of economic and ecological valuation of resources is required.

The valorisation of nature has always occurred in human history when ecosystems and landscapes were modified through human resource use, for example through colonisation and forest clearing for agricultural, industrial and urban land use. As long as these processes were of limited scope and did not significantly modify the global ecological system (before the industrial revolution), the question was not practically relevant. With the globalising economy, the majority of ecosystems currently are strongly influenced by human activities. In the global commons as well, especially in the oceans, more and more resources are being commodified, which implies further and rapid intensification of natural resource use and deterioration of the global environment. In the late phase of development and economic globalisation that overlaps with the sustainability process, the economic valorisation and commercialisation of natural resources is extended to ecosystems and their services with the forms of economic growth discussed as "green growth" and "blue growth" (see Chaps. 5 and 6).

In the interdisciplinary economic and ecological theory of Norgaard, the problems associated with the economic valuation and appropriation of nature in the modern economy are translated into a critical theory of modernity. Norgaard (1984) developed the concept of "coevolutionary development" as a critical analysis of forms of co-evolution of nature and society that are seen as beneficial by humans. The co-evolution concept became part of his theory of the unsustainability of the modern economy and society (Norgaard 1994) that is one of the sources of sustainable development as social-ecological transformation. According to this theory the promise of modernity (economic development and rationality create progress, material abundance, peace and justice) has turned into a nightmare. Greater inequality, violent conflicts and accelerated depletion of the stock of natural resources are now challenges of sustainable development (Norgaard 1994: 1f). These ideas became part of a co-evolutionary approach in ecological economics (Kallis and Norgaard 2010).

3.4.2 Limits to Growth: Naturally and Socially Caused Scarcity of Natural Resources

The limits to growth debate can be understood as a critical response to the economic globalisation and valorisation of nature that has been taking place since the second half of the twentieth century. But it dates back to older economic debates in classical political economy of the nineteenth century, where the term economic growth was not always clearly specified as it is today (as GDP growth). The traditional debate about economic growth and scarcity referred to natural resources, their availability and limits, but not necessarily and only in markets. This debate was about physical and ecological factors: population growth, growth of subsistemce production, and overuse of the natural resource base. Scarcity of food resources can be caused through economic and market processes, but historically, even in the modern market economy, they are caused by the human appropriation of nature that results in ecological scarcity and overuse of resources. The Malthusian scarcity referred to a form of ecological scarcity, although among its causes are economic factors. The forms of naturally and socially or economically caused scarcity are often connected. The Malthusian and later Darwinian thinking about scarcity and limits of natural resources was based on the ecological knowledge of the time. With the development of classical political economy it became a more refined and critical discussion about economic growth, especially by Ricardo, Mill and Marx (see Chaps. 5 and 6).

From the *early political-economic controversies about growth*, scarcity and limits to the Neo-Malthusian growth debate in the new environmental discourse since the 1970s passed long time in which the growth-critical debate became one about global limits and the total resource base on the earth. The new debate about growth and scarcity is still oscillating between ecological and economic phenomena and controversial arguments. It is no longer limited, as the original Malthusian debate was, to population growth and subsistence production, but includes the global population and global use of natural resources. It is unfolding with the discussion of the question asked in ecological economics: what is growing when the economy grows? The consequences of limits to growth are more specifically discussed in the sustainability discourse with the competing concepts of "sustainable growth" and "degrowth" (Chap. 6).

Mills question, what will be when growth in the industrial society has come to an end, is still insufficiently answered in the sustainability discourse. Economists have different views about economic growth, and they mean different things when they talk about growth, its causes and its consequences. The discussion of economic growth in the sustainability process reveals the problem with the arguments for growth: they are differentiated in many models and explanations that cover partial forms and components of economic growth, but economic growth is rarely analysed in the broader context of the use of natural resources. Monetary growth and physical growth (of natural resource use) in the economy are blended in unclear ways and need to be clarified in the sustainability discourse with regard to the *risks of economic growth and resource depletion.*

When economic knowledge becomes part of inter- and transdisciplinary knowledge practices and is combined with knowledge from other disciplines, the question of how to use disciplinary terminologies in interdisciplinary practices arises.

What terminology can be retained from economics, and for which specific purposes and cases of natural resource use should ecological terminology be applied? How can the two terminologies be combined? This debate includes a variety of economic and ecological terms and their combinations, such as risks and systemic risks, vulnerability, resilience and sustainability.

3.5 New Interdisciplinary Analyses of Nature–Society Interaction

Social ecology—theoretical analysis of natural resource use in modern society: The concept of metabolism (as complex flows of energy, material and information in the "anthroposphere" or humankind's sphere of life) includes the uptake, transport and storage of substances, the chemical transformations within the organism and the forms of waste (Baccini and Brunner 1991: 1). In a more specific theoretical form, it is represented in social ecology by the concept of societal metabolism (Box 3.4).

The societal metabolism of pre-industrial agricultural societies in modern Europe can be described following Krausmann and Schandl (2006: 286f; Sieferle

Box 3.4 Modes of Production and Societal Metabolism

A series of theoretical concepts are available from economics, political economy, sociology, world system analysis, cultural anthropology and other subjects for conceptualising, analysing and assessing the modes of production and the societal metabolism in order to describe the complex societal organisation of natural resource use. In interdisciplinary ecological research on human, social and political ecology, the ecological components of modes of production, or socio-metabolic regimes, have been studied. Research on the systemic mechanisms has raised additional questions connected with the modes and relations of production, or relations of power and authority. These include the forms of power institutionalised as components of property systems in coupled economic and political systems. The critical discourses of the "tragedy of the commons" (Hardin 1968) and the "tragedy of enclosures" (Gonner 1912; Martinez-Alier 2001) refer to conflicts within given relations of production and authority. Knowledge about these systemic mechanisms is available, but not much of it used in the sustainability discourse.

(*continued*)

Box 3.4 (continued)

The core of an interdisciplinary social-ecological analysis of historically specific modes of production in human history is the concept of societal metabolism for analysing the dependence of economic production on historically specific relations between society and nature (specified in material and energy flows between nature and society or ecosystems and social systems). For agricultural and industrial societies, an exemplary analysis of social-metabolic regimes was done by Krausmann and Fischer-Kowalski (2010), who conclude that technological innovation and geo-engineering (for example through carbon capture and storage: IPCC 2005), will not be enough to solve the global environmental problems that result from the close coupling of economic growth and throughput of natural resources in the modern economy. Finally growth will reach the physical limits of the earth system which enforces solutions as they are sought in the sustainability transformation: decoupling the ways of life and consumption, the quality of life from increasing material and energy use (Krausmann and Fischer-Kowalski 2010: 24; Krausmann and Schandl 2006). The combination of science and technology as an effective and universal solution to all the environmental problems that must be dealt with in the sustainability process is critically analysed and exemplified by Burns and Schaefer Caniglia (2016) as a history of progress, paradoxes and unanticipated negative consequences. Technology is not neutral; the progress and wealth it has created, especially in modern society, has been at the price of creating further problems such as global environmental change. As with other factors of development and change, technologies do not work in isolation. In order to understand the effects of technological processes of change and innovation they need to be analysed in their connection with other social, economic and ecological changes, their intended and unintended effects, their dependence on levels of economic and social development, and the size and scale of technologies. Technologies coupled with large-scale production, where technological innovations happen with cultural support and legitimation, are culturally framed and "normalised", silently and uncritically accepted as forms of the modern economy and society, even when they create economic and ecological risks. The challenge for further development will be to find technologies which respect the principles of the functioning of ecosystems (Burns and Schaefer Caniglia 2016: 83ff).

Sources: mentioned in the text

et al. 2006): the limits to economic growth in socio-metabolic regimes of agricultural societies are dependent on the energy sources available in agro-ecosystems and forest ecosystems, which provide nearly all available primary energy and also large parts of the material resources—biomass for human food and feed of farm animals and, as the dominant energy source, wood. In the pre-industrial societies of Europe the use of wind and water was limited, as well as the use of coal as fossil energy source. The economically used energy is dependent on the controlled flow of solar energy in agricultural production and the energy through human labour and through the labour of farm animals. The economically disposable energy depends mainly on the rate of renewability of the vegetation, which can be measured in terms of net primary production of ecosystems and the human appropriation of this net primary production (HANPP; see Table 3.1). The ecological limits of growth in this metabolic regime are dependent on the controlled use and harvest of solar energy in plant production, with an elastic boundary varying with the ecological and social forms and conditions of agricultural production. The final ecological limits at global scale, as discussed currently, cannot be transgressed in this mode of production. Under these conditions the agricultural mode of production is ecologically sustainable. Sustainability is determined by the limited availability of agricultural land, the (in terms of energy) stationary state economy and the lack of technological innovations (such as synthetic fertilisers and agrochemicals). A somewhat simplified characterisation of the environmental sustainability of pre-industrial agricultural societies is of economic poverty, the limited role of markets and the dominance of subsistence production. The absence of exponential economic growth prevented the global spread of resource use problems and environmental pollution that came with the *societal metabolism of industrial society*. Its core process is the decoupling of the use of material and energy resources from the local limits of resource availability, through the building of the modern world system and economic globalisation. Under these conditions the sustainability process began as a race against the planetary boundaries of resource use in the late phase of the Anthropocene, as the industrial society is called now.

Important forms of new and integrated knowledge for the sustainability process are provided with the ecological indicators that integrate ecological, social and economic knowledge (Table 3.1). Such indicators and indices are used in the scientific and political sustainability discourses: indices of environmental vulnerability (EVI) and sustainability (ESI) of carrying capacity, of the human appropriation of net primary production, of planetary boundaries and of ecological footprints. The examples show different facets of (non-)sustainability of human

Table 3.1 Ecological indicators—natural limits and sustainability

Name	Definition	Interpretation	Sources
Carrying capacity	Maximum number of organisms of a species an environment/ecosystem can sustain indefinitely without negative consequences	Carrying capacity of a species is defined in ecology as the maximum number of individuals of that species; because consumption levels of humans vary widely, Rees has suggested that carrying capacity of humans should be defined by consumption levels	Sayre 2008
Ecological footprint	Biologically productive area of the earth necessary to support an individual or a population (of a city, region, country, the earth); includes the area for absorption of waste (calculated in global hectares)	The ecological footprint can be compared with the biocapacity of an ecosystem (the capacity to produce biological materials and to absorb waste: the counterpart on the supply side when the footprint is seen as the demand); the concept of the footprint works with simplifications; additional indicators are necessary to measure environmental sustainability	Wackernagel et al. 2019
SPI (Sustainable Process Index)	The material and energy flows necessary for a product or a service, including the whole life cycle from production to consumption and reduction or absorption	Similar to the ecological footprint but more complex (including direct use of surface, use of renewable, non-renewable and fossil resources, emissions in air, water, soil)	Narodoslawsky and Krotscheck 1995
EPI (environmental performance index) and ESI (environmental sustainability index)	EPI measures environmental performance of a state/country to compare it with other countries; 16 indicators are used to measure health impacts, air quality, water quality, wastewater treatment, state of agriculture, forestry, fisheries, biodiversity and habitat, climate and energy; ESI was an earlier variant of the EPI, measuring the overall progress to sustainability of a country with 21 indicators	The variables for the calculation of EPI have changed over time, making the comparison of EPI measured in different years complicated; the environmental performance depends heavily on the wealth of a country: developed/industrialised countries have a high EPI, poor countries have a low EPI	Yale Center for Environmental Law and Policy 2018

Name	Definition	Interpretation	Sources
EVI (environmental vulnerability index)	EVI uses 50 indicators to measure the relative severity of environmental pressure (exposure to a series of environmental risks) in a country	EVI in contrast to EPI: most developed or industrialised countries are highly or extremely vulnerable to environmental risks	Environmental Vulnerability Index 2013 (South Pacific Applied Geoscience Commission and United Nations Environment Program)
HANPP (human appropriation of net primary production)	HANPP is a social-ecological indicator measuring the part of biomass production (net primary production of an ecosystem) which is appropriated by human society	The HANPP quantifies the intensity of human use of the biomass for the purpose of assessing the environmental sustainability of human modification of ecosystems (through agriculture and other forms of land use); the analysis by Erb et al. (2009) shows that the efficiency of land use has increased in recent decades, but in absolute terms the global HANPP has continued to increase, implying high levels of pressure on ecosystems	Erb et al. 2009
Planetary boundaries	Nine planetary boundaries are specified which should not be transgressed through human use of natural resources (to avoid abrupt and catastrophic environmental changes in the earth system)	Three of the nine boundaries have already been transgressed: climate change, biodiversity loss and biogeochemical change; the other boundaries include ocean acidification, land use, freshwater, ozone depletion, atmospheric aerosols and chemical pollution. The concept of planetary boundaries is controversial but has received widespread attention	Rockström et al. 2009

Sources: mentioned in the table

resource use, but additional indicators and combinations of economic and social indicators can improve the knowledge practices in the sustainability process.

A new hypothesis and interdisciplinary concept—the Anthropocene: The connected economic and ecological changes of the earth system are described with the term "Anthropocene" (Costanza et al. 2007) for the epoch of industrialisation, beginning about 250 years ago. It is assumed that for the first time in their history humans are changing the total earth system through economically organised natural resource use, with the help of scientific knowledge and technological innovations. The negative consequences for all of society, the economy and nature are already perceptible. The concept of the Anthropocene, not yet an elaborated theory, is of natural-scientific origin: it is an attempt to describe human society and history as part of the long geological processes of change on the surface of the earth, in the sphere where life exists ("biosphere"). The social-ecological change in the Anthropocene can be specified in ecological indicators that show the global effects of human modification of the earth by measuring the human appropriation and throughput of energy and material resources in social systems in late modern society (Table 3.1). The indicators show that humans are appropriating an ever greater part of the natural resources available on earth. Meanwhile the resource flows in social systems (resources appropriated and consumed by humans) approach the level of resource flows in ecosystems (that produce the natural resources on the earth). Various ecological indicators show the increasing intensity of natural resource use by humans and provide arguments for enforcing the reduction of resource use to achieve environmental sustainability.

The ecological changes that are the focus of the concept and hypothesis of the Anthropocene are global environmental and social change processes caused by the modification of ecosystems by humans through resource use in agriculture, industry and other economic activities. Human activity has impacted all parts of the earth system, even these not inhabited by humans—the Arctic and Antarctic areas, the atmosphere, the oceans and the deep sea. As an ecological description of the modifications of natural ecosystems by humans and the creation of novel ecosystems, the term Anthropocene is not controversial. It becomes controversial with its analysis of the causes of the changes and its simplifications regarding the social subjects responsible. The social subject of action and change is insufficiently described as humankind, ignoring manifold social, political and economic differences between actors, groups and institutions. Responsibility is attributed in more or less similar form to all human beings, although the processes of global environmental change were caused through culturally, socially and economically specific and limited processes in certain countries: more through the systemic organisation of the modern economy and society, through industrialisation and the modern capitalist world system, than through subjective or individual changes of resource use and consumption. Differentiations of worldviews, views of nature and humans, cultural and social differentiations do not count for much in the concept of the Anthropocene, which adopts a specific form of normativity with questions of whether humans are now overwhelming the great forces of nature.

The controversy over the Anthropocene is about the subject of ecological modifications in the last phase of modernity: is it humankind in general, or is it a historically specific economic systemic, mode of production and social organisation of natural resource use in the form of the modern world economic system? The critical counterpart of the concept of the Anthropocene, articulated in the term "capitalocene" (Moore 2017), assumes systemic causes of the ecological crisis. One can find arguments for both variants, but both generalise one aspect in their explanations: either that of human action (Anthropocene), or that of societal system (capitalocene). With a theoretically refined reasoning, both components of social-ecological systems and human subjects would be required to describe the problems: the historically and geographically varying forms of coupled social-ecological systems and human resource use.

The problems of human or anthropogenic changes to nature are a consequence of processes of intensification of human use of natural resources that began earlier in human history. A decisive transformation was the agricultural revolution about ten thousand years ago. With that began the development of agro-ecosystems, and further changes in land use through settlement and urbanisation that began in the Near East and in Europe soon thereafter. The modification of ecosystems is completed with the modification of the earth system through industrialisation and global environmental change. Throughout modernity the problems of environmental destruction have been recognised and scientifically analysed, but they were often ignored politically or forgotten about for long periods.

3.6 Discussion and Conclusions

Opening economics up to allow for interdisciplinary discussion and knowledge exchange is a difficult project, although a necessary one for the sustainability discourse. Interdisciplinarity is not strongly supported in mainstream economics or environmental economics, which have developed by maintaining a disciplinary identity and tradition. The use of economic knowledge in the broader sustainability debate, furthermore, brings a discussion of the limits of economic knowledge and the value bases of economics. Ludwig (2001) criticised economics for its inability to deal with the new situation of governance of complex systems, referring to Kaysen (1968) and Nelson (1987), who argue that economic policy recommendations are value-loaded, selective and committed to a competitive market ideal, thus becoming part of an ideology of economism that suppresses all other values in favour of economic ones and reduces the value of nature to its immediate utility for economic purposes (Ludwig 2001: 759). The situation changes gradually with the development of critical forms of environmental and ecological economics that attempt to critically reflect on their normative premises and assumptions. Burns and Schaefer Caniglia (2016: 89) describe the drama of economics as something that can happen to any discipline: its central ideas "*can come to be dumbed down by common usage and unquestioning belief… As the discipline of economics seeped into the common consciousness in the culture of modernity, a number of ideas came to be accepted as truisms.*"

Understanding complex systems and processes such as climate change or sustainable development requires a plurality of knowledge sources—combinations of different approaches and knowledge that cannot be synthesised in a single and coherent theory or approach, and that do not allow the design of management approaches for natural resources as they are traditionally understood, subjecting resource use to the criteria of economic rationality. Ludwig's conclusion is that economic knowledge must be combined with other forms, and economics must cooperate with other disciplines, with practitioners and interested parties in participatory forms "on an equal footing". This is similar to the situation described as post-normal science by Funtowicz and Ravetz (1993), and to the transdisciplinarity discourse. It can be summarised as the requirement to develop a meta-economic terminology for discussing the integration of ecological and economic knowledge. But it is only one step forward; participatory knowledge practices alone cannot solve the many problems of knowledge integration that require epistemological reflection and methodological improvement.

Conclusions: The rationale of the future sustainability discourse can be described as analysing and assessing the socially structured processes that have the effect of blocking or supporting forms of sustainable development; "*human choices are at the root of everything*", *but they are* "*given effect by larger organizational structures*" (Perrow and Pulver 2015: 61). The authors describe organisations as having the contradictory properties of independent actors, institutional isomorphism and collective rationality in a market-critical perspective. A dialogue between the disciplines and specialised researchers is one precondition of the sustainability discourse; another one is the assessment of economic knowledge with the help of other, interdisciplinary knowledge.

Many of the questions of interdisciplinary embedding of economic knowledge in the sustainability discourse and process touched on in this chapter have not yet been sufficiently discussed; temporary answers need to be found in the continuing processes of knowledge creation for sustainability. The limits of specialised economic knowledge for sustainable development that became evident in the course of the discourse are visible in the controversial themes to be discussed in the following chapters: the valorisation and commercialisation of natural resources, the natural limits of economic growth and the economics of ecosystem services. With them appears the question of the transformation of the growth-based economy to a non-growing economy.

Summary of Chap. 3: Inter- and Transdisciplinary Knowledge Integration
In the knowledge chain from production to integration, dissemination and application of knowledge, the core process on which the success of sustainability governance and the global transition to sustainability depends has become the inter- and transdisciplinary combination and integration of knowledge. Epistemologically and methodologically such integration is difficult; it is neglected in the sustainability process, where knowledge integration and application are often reduced to problems in the policy and practice of sustainable development, where power relations determine what knowledge is applied.

In the sustainability discourse, economic knowledge needs to be combined and integrated with other knowledge from ecological and environmental research:

1. From ecological research, two processes are important: (a) material and energy flows in ecosystems to maintain their functions and services, and (b) the societal metabolism analysed in social ecology, including the material and energy flows between nature and society (the "physical economy" that connects humans with nature). Connected with the analyses of these flows are analyses of environmental pollution, global environmental change, overuse of natural resources, "limits to growth" and "planetary boundaries".
2. Knowledge from social-scientific research is important to understand the institutional transformation of natural resource use, the systemic organisation of the economy, the social structures and the division of labour in modern society. This knowledge is often ignored in ecological research and in the sustainability process.
3. Additional knowledge from interdisciplinary ecological research in human, social and political ecology or sustainability science helps to connect the processes of knowledge production and application in the sustainability process. Knowledge transfer and application invoke difficult problems of the entanglement of knowledge and power that are visible in controversial themes in economic research.

Appendices: Further Information and Material

Learning Exercise 1: Discussion of Theoretical Concepts for Interdisciplinary Sustainability Research

Discuss individually or in groups the combined theoretical concepts described in Boxes 3.3 and 3.4 for modern society, modes of production and societal metabolism. The aim is to find out why these concepts are relevant for the sustainability discourse and how they can be used (as theoretically defined, or in simpler forms and as descriptive terms).

The boxes can be used as introductory texts for the discussion. No additional reading is required for this discussion. Learning exercise 3 provides the opportunity to discuss in detail a specific concept of society that is relevant for sustainability discourse: the concept of "macrosociety" conceived by Marchalek. The following chapters provide further knowledge to clarify the concepts and their application in the sustainability discourse.

Learning Exercise 2: The Interests of Future Generations

The discussion about obligations or possible forms of action (and their justification) in the interest of future generations in environmental and sustainability governance is the domain of environmental ethics or intergenerational ethics. In the controversial debates on the responsibility of present generations for future generations, new ideas about how to represent politically the interests of future generations have been suggested. The introductions to environmental ethics by Attfield (2003) and DesJardins (2006) discuss the ethics of sustainable developmental with regard to responsibilities to future generations that derive from the influence of current resource use practices on future generations. It is not necessary to read these books; the arguments are summarised in the following paragraphs (where you will also find additional suggestions for discussion and reading). As a short introduction to environmental ethics you can read the recently published article "Environmental ethics" by Ott (2020), available online.

1. *The ethical discourse of future-related responsibilities* (Attfield 2003: 96ff): "Future generations" refers to all future generations for which action or inaction in the present makes a difference to their living conditions, not only that of the near future (of our children and grandchildren or of the next generation). Sustainable development aims to maintain the natural resource base and the

ecosystems functioning for the continuity of human life. The ethical reasoning for sustainable development aims to maintain the environmental conditions for a decent life for future generations: maintaining the stocks of renewable resources, maintaining ecosystem services and reducing environmental pollution to prevent harm to future generations of living beings, not only humans. To protect the interests of future generations in this sense does not require knowing the people and their concrete interests, only their basic needs; protection requires, for example, avoiding uncontrollable human-made risks such as nuclear or bacteriological wars, not modifying or destroying ecosystems such that they cannot deliver their life-supporting services, and not using risky technologies with unforeseeable consequences that would pose risks and cause damage to future generations. Future-oriented sustainability policies imply responsibilities for negative changes (harm to future generations, catastrophes) and for positive changes (to maintain or improve the conditions and the quality of life). The interests of generations in the distant future can be seen in an analogy to spatially distant people; most ethicists argue that we have responsibilities for them, not only for the ones we know or that live near us. With the current anthropogenic climate change the risk is that it will impact on the conditions of life on earth for many thousands of years. Differentiating between the near and the distant future brings to the fore problems with trade-offs, such that decisions made now can have positive effects for the generations in the near future but negative ones for those in the distant future.

2. *Transferring the ethical debate about rights of future generations into a policy- and economy-related discussion of sustainability*: The discussion of the relations between economics, ecology and ethics of sustainability is differentiated and complicated, as shown in the classic anthology edited by Daly (1980). Howarth's (2007) attempt to simplify the ethical requirements for sustainable development is useful; he defines sustainability operationally for the political and economic processes of sustainable development in three components. (1) Members of future generations are morally entitled to enjoy conditions of life as good as or better than today. (2) Sustaining a non-declining level of welfare for future generations is fraught with difficulties because economic values, preferences, institutions and technologies change over time and cannot be predicted for the long term—economic operationalisation of future welfare remains difficult or practically impossible. (3) Duties to conserve and protect the environment for the utility of future generations do not imply that natural resource stocks need to be conserved permanently, but future generations need to be compensated for resource depletion or environmental degradation. This can be done with the help of a "structured bequest package" (Norton and Toman 1997) in which manufactured capital, technological capacity, environmental re-

sources, and functioning political and social institutions are maintained over time (Howarth 2007: 661f).

Suggestions for further reflection and discussion: On the internet you may find many articles and texts about the interests of future generations. Read the "Declaration on the Responsibilities of the Present Generations for Future Generations" from 12 November 1997 by UNESCO, and one or two texts that specify the possibilities of future-oriented action to deal with the ethical problems in sustainable development, for example Nesbit and Illés 2015 or Anderson 2018. Discuss (individually or in groups) how you would define and specify the responsibilities for future generations? How can resources be shared fairly between present and future generations? How and how far do the responsibility for humans and for nature converge? Who are the actors responsible for future generations, and what concrete actions can they take for future generations?

Questions for in-depth reading and discussion: How can ethical principles of sustainable development be introduced and applied in the policy and governance processes (see sources mentioned above: Attfield 2003, DesJardins 2006, Ott 2020)? Is a global ethic necessary or can sustainability be achieved with different and culturally specific ethics? Can the "fair earth share" help to achieve intra- and intergenerational solidarity in resource use? How can the problems of "discounting the future" (Mathis 2011: 165ff) and the "nonidentity problem" (Boonin 2014) regarding future generations be dealt with in practice?

Learning Exercise 3: Development of Macro-societies as Exceptional Forms in Human History

Discuss the following ideas of Marchalek and their relevance for an interdisciplinary social-ecological analysis and explanation of modern society and resource use problems in the global economy:

The development of large societal, political and economic systems is a central topic in cultural anthropology, but the reasons they developed are not easily found in anthropology; they require further knowledge from the social and natural sciences. In anthropology large societal systems appear as more unstable and unsustainable than small societal systems (Bodley 1994). Macro-societies are exceptional forms in human society that developed relatively recently, around four thousand years ago (Marchalek 1992).

Such societies do not develop simply through population growth but are based on specific agricultural and industrial modes of production. Marchalek explains the rarity of macro-societies among humans and other animal species through a highly improbable coincidence of factors. Four types of constraints limit their development: organismic, ecological, economic and sociological (Marchalek 1992: 39ff). Organismic constraints include the biological factors of reproduction capability of species; ecological constraints include the size and carrying capacity of ecosystems that determine the availability of life-supporting resources such as food and raw materials; and economic constraints include the forms of division of labour and the costs and benefits of organising large groups. Whereas these are only briefly discussed by the author, the sociological constraints are the focus of his analysis, including (1) the ability of individuals to cooperate impersonally (outside kinship groups), (2) the division of labour between members of different social groups and classes (not gender groups) and (3) specialised forms of labour that need to be coordinated and integrated (historically through hierarchic forms of organisation and bureaucracies).

This interdisciplinary, natural- and social-scientific explanation of societal development is controversial in disciplinary social-scientific explanations of societies, where the organismic and ecological factors are excluded. The sociological explanation of societies usually begins with the division of labour, cooperation and social organisation, where natural-scientific explanations would be rejected as ahistorical and simplistic, as biological or ecological reductionism that applies false explanations to human societies. This controversy continues. Marchalek's theory brings some plausible and sociologically meaningful explanatory components to the discussion. It is compatible with other explanations of modern society that are discussed in the social sciences: its historically exceptional development path, its late emergence in human history, its dependence on cultural and social factors and capacities such as knowledge, culture and technology. It is a simplification by way of interdisciplinary generalisation and abstraction, but it helps in analysing the social, economic and ecological consequences of the development of the modern global society and economy. Such analysis is developing in theoretically elaborate and more detailed forms with world system theory (Wallerstein) and the connected world ecology theory (Moore).

Further Reading Suggested: Deepening, Thematically Specialised

Interdisciplinary knowledge integration: Klein (1990)—introduction and overview; Barry et al. (2008)—the logics of interdisciplinarity; Jacobs and Frickel (2009)—critical review of interdisciplinarity; Nowotny et al. 2003—transdisciplinary knowledge practices.

Sustainability: Leal Filho (2018), Leal Filho et al. (2018).

Scientific journals (international, peer reviewed) with research relevant for the themes of this chapter: "Annual Review of Environment and Resources", "Consilience: The Journal of Sustainable Development"; "Current Opinion in Environmental Sustainability", "Environmental Science and Technology", "Economics, Management and Sustainability" "Environmental Research Letters", "Environmental Innovation and Societal Transitions" "Frontiers in Ecology and the Environment", "Global Environmental Change", "International Journal of Sustainable Development", "International Journal of Sustainable Economy", "Journal of Biodiversity and Conservation Research", "Science of the Total Environment", "Sustainability", "Sustainability Science", "Sustainability: Science, Practice and Policy", "The Journal of Environmental Education"

References

Altvater, E. (1992). *Der Peis des Wohlstands oder Umweltplünderung und neue Welt(un) ordnung*. Münster: Westfälisches Dampfboot.

Anderson, V. (2018). *Protecting the Interests of Future Generations*. CUSP – Centre for the Understanding of Sustainable Prosperity. CUSP Working Paper No. 14. University of Surrey, Guildford, UK.

Attfield, R. (Ed.). (2003). *The Ethics of the Environment*. London and New York: Routledge.

Baccini, P., & Brunner, P. H. (1991). *Metabolism of the Anthroposphere*. Berlin et al.: Springer.

Barkin, J. S. (2006). Discounting the Discount Rate: Ecocentrism and Environmental Economics. *Global Environmental Politics, 6*(4), 56–72.

Barry, A., Born, G., & Weszkalnys, G. (2008). Logics of Interdisciplinarity. *Economy and Society, 37*(1), 20–49.

Becker, U. (1988). From Social Scientific Functionalism to Open Functional Logic. *Theory and Society, 16*(6), 865–883.

Beckert, J. (2007). *The Great Transformation of Embeddedness: Karl Polanyi and the New Economic Sociology*. MPIfG Discussion Paper 07/1. Max Planck Institute for the Study of Societies, Cologne, Germany.

Bennett, J. W. (1976). *The Ecological Transition: Cultural Anthropology and Human Adaptation*. New York: Pergamon Press.

Benson, M. H., & Craig, R. K. (2014). The End of Sustainability. *Society and Natural Resources, 27*, 777–782.

Bicchieri, C. (2006). *The Grammar of Society: The Nature and Dynamics of Social Norms*. New York: Cambridge University Press.

Bodley, J. (1994). A Culture Scale Perspective on Human Ecology and Development. *Advances in Human Ecology, 3*, 93–112.

Boonin, D. (2014). *The Non-Identity Problem and the Ethics of Future People*. Oxford: Oxford University Press.

Boulding, K. E. (1965). *The Meaning of the 20th Century: The Great Transition*. New York: Harper & Row Publishers.

Brockington, D. (2012). A Radically Conservative Vision? The Challenge of UNEP's Towards a Green Economy. *Development and Change, 43*(1), 409–422.

Brown, T. C. (1984). The Concept of Value in Resource Allocation. *Land Economics, 60*(3), 231–246.

Bruckmeier, K. (2016). *Social-Ecological Transformation: Reconnecting Society and Nature*. London: Palgrave Macmillan.

Bruckmeier, K. (2019). *Global Environmental Governance: Social Ecological Perspectives. Cham*. Switzerland: Palgrave MacMillan and Springer Nature.

Buchanan, J. M., & Brennan, G. (1985). *The Reason of Rules: Constitutional Political Economy*. Cambridge: Cambridge University Press.

Buchanan, J. M., & Tullock, G. (1962). *The Calculus of Consent: Logical Foundations of Constitutional Democracy*. Ann Arbor: University of Michigan Press.

Burkett, P. (1999). *Marx and Nature: A Red and Green Perspective*. New York: St. Marin's Press.

Burns, T. J., & Schaefer Caniglia, B. (2016). *Environmental Sociology: The Ecology of Late Modernity*. Norman, OK: Mercury Academic.

Colyvan, M., Linquist, S., Grey, W., Griffiths, P., Odenbaugh, J., & Possingham, H. P. (2009). Philosophical Issues in Ecology: Recent Trends and Future Directions. *Ecology and Society, 14*(2), 22. [Online]. Retrieved from http://www.ecologyandsociety.org/vol14/iss2/art22/.

Costanza, R., Graumlich, L., & Steffen, W. (Eds.). (2007). *Sustainability or Collapse? An Integrated History and Future of People on Earth*. Cambridge, MA and London, UK: MIT Press.

Cudworth, E. (2005). *Developing Ecofeminist Theory: The Complexity of Difference*. New York: Palgrave Macmillan.

Daly, H. (Ed.). (1980). *Economics, Ecology, Ethics*. San Francisco: Freeman and Company.

DesJardins, J. R. (2006). *Environmental Ethics. An Introduction to Environmental Philosophy*. Belmont, CA, et al.: Thomson Wadworth.

Dieng, B., & Pesqueux, Y. (2017). On 'Green Governance'. *International Journal for Sustainable Development, 20*(1/2), 111–123.

Douai, A. (2009). Value Theory in Ecological Economics: The Contribution of a Political Economy of Wealth. *Environmental Value, 18*, 257–284.

Dunlap, R. E., & Catton, W. R. (2002). Which Function(s) of the Environment Do We Study? A Comparison of Environmental and Natural Resource Sociology. *Society and Natural Resources, 15*, 239–249.

Environmental Vulnerability Index. (2013). Retrieved March 14, 2020, from www.vulnerabilityindex.net/category/indicators/.

Erb, K.-H., Krausmann, F., Gaube, V., Gingrich, S., & Bondeau, A. (2009). Analyzing the Global Human Appropriation of Net Primary Production – Processes, Trajectories, Implications. An Introduction. *Ecological Economics, 69*, 250–259.

Felt, U., Fouché, R., Milelr, C., & Smith-Doerr, L. (Eds.). (2017). *The Handbook of Science and Technology Studies* (4th ed.). Cambridge, MA: MIT Press.

Fernández Pinto, M. (2019). Scientific Ignorance: Probing the Limits of Scientific Research and Knowledge Production. *THEORIA, 34*. https://doi.org/10.1387/theoria.19329.

Fischer-Kowalski, M., & Haberl, H. (Eds.). (2007). *Socioecological Transitions and Global Change: Trajectories of Social Metabolism and Land Use*. Edward Elgar.

Funtowicz, S. O., & Ravetz, J. R. (1993). Science for the Post-Normal Age. *Futures, 25*(7), 735–755.

Funtowicz, S. O., & Ravetz, J. R. (1994). Emergent Complex Systems. *Futures, 26*(6), 568–582.

Gaard, G. (2011). Ecofeminism Revisited: Rejecting Essentialism and Re-Placing Species in a Material Feminist Environmentalism. *Feminist Formations, 23*(2), 26–53.

Gaard, G., & Gruen, L. (1993). Ecofeminism: Toward Global Justice and Planetary Health. *Society and Nature, 2*, 1–35.

Ginsburg, V. A., & Throsby, D. (Eds.). (2006). *Handbook of the Economics of Art and Culture* (Vol. 1). North Holland.

Gonner, E. C. K. (1912). *Common Land and the Inclosure*. London: Macmillan and Co..

Granovetter, M. (1985). Economic Action and Social Structure: The Problem of Embeddedness. *American Journal of Sociology, 91*, 481–510.

Granovetter, M., et al. (2004). Polanyi Symposium: A Conversation on Embeddedness. *Socio-Economic Review, 2*, 109–135.

Gross, M., & McGoey, L. (Eds.). (2015). *International Handbook of Ignorance Studies*. Abingdon and New York: Routledge.

Habermas, J. (1991). Erläuterungen zur Diskursethik. Frankfurt a.M.: Suhrkamp.

Haila, Y., & Heininen, L. (1995). Ecology: A New Discipline for Disciplining? *Social Text, 42*, 153–171.

Haila, Y., & Levins, R. (1992). *Humanity and Nature: Ecology, Science and Society*. London: Pluto.

Hardin, G. (1968). The Tragedy of the Commons. *Science, 162*, 1243–1248.

Hartmann, T. (2012). Wicked Problems and Clumsy Solutions. *Planning Theory, 11*(3), 221–241.

Howarth, R. B. (2007). Towards an Operational Sustainability Criterion. *Ecological Economics, 63*, 656–663.

IPCC. (2005). *Carbon Dioxide Capture and Storage*. Cambridge et al.: Intergovernmental Panel on Climate Change, Working Group III. IPCC Special Report.

Ives, C. D., Abson, D. J., von Wehrden, H., Dorninger, C., Klaniecki, K., & Fischer, J. (2018). Reconnecting with Nature for Sustainability. *Sustainability Science, 18*, 1389–1397.

Jacobs, J. A., & Frickel, S. (2009). Interdisciplinarity: A Critical Assessment. *Annual Review of Sociology, 35*, 43–65.

Kallis, G., & Norgaard, R. (2010). Coevolutionary Ecological Economics. *Ecological Economics, 69*(4), 690–699.

Kaysen, C. (1968). Model-Makers and Decision-Makers: Economists and the Public Process. *Public Interest*, 80–95.

Klein, J. T. (1990). *Interdisciplinarity: History, Theory, and Practice*. Detroit: Wayne State University.

Köhler, J., Geels, F. W., Florian Kern, F., Markard, J., Onsongod, E., Wieczoreke, A., et al. (2019). An Agenda for Sustainability Transitions Research: State of the Art and Future Directions. *Environmental Innovation and Societal Transitions, 31*, 1–32.

Koren, I., & Krishna, C. M. (2007). *Fault-Tolerant Systems*. San Francisco, CA: Morgan Kaufmann.

Krausmann, F., & Fischer-Kowalski, M. (2010). *Gesellschaftliche Naturverhältnisse: Energiequellen und die globale Transformation des gesellschaftlichen Stoffwechels.* Social Ecology Working Paper 117. Vienna: Institute of Social Ecology.

Krausmann, F., & Schandl, H. (2006). Der soziale Metabolismus der Industrialisierung: Die Überwindung der energetischen Schranken des agrarischen Wirtschaftens. *GAIA, 15*(4), 285–293.

Kuhn, T. (1962). *The Structure of Scientific Revolutions.* Chicago: University of Chicago Press.

Lansing, J. S. (2003). Complex Adaptive Systems. *Annual Review of Anthropology, 32*, 183–204.

Leal Filho, W. (Ed.). (2018). *Handbook of Sustainability Science and Research.* Palgrave Macmillan.

Leal Filho, W., Marans, R. W., & Callewaert, J. (Eds.). (2018). *Handbook of Sustainability and Social Science Research.* Cham, Switzerland: Springer Nature.

Leary, H., & Walker, A. (2018). Meta-Analysis and Meta-Synthesis Methodologies: Rigorously Piecing Together Research. *TechTrends, 62*, 525–534.

Loubser, J., Baum, R., Effrat, A., & Meyer Lidz, V. (Eds.). (1976). *Explorations in General Theory of Action: Essays in Honor of Talcott Parsons.* 2 Vols. New York: The Free Press.

Ludwig, D. (2001). The Era of Management Is Over. *Ecosystems, 4*, 758–764.

Malhi, Y., Franklin, J., Seddon, N., Solan, M., Turner, M. G., Field, C. B., et al. (2020). Climate Change and Ecosystems: Threats, Opportunities and Solutions. *Philosophical Transactions of the Royal Society B, 375*(1794). https://doi.org/10.1098/rstb.2019.0105.

Marchalek, R. (1992). The Evolution of Macrosociety: Why Are Large Societies Rare? *Advances in Human Ecology, 1*, 33–64.

Martinez-Alier, J. (2001). Ecological Conflicts and Valuation: Mangroves Versus Shrimps in the Late 1990s. *Environment and Planning C: Government and Policy, 19*(5), 713–728.

Mathis, K. (Ed.). (2011). *Efficiency, Sustainability, and Justice to Future Generations.* Dordrecht et al.: Springer, Science+Business Media B.V.

Moore, J. W. (2017). The Capitalocene, Part I: On the Nature and Origins of Our Ecological Crisis. *The Journal of Peasant Studies, 44*(3), 594–630.

Nakamori, Y. (Ed.). (2016). *Knowledge Synthesis: Western and Eastern Cultural Perspectives.* Tokyo et al.: Springer Japan.

Narodoslawsky, M., & Krotscheck, C. (1995). The Sustainable Process Index (SPI): Evaluating Processes According to Environmental Compatibility. *Journal of Hazardous Materials, 41*(2+3), 383–397.

Nelson, R. H. (1987). Sustainability, Efficiency and God: Economic Values and the Sustainability Debate. *Annual Review of Ecology and Systematics, 26*, 135–154.

Nesbit, M., & Illés, A. (2015). *Establishing an 'EU Guardian for Future Generations'.* Report and Recommendations for the World Future Council. Institute for Environmental European Policy, London.

Nesshöver, C., Assmuth, T., Irvine, K. N., Rusch, G. M., Waylen, K. A., Delbaere, B., et al. (2017). The Science, Policy and Practice of Nature Based Solutions: An Interdisciplinary Practice. *Science of the Total Environment, 529*, 1215–1227.

Norgaard, R. (1984). Coevolutionary Development Potential. *Land Economics, 60*(2), 160–173.

Norgaard, R. (1994). *Development Betrayed: The End of Progress and a Coevolutionary Revisioning of the Future*. London and New York: Routledge.

Nowotny, H. (2003). Science in Search of Its Audience. *Nova Acta Leopoldina, 87*(325), 211–215.

Ostrom, E. (2009). A General Framework for Analyzing Sustainability of Social-Ecological Systems. *Science, 325*, 419–422.

Ott, K. (2020). Environmental Ethics. In T. Kirchhoff (Ed.), *Online Encyclopedia Philosophy of Nature*. https://doi.org/10.11588/oepn.2020.0.71420.

Park, C. S. (2003). On Replacing Labor as the Substance of Value: Early and Recent Arguments. *Science & Society, 676*(2), 160–172.

Perrow, C. (1984). *Normal Accidents*. Basic Books.

Perrow, C., & Pulver, S. (2015). Organizations and Markets. In R. E. Dunlap & R. J. Brulle (Eds.), *Climate Change and Society: Sociological Perspectives* (pp. 61–92). New York: Oxford University Press.

Polany, K., Arensberg, K., & Pearson, H. (1957). *Trade and Market in the Early Empires*. New York: The Free Press.

Polanyi, K. (1944). *The Great Transformation*. New York: Rinehart.

Proctor, R., & Schiebinger, L. (2008). *Agnotology: The Making and Unmaking of Ignorance*. Stanford, CA: Stanford University Press.

Ravetz, J. R. (1987). Usable Knowledge, Usable Ignorance, Incomplete Science with Policy Implications, Knowledge. *Creation, Diffusion, Utilization, 9*(1), 87–116.

Rockström, J., Steffen, W., Noone, K., Persson, A., Chapin, F. S., III, Lambin, T. M., et al. (2009). Planetary Boundaries: Exploring the Safe Operating Space for Humanity. *Ecology and Society, 14*(2), 32. Retrieved from http://www.ecologyandsociety.org/vol14/iss2/art32/.

Ross, E. B. (1998). *The Malthus Factor: Poverty, Politics and Population in Capitalist Development*. London and New York: Zed Books.

Sagan, S. D. (2004). The Problem of Redundancy Problem: Why More Nuclear Security Forces May Produce Less Nuclear Security. *Risk Analysis, 24*(4), 935–946.

Sarewitz, D. (2004). How Science Makes Environmental Controversies Worse. *Environmental Science & Policy, 7*(5), 385–403.

Sayre, N. F. (2008). The Genesis, History, and Limits of Carrying Capacity. *Annals of the Association of American Geographers, 98*, 120–134.

Sieferle, R. P., Krausmann, H., Schandl, V., & Winiwarter, V. (2006). *Das Ende der Fläche: Zum gesellschaftlichen Stoffwechsel der Industrialisierung*. Wien: Böhlau.

UNEP. (2011). *Towards Green Economy: Pathways to Sustainable Development and Poverty Eradication*. Geneva: United Nations Environment Programme.

Vlachou, A. (2002). Nature and Value Theory. *Science & Society, 66*(2), 169–201.

Wackernagel, M., Lin, D., Evans, M., Hanscom, L., & Raven, P. (2019). Defying the Footprint Oracle: Implications of Country Resource Trends. *Sustainability, 11*(7), 2164. https://doi.org/10.3390/su11072164.

Wagner, A. (1999). Causality in Complex Systems. *Biology and Philosophy, 14*, 83–101.

Wagner, A. (2005). *Robustness and Evolvability in Living Systems*. Princeton: Princeton University Press.

Wagner, A. (2011). *The Origins of Evolutionary Innovations*. Oxford et al.: Oxford University Press.

Wallerstein, I. (1991). Beyond Annales. *Radical History Review, 49*, 7–15.

Wallerstein, I. (1995). What Are We Bounding, and Whom, When We Bound Social Research? *Social Research, 62*(2), 839–856.

Warren, K., & Erkal, N. (Eds.). (1997). *Ecofeminism: Women, Culture. Nature*. Bloomington; IN: Indiana University Press.

Yale Center for Environmental Law and Policy. (2018). EPI Report 2018. Retrieved March 14, 2020, from https://epi.envirocenter.yale.edu/epi-report-2018.

Zylstra, M., Esler, K., Knight, A., & Le Grange, L. (2019). Integrating Multiple Perspectives on the Human-Nature Relationship: A Reply to Fletcher 2017. *The Journal of Environmental Education, 50*(1), 1–10.

Part II

Economic and Ecological Knowledge in the Sustainability Process

Economics Outright: Management of Natural Resources

4

This chapter describes the use of economic and ecological knowledge in natural resource management for sector-specific areas such as agriculture, fisheries and forestry, and for integrated or ecosystem-based management. In the management and use of biological or living resources such as plants and animals it is evident that ecological knowledge and criteria need to be used and combined with economic knowledge and criteria. Furthermore, other forms of resource use, of physical, mineral and energetic resources, also require knowledge about ecosystems and their maintenance and the application of ecological criteria. The practical challenge is to find adequate combinations of criteria for the many natural resources of different types and quality. Natural resource economics as economics in the strict sense of the term, of managing scarce resources with multiple criteria, requires a balancing of ecological and economic rationalities.

Figure 4.1 illustrates the forms of natural resource management, their interrelations, and their ecological and economic criteria.

The processes of using and managing natural resources—material and energetic—are economic ones where economic knowledge needs to be matched with ecological knowledge in the routine processes of resource management. Natural resources are used in all economic processes in some way and to some degree: as means or objects of trade and exchange, production, distribution and consumption. Not all economic activities use or produce living resources as in the first sector of the economy (hunting, agriculture, fisheries and forestry); for example the extracting and processing industries use energetic and material resources. The economic processes where natural resources are produced and processed can be called eco-

K. Bruckmeier, *Economics and Sustainability*,
https://doi.org/10.1007/978-3-030-56627-2_4

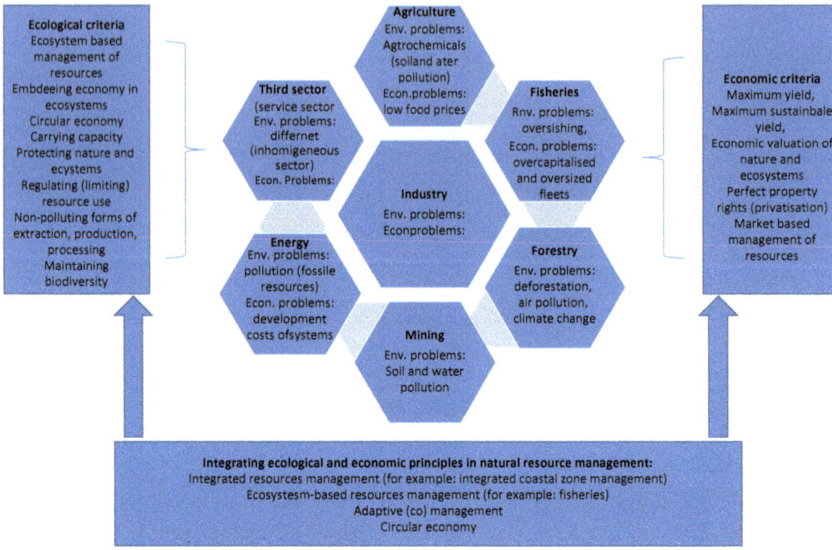

Fig. 4.1 Natural resource management: ecological and economic criteria

nomics outright, meaning: economics, where ecological knowledge and ecological limits are observed in economic production, use and management of resources. The economic objectives of resource management are subordinated to those of sustainability in the sense of intra- and intergenerational solidarity as guidelines for long-term development (Binswanger and Chakraborty 2000: 3; see also: Dale and Huber 2001 for land use management; Cooney 2004). Natural resource economics in this sense (Conroy and Peterson 2014; Anderson 2019) deals with the integration of economic processes with processes in ecosystems. This integration is described as integrated management or ecosystem-based management (see Sect. 4.2.2). For most of human history natural resource use was practised without modern economic knowledge, on the basis of the practical knowledge and experience of the resource users, often in sustainable forms. The forms of non-scientific knowledge in resource management are rediscovered in sustainable resource management.

The economy, in the sense of using and processing natural resources, comes close to the ancient Greek term "oikonomika" in the writings of Aristotle, meaning the administration and resource use in a house or household. This definition of economy is also adopted in ecology, where parallels between the "economy of nature" and the "economy of society" are drawn (see below). The other form of the economy, more similar to the modern market economy, is referred to by Aristotle as "chrematistics" or the monetary economy. The undifferentiated use of the term economy for both forms causes much of the current confusion, including in environmental economics and the sustainability discourse. The global economy functions as a chrematistic market economy, but the other economic functions cannot be ignored. It is necessary to note the various functions of resource management, chrematistics and the resource flow between the economy and nature—different forms of the economy that need to be integrated.

4.1 Economics, Natural Resource Management and Sustainable Development

Competing definitions of economics, as the science of the economy or otherwise, influence resource management, environmental economics and the sustainability discourse. Controversies about the definitions and views of economics and the economy illustrate the normative and value-based views that influence scientific thinking and research. Such controversies make interdisciplinary cooperation and knowledge exchange more difficult: which knowledge is important for natural resource management? The differentiation between social, economic and environmental sustainability in the dominant policy approaches does not help much in the clarification: regarding the use of economic knowledge in the sustainability discourse, two conflicting views of "economising ecology" or "ecologising the economy" clash. The first view assumes that the principles and institutions of the modern market economy that direct the appropriation, distribution and use of natural resources are also the universal principles for managing natural and other resources sustainably (understanding economics as the science of allocating scarce resources to alternative uses). The second view assumes that economic principles and institutions alone cannot rule the use of natural resources; they need to be embedded in the broader systems of society and nature, which requires prioritising the knowledge about social and ecological systems and, in natural resource management, prioritising ecological knowledge.

4.1.1 Controversies in Economics about the Meaning of Economics and the Economy

The controversial discussion about different approaches to economics and different understandings of the modern economy (Box 4.1) influences the sustainability discourse. It appears as a controversy between a disciplinary view that is dominant in conventional economics and environmental economics and an interdisciplinary view found in ecological economics.

In economics in the sense of *neoclassical and orthodox economics* (see Box 4.1), ecological criteria of resource use are neglected. In this view of economics as a specialised science, the economy appears as a closed system where economic principles determine resource management, ignoring the coupling between economics and ecosystems. The question of whether the economy as a social system of resource use can be decoupled from nature and ecosystems marks a cleavage between orthodox traditions of economics in the established academic sciences (with the assumptions of decoupling and substitution of natural capital through other capital forms) and the traditions of critical economics (see Chap. 5). In critical, heterodox forms of economics (political economy, institutional economics, development economics, feminist economics and ecological economics), the necessity of embedding the economy in social and ecological systems and the assumption of limits to growth are guiding ideas (see Chap. 6). Sustainable development as social and ecological transformation cannot be investigated in specialised science and realised in policy or governance practices with fragmented knowledge from specialised research, or with a single approach or theory. In the sustainability discourse and practice it is necessary to integrate different forms of economic knowledge.

Box 4.1 Different Views and Definitions of Economics and the Economy

In the social-scientific and interdisciplinary discourse the following views of the modern economy as a system can be found: (a) the orthodox view in economics is that of the economy as a market-based autonomous system that functions according to the principles and rules described in economics; (b) in sociology, the economy is seen as a system in the sense of a differentiated societal subsystem of modern society that forms, together with other specialised systems of politics, education and science, the overall system of modern society (the dominant view in functionalist sociology: Parsons and Smelser 1956; Luhmann 2012–2013); (c) in critical political economy, world system theory and world ecology, the historically specific system of the economy is the modern world economic system based on the capitalist mode of production (Wallerstein 1979; Moore 2015); (d) in interdisciplinary ecology and sustainability research, the economy is seen as part of a network of coupled,

(continued)

Box 4.1 (continued)

complex and adaptive systems in nature and society (Bruckmeier 2016). All variants except the first one adopt an interdisciplinary perspective in trying to understand the complex social reality, and the last two also its interaction with the biophysical reality.

In the academic economic discourse there is a clash between different views of economics as science that are often inexactly classified as *orthodox and heterodox approaches of economics*. Lee (1989, chap. six) describes orthodox economics (he includes classical and neoclassical variants) as based on a mechanistic paradigm of causation, on exponential growth and ecologically insensitive technology—a view of the economy that ignores the laws of thermodynamics and the principles of ecology in complex ecosystems.

Neoclassical economics (Morgan 2016) remains influential, although it has developed and adopted new concepts; it is an approach that analyses the determination of prices, outputs and income in a market economy through supply and demand (it assumes that economic actors aim to maximise utility according to rational choice theory; the methodology of marginal utility analysis is used to identify the value or price of goods). The market economy is the modern capitalist economy (although the term is not used in neoclassical economics). The combination of neoclassical microeconomics and Keynesian economics is called the "neoclassical synthesis" and is seen as mainstream economics today. This widespread view is contested, however, especially by Colander (2000: 130), who sees the current approach to economics as eclectic rather than as a form of neoclassicism. He argues from a historical perspective on the development of economics, where different approaches compete with each other. However, in his description of the principles of neoclassical economics and modern economics, the differences are limited: modern economics appears as more developed and differentiated but using similar principles. More relevant for the discussion of orthodox and neoclassical economics is Albert's (1967) critique of neoclassical economics as "model Platonism" (the model constructions in economics are immune to experience, empirical knowledge and falsification). Similarly, Rosenberg (1992, 1994) criticises neoclassical economics as lacking empirically based theories and empirical research, as only constructing complex mathematical models with little relevance for social and economic practice and reality.

(continued)

Box 4.1 (continued)

For the study of economic processes of natural resource use and management, two contrasting views of conventional disciplinary (orthodox) and critical interdisciplinary (heterodox) economics are described by Hall and Klitgaard (2012: 8f): conventional economics studies "the allocation of scarce resources among alternative choices", while critical interdisciplinary institutional economics studies (according to Polanyi) "how groups of humans transform nature to meet their needs". Polanyi called this view "substantive" because it derives from human dependence upon nature and other humans. Institutional and ecological economics as variants of heterodox economics study economic processes that are framed, structured and guided by different clusters of social institutions, including social cooperation, not only markets (Bromley 1991, 1992, 2014). To define the modern economy only as the market economy is misleading as it ignores the many social influences on economic processes and assumes that markets are autonomous and self-regulating institutions.

Sources: mentioned in the text

Polasky et al. (2019) summarise the discussion about the role of economics in analysing environmental and sustainability problems. They conclude that interdisciplinary cooperation and knowledge exchange are insufficient: the core of knowledge about sustainable development is provided by ecology and natural sciences, whereas "*the center of gravity in economics remains far removed from the challenge of sustainable development*" (Polasky et al. 2019: 5234). The insufficient interdisciplinary cooperation continues with the adoption of the growth-based "green economy" in global sustainability policy by the powerful political and economic elites, supported by ideas from orthodox economics and environmental economics (Heynen et al. 2007; Brockington 2012). The building of an interdisciplinary knowledge culture for sustainability research requires combining and integrating economic knowledge with that of interdisciplinary subjects such as human and social ecology, sustainability science, or the more recently developing transformation science and transdisciplinary fields connecting research on complex systems, sustainability and the transformation of society and the economy (Swilling 2016; Biggs et al. 2018; as part of a developing transformative science: see Chaps. 8 and 9). The process of building an interdisciplinary science for sustainability began with the opening up of ecology to the input of interdisciplinary knowledge from the natural and social sciences (Scoones 1999), elaborated more systematically in human, social and political ecology (Becker and Jahn 1999; Fischer-Kowalski and Haberl 2007; Bruckmeier 2019) and ecological economics (Martinez-Alier and Muradian 2015; Spash 2018).

In natural resource use, and for sustainable development, it is not possible to abstract from the social, material or energetic nature of resources, which are not used in a social vacuum. When natural resources are reduced to tradeable goods on markets, their valuation and use becomes distorted, creating unintended and unwanted effects that are analysed in social and ecological research on the environment and ecosystems. Scarcity of natural resources has become the main theme in the sustainability process, but the economically defined scarcity measured in monetary forms does not sufficiently reflect ecological scarcity, where natural limits to resource use need to be taken into account. Social scarcity as rooted in the institutional structure of an economy (for example structural scarcity through unequal access to natural resources) differs from natural or ecological scarcity, which is rooted in the physical and biological characteristics of ecosystems and the biosphere that determine the limits to growth (Lee 1989). Market prices alone do not adequately reflect the scarcity of natural resources, as they are influenced by a series of social and economic factors and criteria among which asymmetric power relations and regimes of appropriation and accumulation are relevant. This argument was first developed in institutional economics, where power relations were introduced as a variable in the analysis of price finding, drawing attention to the oligopolistic and monopolistic closure of markets that has reached extreme levels in the globalising economy.

The different scarcity problems that must be dealt with in the sustainability discourse include the systemic problems and effects of the economy as a social system: scarcity in terms of the global scarcity of natural resources, and scarcity problems created through the unequal distribution and appropriation of resources in the global economic system. These problems appear in another form in the sustainability discourse, as that of intra-and intergenerational solidarity. Connected to the scarcity questions is that of economic rationality. The discussion of rational behaviour in neoclassical and behavioural economics struggles with the construction of individual and collective rationality. Alternatives to rational choice theory have arisen with the construction of collective rationality, which analyses economic behaviour in groups or organisations (Newell and Simon 1972; Simon 1979, 1995), where the concept of bounded rationality emerged. This concept proposes two kinds of limits to rationality: (a) procedural limits in the approach to decision-making (resulting from context) such as organisational limits, lack of time and information, using insufficient or inadequate knowledge, insufficient or unclear rules, and inadequate use and reflection of value premises in decision processes; and (b) substantial limits resulting from the cognitive limits of information processing by decision-makers (Jones 1999: 299ff). Such limits of economic rationality and decision-making influence the possibilities to find solutions for the problems caused by economic growth, overuse of resources, pollution of the environment and destruction of ecosystems. The construction of refined or situation-specific models and the use of ever more data for economic and political decision-making (with the help of modern information technology) cannot solve the decision problems. The value basis of decisions and the guiding normative assumptions influence the results of decision-making as well as the power relations in the economic system.

With the two concepts of social and natural scarcity (with further differentiations) more variables and assumptions need to be specified in *economic models* constructed theoretically or with empirical knowledge (de Marchi and Blaug 1991; Caldwell 1994). Interdisciplinary integration of economic knowledge with other social and ecological knowledge creates possibilities of learning and modelling of the social and biophysical complexity that must be dealt with in the sustainability process. The practical question is, how many variables can be accounted for in a model and in economic and political decisions? The discussion in the research on sustainable resource management (Agrawal 2003, Ostrom 1990) shows that the limits of modelling the complexity of social and ecological systems for decision-making are quickly reached. The idea of including ever more data in computer-based modelling does not always represent a better approach to complexity but can also result in more artefacts: modelling is based on guesses, albeit informed ones. The schism between "economising nature" in neoclassical environmental economics and "ecologising the economy" (and economic thinking) in ecology and ecological economics that complicate the sustainability discourse cannot simply be resolved by improved economic and ecological modelling. Overcoming the controversies will require interdisciplinary research and knowledge integration.

4.1.2 Ecology as the "Economy of Nature"

Economic knowledge is used in ecology in the construction of two types of economy, the "economy of nature", analogising ecosystems with economic systems, and the "economy of society". Similarly, the concept of markets is used in biology by analogy with economic markets, and the comparison of rational economic decision-making of humans with adaptation through natural section in biological evolution (Hammerstein and Hagen 2005). Such forms of concept transfer between biology or ecology and economics, made possible through the abstract and simplified use of the concepts, ignore the differences between ecosystems and social systems. These differences are described in theoretical analyses in sociology and ecology. Couplings of social and ecological systems are couplings of two fundamentally different types of systems.

Ecology is called the "economy of nature" by Ricklefs (1993, see Box 4.1), using a term from historical descriptions of ecology, for example by Haeckel in 1870, and from earlier ecological thinking, especially in the work of Linné (2010), where it is connected to ancient ideas of natural harmony or order of the cosmos (Trepl 1994; see Chap. 2). Economy of nature utilises the old meaning of economy (from Aristotle) as the organisation of human resource use in the administration of a house or household. In analogy to that concept, economy of nature can be seen as "nature's householding", or the use and management of natural resources. The analogy does not work well for ecological processes; nature and society differ greatly in function, organisation and reproductive processes. In the environmental history of Worster (1979), the economy of nature is not seen as a consistent and coherent concept; rather, it differs across the many views of ecology historically, with the approach of Darwin and Haeckel being dominant in the development of ecology as a biological science (Box 4.2).

Box 4.2 The Economy of Nature

Ricklefs (1993) describes the economy of nature by analogy with the economy as a social system: "everything on the surface of the earth is interconnected in the same way that human enterprises are interrelated and defined by economic principles" (Ricklefs 1993: 2). This concept of an economy of nature does not say much about the relations between the economy in modern society and ecology as an economy of nature, but he concludes from his assertions about the analogy between ecosystems and economic systems that present-day environmental problems require the "intelligent application of general principles of ecology" in political, economic and social contexts (Ricklefs 1993: 22); this can be seen as a reformulation of the idea of sustainable development.

The economic principles are not discussed more concretely by Ricklefs. If this were to be done, it would be necessary to explain the differences between

(continued)

Box 4.2 (continued)

the two forms of economy: the societal economy varies widely in the course of human history, and it has existed in different historically specific systems and for short time spans as ecosystems, rather than being homogeneous, stable and long-lasting as with ecological processes in nature. To compare nature's and society's economies, the similarities of ecological and economic processes need to be described, as is attempted in the ecological and sustainability discourses. Essential principles in the economy of nature are the circulation of materials and the continually renewed processes of production, consumption and reduction in ecosystems; these differ from the historically varying forms economic production. The modern economy comprises a variety of institutions, structures, processes and forms of behaviour which cannot be analogised with nature or ecosystems. Nature does not know markets, money, capital, interest rates, property rights, enterprises, governments and states. The concept of resources as understood by the modern economy and its enterprises would not make much sense in ecosystems (except for living nature, plants and animals in agriculture or forestry, where production is a modified form of production in ecosystems). The basic processes that seem to support the analogy are those of material and energy flows that exist in ecosystems and social systems, and flow between ecosystems and social systems.

In enterprises in the societal economy, a great deal of the waste that is produced is not or cannot be used as material in economic processes or recycled. This can be described as an ecological deficit of the modern societal economy: it has no reduction function such as ecosystems have. From this deficit derives an ecological critique of the dysfunctionality of the modern economy. The critique is not theoretically based and empirically validated, but is based on an abstract analogy or metaphorical comparison. The difficulties with a closed circular economy become visible when recycling and waste production are analysed more concretely. Effective recycling is not possible in the societal economy: "recycling is downcycling", meaning that after several cycles the recycled materials become waste that can no longer be used in the economy and is therefore given back to nature. Certain forms of waste from the modern economy, including toxic and non-biodegradable waste, cannot be absorbed and reduced in the reduction function of ecosystems. These wastes remain in nature as residues of the human economy, creating dysfunctional absorption processes. An example is plastic, which

(continued)

Box 4.2 (continued)
has polluted the oceans and entered the food chain. Furthermore, reduction in ecosystems is bound to the periodical ecological processes of growth and primary production. In the societal economy only some branches (those where living resources are produced, such as hunting, agriculture, aquaculture, fishery, forestry) are closely connected to ecological principles of growth and periodical cycles.

Source: mentioned in the text

Ricklefs's analogy between enterprises and ecosystems gives a vague impression of how he sees the relations between ecology and the economy, or between the economy of nature and the economy of society. The analogy does not work well to explain the different principles of the functioning of ecosystems and enterprises. To describe the relations between economic organisations and ecosystems more adequately, the application of economic principles needs to be consistent with the application of ecological principles in natural resource use. Economic systems are part of larger systems: society and nature. Pollution and destruction of the environment through the industrial economy cannot be prevented by organising enterprises in a similar way to ecosystems based on a simplified concept of a circular economy. The problems need to be solved by preventing pollution, which requires observing ecological principles and reducing emissions; but in doing so, enterprises would not become similar to ecosystems.

In the systems ecology of Odum (1997a), where ecology is understood as a bridge between science (that of ecology and nature) and society, the comparison of ecosystems with social systems is specified (see Box 4.3). But thinking in analogies and transferring concepts between the two disciplines of ecology and economics risks distorting and misusing the specific concepts of both disciplines and drawing misleading conclusions from comparisons between nature and society, as in Odum's concept of "dual capitalism", which does not account for the functions and dysfunctions of the modern capitalist economy and simplifies the term capitalism.

Box 4.3 An Ecological Discussion of the Interaction Between Nature and Society
Odum (1997a), mainly in the epilogue of his textbook, discusses some ideas about the interaction between modern society and nature. He differentiates between economic growth and economic development (Odum 1997a:306ff),

(continued)

Box 4.3 (continued)

adopting ideas from the Brundtland Report and the subsequent sustainability discourse. He does not systematically analyse the relations between nature and society, or the economy and ecology; he mentions only the differentiation between (quantitative) economic growth and (qualitative) economic development introduced by Goodland et al. (1991). Odum tries to develop the critical discussion about economic growth with the concept of "dual capitalism", referring to ecological economists. He suggests that a business or industry should not only consider the market possibilities for new products or services, but should invest in resource-efficient, pollution-reducing and material-reusing production processes and the internalisation of externalities as waste and pollution. The burden in the form of costs should not be shifted to the taxpayers (Odum 1997b, 2012). The concept of dual capitalism is seen as the bridge between nature and society. The comparison of society and nature is reduced to the basic idea that in ecological economics and other forms of interdisciplinary ecology, two valuation principles should be used for market and non-market goods: economic and monetary valuation criteria, and ecological criteria for natural resources. The thinking remains abstract, using formulations such as giving "equal consideration to market capital and natural capital", referring to Daly's discussion of operationalising sustainable development (Odum 1997a: 314). For the integration of the economy and ecology, the focus should be the coupling of social and ecological systems, or the interface between economic and ecological processes (Odum 1997a: 314). The vagueness of Odum's notion of dual capitalism shows how difficult it is to work with the terminology of economics in ecology, or the other way round. Instead of interdisciplinary integration of concepts and knowledge, one often encounters questionable analogies, metaphorical reasoning, or epistemologically and methodologically weak concept transfer.

The few and vague ideas discussed by Odum and in natural resource management do not create sufficient understanding of the integration or the dysfunctional interaction of the modern economy and nature. What can ecological, economic and social-scientific research provide in terms of new and better knowledge to integrate economic and ecological processes under the condition of a globalising modern economy, organised in complex multi-scale processes? Odum's term dual capitalism does not provide more theoretical, more interdisciplinary, more critical or more concrete knowledge and

(continued)

Box 4.3 (continued)

critique; rather, the analysis becomes confusing. Playing with the term capitalism, without a theoretically specified concept that can be applied to nature as well as for society, results in analogies, metaphorical terminology and doubtful comparisons. Which principles of the capitalist mode of production analysed in political economy can be used to describe the function of nature's economy operating in ecosystems? The complexity of natural processes differs greatly in principle and function from that of social or economic processes (see Chaps. 5 and 6). Only trivial similarities or misleading interpretations can be found through simple concept transfer without interdisciplinary knowledge integration. The misleading concept of dual capitalism implies that with ecological knowledge, the future of the modern economy can be conceptualised as a capitalist one. But nothing about the mode of production, the structures and processes, the asymmetric power relations, the dysfunctionalities and crises of the global financial economy is analysed by Odum with this concept. Similar misunderstanding is possible with the term "natural capitalism" (Hawken et al. 1999). Hawken later clarified that the term referred to Schumacher's concept of "natural capital", not to the modern capitalist economy or its justification, as many readers of the book had thought.

Source: mentioned in the text

One of the conclusions to be drawn from this discussion of "nature's economy" and "modern society's economy" is the simple one found in the basic argument of all ecological debates about sustainable resource use: the modern market economy should not only be embedded in or regulated by society (culturally or politically); it should also be embedded in ecosystems, and the interaction between ecosystems and economic systems should be regulated. The opposite approach, which would be to integrate all nature in markets and commodify all natural resources and ecosystem services, is only possible with the neoliberal or market ideology and orthodox economics that guide the economic globalisation process, based on a variant of the dominant Western worldview or human exceptionalism paradigm (Catton and Dunlap 1978): the resources on earth are unlimited and human science and technology always find solutions to resource use problems through innovations. In such forms of "economising ecology" several conditions are ignored: (a) the limits of scientific knowledge that appear in the analysis of complex systems; (b) the social limits of growth and its consequences that appear in the institutional failure of the economy and of policy (of the property and accumulation regimes, the market failure, and the power- and interest-based distortions of policies); and (c) the conditions of functioning of ecosystems and the limits of natural resources or global boundaries.

With the new interdisciplinary research in ecology and new knowledge about limits of natural resources, it seems possible to come to more meaningful comparisons between ecological principles in nature and economic principles in human society. The ideas of Ricklefs and Odum can be developed further, corrected or improved, through the discussion in ecological and social-ecological economics that developed in parallel to the growth-critical ecological debate (see Chap. 6). Ecological and other critical forms of economics introduce ecological knowledge, concepts and value criteria in economic analyses. The basic idea of integrating economic and ecological thinking, research and knowledge by ecologists and ecological economists is compressed in the formulation of the dependence of economic processes on processes in ecosystems. Somewhat differing interpretations of the dependence of the human economy on nature are given by Georgescu-Roegen (1975, thermodynamic laws), Daly (1994, steady state) and Boulding (1966, economy of the spaceship earth), as classical authors of ecological economics, and by the ecologists Hardin, Odum and Noorgard (Costanza et al. 2010). A summarising discussion of the dependence of the economy on nature by Hardin (1991) includes the following main points: (a) economic processes of natural resource use are dependent on the physical laws of thermodynamics; (b) many resources on the earth are limited, and their limits are dependent on the consumption levels of a growing global population; (c) limits of resources are given by the carrying capacity of ecosystems.

"To stay within the limits of *the carrying capacity* of ecosystems" (see Glossary) is the main ecological criterion for natural resource management to specify the limits of resource use in economic processes of production and consumption. For that purpose the carrying capacity concept needs to be modified for humans; it cannot be assessed in terms of the maximum number of humans an ecosystem can feed, as is done for other species. Other animal species and plants consume resources in more equal, biologically determined quantities. Humans, however, do not consume resources at the level of the biological minimum to survive, but in extremely differing quantities determined by the social organisation of society and variations in access to resources, economic wealth and ways of life. Therefore, Rees has suggested calculating the carrying capacity for humans not in numbers of people, but in quantities of "loads" that include the varying levels of per capita consumption (Rees 1996: 197).

Additional ideas, concepts and indicators developed in the ecological and sustainability discourse to specify the forms of sustainable resource use and ways to achieve it include circular economy principles (Geissdoerfer et al. 2017), reducing or reusing waste, saving material and energy resources, and regulating the modern economy using combined principles and assessment methods. Such regulation cannot take the form of self-regulation and self-correction of markets, which are systems with limited information use—that is, reactive systems that are not able to use complex knowledge but only react to signals in the form of prices. The principles of markets, which make short-term attempts to balance supply and demand, are criticised by Hardin and other ecologists and ecological economists. Hardin, however, did not make a coherent argument in his critique of overuse of natural resources; previously, in an article titled "The Tragedy of the Commons", he proposed the privatisation of natural resources to reduce environmental damages and overuse, accepting markets as a regulatory mechanism. Privatisation as a main feature of the modern market economy has not prevented environmental destruction, however, but instead has stimulated it when combined with the "growth engine" of the capitalist economy.

Discussing environmental sustainability from interdisciplinary perspectives requires specifying the theory–practice relationship for the use of economic knowledge: in which forms and through which channels is knowledge transferred from scientific research to the practices of the sustainability process? These questions reveal the open boundaries of economics as a discipline—where is it still economics, and where is it already something else or more interdisciplinary, taking into account a variety of social and ecological conditions of resource use, as, for example, in management science and in the policy-guided practices of natural resource management? Different and competing criteria and interpretations in resource use practices, and specific criteria for sustainable resource management, are required to answer this question. Resource management is no longer pure economics, even in the core themes of management science—the organisation and regulation of economic production, of structuring and processing and of natural resource use in a variety of possible forms of organisation. Finally, resource management dissolves into knowledge practices based on inter- and transdisciplinary knowledge integration with multiple criteria for resource use.

The first steps towards a more critical and interdisciplinary analysis of problem-solving processes were taken in management science with the research of Simon and others (Simon 1995) beginning in the 1950s, resulting in a science of problem-solving through the structuring of decision-making processes according to the premises and conditions in a specific situation; in economic decisions this primarily took the form of decisions in organisations of enterprises. This resulted in the limitation of economic rationality to forms of "bounded rationality", and it is convincing as far it concerns the knowledge used in decision-making processes; however, it quickly reaches a dead end when the systemic constraints and limits of decision-making in terms of political or economic power relations, asymmetric power relations, structural power and the potential to destroy the environment and overuse the natural resource base are ignored. A more critical discussion of problem-solving in the environmental discourse includes different disciplinary perspectives, for example Miller's (1999) analysis of environmental problem-solving in relation to psychosocial barriers to adaptive change, the research of Ostrom and others about institutional changes in natural resource use (National Research Council 2002), and the research about complex adaptive systems (see Chap. 3), where the conditions for combined economic and ecological rationality have been successively stated more precisely and summarised for a variety of situations. This research has broadened the perspectives of problem-solving by accounting for the interplay of individual, social, political and economic dynamics that influence and shape collective problem-solving processes, re-conceptualising these as interinstitutional, pluralistic and multi-agent processes.

Preliminary conclusions from the debate on natural or ecological limits and attempts to formulate an economy of nature as a guiding idea in terms of principles and practices of natural resource management include the following:

1. *Whatever the differing and competing ideas, the theoretical and normative premises in the economics of resource management are that they need to be combined with ecological principles and specified and differentiated for a variety of practices of natural resource management.* Neither economic activities in primary production (hunting, agriculture, fisheries, forestry) nor those in secondary, artisanal or industrial production and processing of resources can be organised only in monetary and market terms, according to economic principles of rationality and profit. The economy of nature is not clearly understood when it is formulated as a natural system of production, processing and resource use in ecosystems, parallel to that of the societal economy (in a superficial analogy to the idea of circularity in ecosystems, to criticise the modern economy for its lack of a decomposition function). It is more adequate and open to theoretical interpretation to see the economy of nature as a guiding idea or regulative principle to deal with ecological limits in human production, distribution and consumption under conditions of the modern economy.
2. *Accounting for natural limits requires conceiving the modern economy not as an autonomous system, independent or decoupled from nature and ecosystems, but as part of ecosystems and as a form of coupled social-ecological systems.*

Ecologists such as Odum and Hardin and economists such as Polanyi, Daly and others have argued that this requires the containment and societal control of markets with a limitation of resources that can be allocated through markets. Complementary economic principles and institutions beyond markets need to be (re-)established in the economies of the future for the building of systems of non-market-based transactions of resources: reciprocity, redistribution and sharing, supported by normative principles of solidarity, fairness and equity. In the sustainability transformation such forms of differentiation of the economy can become guiding ideas to reduce exponential economic growth by developing socially, economically and ecologically sustainable levels of resource use within the limits given by the carrying capacity of the global ecosystem.

4.2 Resource Economics as Management Science: Organisation and Knowledge Practices

How the economy of nature and the economy of society interact is not systematically analysed in economics nor in environmental economics. Ecological economics is opening up such analyses with the help of interdisciplinary knowledge to understand the coupling of social and ecological systems and the interaction between the *social and the biophysical lifeworld* of humans (see Glossary).

The building of an ecologically integrated economy can be described in the forms and principles of natural resource management from two perspectives: (a) economically differentiated by sector (in simplified form; first, second and third sectors) or (b) ecologically differentiated by type of natural resource use—water, land, air, living and non-living resources, and mineral resources. The sector-specific view of the economy needs to be complemented by theoretically based analyses of the systemic organisation of the economy in modes of production and the socio-metabolic regimes regulating the resource flows between economic systems and ecosystems as they are found in interdisciplinary and holistic approaches to resource management.

4.2.1 Sector-Specific Resource Management: First, Second and Third Economic Sectors

Natural resource management is paradigmatically discussed for the primary economic sector as it is called in conventional economic terminology—fishery, agriculture, forestry, hunting and wildlife management, land and water use and management. Industry also uses natural resources, but mostly sources others than the living resources used in the first sector. Industry's organisation, technology and integration in the global economy and market processes differ from those of primary sector activities, which were long organised with little or no integration in the market economy. Agriculture remained to a large degree local, small-scale and subsistence production in the modernising economy, in alternatives to a market-based economy, developing without industrial forms of specialisation and division of labour (see Harrison 1977). Even in European countries, the primary production activities in the first sector were not fully commercialised, capitalised, industrialised and integrated in global markets until the second half of the twentieth century. This gave rise in economics to a discussion about a historically specific "peasant mode of production" (Harrison 1975, 1977) and family farming that followed the principles of simple commodity production, where large parts of the production were for purposes of subsistence and only surplus production was sold on local markets. Although certain forms of large-scale capitalist agriculture and cash-crop production developed in the colonial economy beginning in the sixteenth century, they remained exceptional in European countries until the second half of the twentieth century when the industrialisation of agriculture gained momentum, supported by the modernisation practices in the common agricultural policy of the European Union (EU).

The model of the primary, secondary and tertiary sectors is the prevailing view on resource use processes in economics as well as in policy and management practice. With it comes a reduction of resource- or species-specific production and management perspectives that do not sufficiently account for the ecological components of the production and processing of resources. The sustainability discourse and the broader environmental discourse were the debates which brought a breakthrough in trans-sectorial and integrated views of resource management, which can be seen as a prerequisite of the further sustainability discourse. The sector-specific analysis of production has the advantage, however, of showing the specific requirements of changing production systems in the transformation to sustainability that work with living natural resources as their only purpose—the domain of the first economic production sector. In recent research on transforming the first and other sectors towards sustainability, the following ideas and principles can be found regarding ecological principles; these principles differ, whereas economic principles of resource management are rather similar regarding the maximisation of utility by resource users.

(1) The agricultural sector—ecological principles of resource management in agriculture: The discussion about sustainable management of agricultural re-

sources and the transformation of modern agriculture towards sustainability began with the environmental movements that developed in the 1960s and 1970s in Western countries. The paradigmatic case of environmental problems was the use of pesticides, beginning with the classical analysis of the consequences of pesticide use in "The Silent Spring" (Carson 1962). Pesticides and agrochemicals that kill insects, birds and other animals have remained an enduring point of controversy regarding modern agriculture and its environmental regulation. In European countries and their environmental policies, modern agriculture since the 1970s has been classified as an environmental polluter and destroyer through the use of pesticides and agrochemicals such as synthetic fertilisers, as well as the development of large-scale and specialised animal production resulting in high quantities of liquid manure which, when used as organic fertiliser, increases through leaching the nitrate pollution of ground and drinking water. The level of nitrates in drinking water has been regulated by the EU since 1980, but less with regard to emissions from agriculture, and in legally binding forms only with regard to the quality of drinking water, which in most areas needs to be purified, and the provision of drinking water through centralised systems and long-distance delivery through pipelines, decoupled from the locally available ground or surface water. Pesticide and nitrate pollution from large-scale, high-tech agriculture has yet to be adequately addressed, resulting in continuous conflicts between agriculture and environmental protection. The second big problem has been the living conditions of farm animals on mass-production, large-scale specialised farms in Europe, especially poultry, pigs and cattle, which is as difficult to regulate as agrochemicals. Further and indirect consequences of large-scale and industrialised farming have contributed to the negative environmental balance. Farmland consolidation and modification of agro-ecosystems to meet the conditions of technically modernised agriculture has had environmentally negative consequences such as increasing erosion of fertile land and the reduction of biodiversity of plants and animals. Mass animal production has made agriculture a significant contributor to greenhouse gas emissions and thus to climate change, primarily through methane emissions from animal husbandry. The high input of fossil and other natural resources for agriculture has contributed to resource scarcity and the pollution of air, water and soils (fuel for tractors and other machines and for the production of agrochemicals and synthetic fertilisers). The integration of the agricultural and industrial sectors of the economy has produced a combination of environmental problems resulting from modern agriculture. The ecological regulation of agriculture and its transformation to sustainability will require either changing modes of production, from conventional to organic agriculture, and other forms of environment-friendly, extensive and small-scale agriculture, or complicated legal and technical regulation of large-scale and intensive farming through governmental agricultural policies.

The large-scale environmental damage caused by modern agriculture has been targeted for decades in Europe through policies for environmental regulation, reduction of use of pesticides and agrochemicals, control and sanctioning, but with limited success; the industrialisation, specialisation and concentration of capital and land in farming is continuing, the pressure for economic growth is increasing, and the power of political regulation is decreasing with the globalisation and deregulation of markets. Voluntary or self-regulation of agriculture has had limited success, mainly in organic farming, which applies ecological principles to production and land use, to some extent in locally specific, extensive and small-scale agriculture. Other forms of voluntary regulation for modernised agriculture are more defensive reactions and "greenwashing", attempts supported by the chemical industry to counter organic farming through "integrated pest management". The pressure to pursue economic growth and profit is growing in modern agriculture; economic growth goes together with industrialisation, specialisation, concentration of capital, further technological modernisation and transforming agriculture into a sector dependent on the agro industry and the international corporations that claim to "feed the world". Farmers become suppliers of raw materials to agro-industrial companies, when they should be producers who balance multiple ecological and economic criteria.

(2) The fisheries sector—ecological principles of fisheries management: In fisheries management the observance of ecological rules in fishing seems simpler than in agriculture: not to overfish the fish stocks. This is a simplification, as the experience with the overfishing of the oceans has shown. European and global fisheries have undertaken policy-supported technological modernisation and concentration of capital in a similar way to agriculture in recent decades, only in a shorter time period. The oceans long seemed to be an unlimited source of food for humankind that could help to solve global nutrition problems, and the possibility of overfishing the oceans appeared unlikely or impossible. However, after a few decades of intensive fishing, in the 1980s most fishing grounds in the oceans were overfished. The rapid technical modernisation and building of large, capital-intensive fishing fleets dependent on growth and overfishing happened in similar forms in many countries. The EU, after subsidising the oversized national fishing fleets in the Common Fisheries Policy, soon had to introduce subsidies for dumping the fleets and to limit catches. The misguided policy of growth based on the assumption that, in a market economy, fishery needs to be profitable resulted for the European countries in a situation where fisheries are declining. In the EU's new strategy of blue growth, fisheries is no longer seen as an important subsector of the future maritime economy. Aquaculture and the exploitation of other maritime resources have priority in the green economy that aims at the sustainable use of maritime resources, but this

encourages the commercialisation and more intensive exploitation of maritime resources. The future of fish production is that of aquaculture in "high-tech" forms that are rapidly expanding and already exceeding the catch of wild fish globally.

Alternatives to growth-based fisheries have been intensively discussed in the past. To prevent overfishing and "fishing down the food-web" (Pauly et al. 1998; Pauly 2019), alternatives include small-scale, locally adapted, species-specific, ecosystem-based, seasonally limited, catch-limited and environmentally certified catch of fish. These ideas for sustainable fisheries were insufficiently practised. The struggle for sustainable fisheries will be more difficult with the advance of new technologies and the decline of local fishing fleets, which have nearly vanished in most European countries and remain mostly in the poor countries of the Global South.

The development of global fisheries in the second half of the twentieth century, its rapid modernisation and growth and its quick decline, shows the failure of growth-driven development and change. It was a "primitive" form of economic growth through technological modernisation and capital-intensive fleets, plundering the oceans with economically effective technologies, quickly reaching the limits of resource use through ecological disasters of overfished fishing grounds and the scrapping of fishing fleets. It became an unintended experiment in unsustainable growth that showed how the limits to growth affect natural resource management. In fisheries the consequences of growth are clearly evident: it has only one kind of resource, fish, that is (in spite of the species-specific ecological and biological differences of the many types of fish) economically seen as a single resource with regard to its purpose as food for animals and for humans. The management of this scarce resource requires refined and combined ecological and economic criteria.

(3) The forestry sector—ecological principles of resource management in forestry: As discussed in Chap. 2, forestry management in Europe has struggled since the Middle Ages with increasing scarcity of wood, local deforestation and overuse of forests. Globally deforestation has been advancing rapidly since the twentieth century and has become an environmental risk due to climate change caused by the loss of climate-stabilising tropical rainforests and boreal forests. The restoration of forests through reforestation becomes less effective when the forests are vulnerable to climate change and other environmental risks, especially air pollution; tropical rainforests, meanwhile, cannot be restored to their original forms as complex ecosystems by reforestation.

Similar to agriculture and fisheries, the ecological principles of forest management are fairly clear: they require protecting the forests and maintaining their stock of natural capital for long-term use, observing the principles of the functioning and reproduction of ecosystems. The economic redefinition of forests as "natural capital" is another questionable example of the transfer of concepts between disciplines through analogies (see Sect. 4.1.2), already terminologically reduced from ecosystems to economic resources and capital. The practice of deforestation thus becomes a form of "economising ecology". Moreover, reforestation becomes in practice a profit-oriented business when fast-growing and economically profitable trees are planted. In the long run global deforestation has not slowed down or stopped; the global forest cover has decreased in the past millennium from about 40% to 20% of the land. Sustainable forest management can no longer be limited to reforestation but must also include managing the forest ecosystems in the context of the earth system and as part of global sustainability transformation. For this purpose one needs to combine various ecological and economic criteria for protecting forests, regulating the use of forest resources, restoring forest ecosystems according to ecological principles, reducing air pollution that damages trees and mitigating climate change.

(4) The mining sector—ecological principles for the extractive economy: The extractive economy that provides the raw materials for other branches of economic production through mining of coal or metals and other resources has always been a risky, environmentally polluting and destructive form of the economy, even prior to its widespread expansion in the era of industrialisation. In the modern world economic system, based on the division of labour between the core countries in Europe and the dependent countries in the periphery or the Global South, extracting industries became a main component of the economy in the colonies and later in the developing countries (Bromley 1999). The consequences have been investigated and discussed in recent decades in terms of the economically and ecologically unequal exchange in the world system and the shifting of burdens for environmental destruction to the least developed countries in the South (Rice 2007). These countries deliver the natural resources and the raw materials of which large parts are exported and used in the European and industrialised countries, perpetuating the economic dependency of the developing countries and their extracting economies with the global divide between rich and poor countries. Another contributing factor is the current shifting of polluting forms of industrial production to the newly industrialising countries, where environmental pollution and destruction are rapidly increasing along with unsustainable forms of economic growth.

A sustainable approach to the extractive economy to limit the pollution and destruction of the environment through mining in the form of polluting soils and waters with toxic residues and waste requires the consideration of multiple economic and ecological criteria, not just technical issues. Even more important factors are the reduction of mining, the saving of material, reuse and recycling. New criteria aimed at overcoming the structural divide and inequality between the Global North and South need to be included to address the central aim of sustainable development as laid out in the Brundtland Report. This will require new approaches to economics based on principles of solidarity, sharing and redistribution of resources. Economic growth in the poor countries of the Global South depends on the sale of their natural resources at low prices, with the benefits going to the importing, affluent countries, whereas the negative externalities and social costs of environmental damage remain in the developing countries. In the discussion of the extractive economy and the global resource flows in the modern world economic system—always moving from the poor to the rich countries—the question of growth and its unequal forms becomes the core problem of sustainable development. The global divide is a mechanism to maintain a socially, economically and environmentally maladaptive system, for which no solutions have yet been found: not through decolonisation, not through political reforms, not through development aid, and not through the valorisation of nature and privatisation of natural resources.

(5) The energy sector—ecological principles of transforming energy systems: The systemic risks created by industrial economies are nowhere more visible than in the global environmental destruction from fossil energy sources on which industrialisation is based, first manifesting in the coal regime and thereafter in the oil regime. Energy systems are historically specific to the different forms of human society and modes of production that have existed in human history—hunter-gatherer societies, agricultural societies, industrial society—and are differentiated within these long periods, becoming less and less sustainable. In the history of humankind there have only been two forms of social-ecological transformation where energy systems changed dramatically: the emergence of agricultural societies (with large-scale deforestation for agricultural production) and of industrial society (where the use of fossil energy sources liberated the development of society and the economy from the limits of locally available land and other resources, which enabled the building of a global society and economy). The history of energy systems (Debeir et al. 1991) shows the complexity of functional and dysfunctional connections and interactions between ecological and social systems. The difference between renewable and non-renewable sources, and the forms and purposes of energy use, are both ecologically relevant. Wind or solar energy can be used for different purposes, not only for electricity generation. Nuclear energy in its non-military application can only be transformed into electricity, and the largest and growing part of the energy that is used is electricity. Multiple criteria of management are required for the energy sector as well, including use of renewable sources and reduced energy consumption (for the transformation of the energy

sector see the example of bioenergy, learning exercise three, described at the end of the chapter).

(6) The industrial sector—ecological principles of industrial resource use: The industrial sector is today decreasing in the economies of the early industrialising European countries; globally seen, however, industrial production is still dominant in terms of resource use and environmental pollution. Industrialisation has never been managed in a non-polluting form. The early industrialising countries in Europe were based on heavily polluting industries, as are the newly industrialising countries in the Global South. The environmentally clean technologies that are being developed work under exceptional conditions in some countries, mainly in Europe, but have not spread globally, in part because the new—expensive—technologies are the privileged domain of the industrialised countries. The industrial economy includes many different branches and sectors, each of which has specific forms of externalities and environmental consequences. The future of the industrial society has become a controversial question both generally and in the sustainability discourse. In these debates the concept of industrialisation becomes blurred, and the new knowledge economy and the digital economy appear as new forms of industrialisation or industrial modernisation. The term is then used to describe technological innovations that happen periodically in the industrial economy, discussed as "long waves" or Kondratiev cycles (Barnett 1998; Devezas 2006; Nefiodow and Nefiodow 2017; Grinin et al. 2016). The debate about industrialisation that is central to the transformation to sustainability is not about technological innovations and the development of economic sectors, but about the phasing out of the industrial society and the transformation to a post-industrial society. This transformation happens in asynchronous forms and with time lags: when the larger part of the global population is just beginning to industrialise, it is being phased out in the early industrialising countries in Europe and the West. Post-industrial society cannot be based on individual countries but must represent a global transformation of the modern world economic system of which it is part. The future global economy needs to work with several economic forms, reducing the dominance of the market economy and building new economic forms based on saving, sharing and redistributing resources.

(7) Beyond agriculture and industry—the "third sector" or the service economy: The service economy in the third sector is heterogeneous, comprising various forms of material and immaterial economic activities for which the term "services" is insufficient to describe their economic functions and their social and ecological consequences. Schools, hospitals, banks, public and private transport systems, recycling systems, the knowledge economy and the digital economy based on information technology and the internet differ in their functioning, principles, resource use processes and consequences. The sustainable management of services that in-

clude material and immaterial resources requires multiple and varying combinations of criteria where social, economic and ecological criteria are balanced.

Problems and limits of sectorial approaches in resource management: Ecological classifications of natural resources do not match with the economic differentiation of sectors; whereas in the first sector (agriculture, fisheries, forestry) the production and use of living resources is dominant, in the industrial and service sectors production and resource use is mainly technological processing and modification of non-living and energetic resources (changes in the first sector come with the new forms of genetic modification of plants and animals, where agriculture becomes more an "industry of living resources"). The first sector of the economy deals directly with the use and the risks of overuse of living resources, plants and animals, and is thus closely coupled to ecosystems and the necessity of observing ecological principles in production, making it a key sector from an ecological perspective. Agriculture is in the industrialised countries a small sector in terms of employment and the economic value of its production, but it uses large amounts of resources for food production, especially land and water; more than 70% of the freshwater resources on the earth are used in agriculture. The advantage of the sectorial perspective and the sector-wise transformation of the economy is that the different sectors, subsectors and branches cause concrete problems and ecological consequences of the use of certain natural resources that require specific forms of ecological regulation and transformation, which are not the same for all sectors and branches. In practice the sustainability transformation of the economy occurs through many different sector-specific transformation processes at local, reginal and national levels. However, the inter-sectoral links in the economy need to be observed in the social-ecological transformation of the global economy, together with the cross-scale links in spatial and temporal terms. The transformation of the global economic system will require the ecological transformation of global components, institutions and regimes; the reduction of large-scale industrial systems, the global division of labour and global value chains; the de-globalisation and re-regionalisation of the economy; and the strengthening of local production and trade. The transformation cannot be completed via piecemeal steps in bottom-up approaches, as has been proposed by many ecological movements. It needs to be a connected transformation from both ends, from the local and the global, which is one of the dilemmas and difficulties of the sustainability transformation.

Trans-sectorial, integrated and holistic approaches to resource management, or the idea of a "circular economy" imitating ecosystem processes of production, consumption and reduction, create paths to more systematic and coherent forms of the ecological transformation of the economy. At the same time, inter- and transdisciplinary knowledge integration and synthesis is developed as a requirement of the sustainability transformation in the management of natural resources.

4.2.2 Interdisciplinary and Holistic Approaches to Natural Resource Management

The holistic resource management approaches described in the following examples of integrated management, ecosystem-based management and adaptive co-management apply similar ideas and are connected to the principles of a circular economy, which is discussed as a fourth form. These management approaches attempt to integrate knowledge from natural- and social-scientific concepts, frameworks and theories of resource use and management and to translate scientific thinking and knowledge into action in environmental policy and governance, thus creating agency for the sustainability transformation. They all take into account the natural and social contexts of natural resource management, although not necessarily both contexts sufficiently. The approaches originate from natural-scientific and ecological research and describe social systems in the functional terminology commonly used in the natural sciences.

(1) Integrated resource management: The abstract and vague term "integrated management" is used in a variety of contexts, including in the scientific literature, thus creating confusion. The term is also used in business management for the purposes of quality management of enterprises, where integration means the creation of a single system to manage a number of aspects such as quality, safety, environment, health and others. Examples include the International Organization for Standardization (ISO) standards for quality management and for environmental management in enterprises (Jorgensen et al. 2006); the authors see integration improving through three levels, from increased compatibility of system elements, to the coordination of generic processes, to an integrated management system embedded in a culture of learning and continuous improvements. Inter-organisational and intersectoral systems of natural resource management, including firm-based models, developed in parallel to these. The term was applied to the early environmental and sustainability discourse in the 1990s in more area- and policy-specific variants, for example as integrated coastal zone management (ICZM), where integration meant the spatial, temporal and trans-sectoral integration of the use and management of the marine and terrestrial resources on the coasts (Cicin-Sain and Knecht 1998). The approach was briefly influential in the environmental policy of the EU, but it was not successful as a means to achieve sustainable resource management. This was due in part to insufficient integration of scientific knowledge as well as its insufficient translation in policy approaches, which furthermore lacked sufficient support from political actors and stakeholders. The approach turned out to be based on false assumptions (Billé 2008), as often occurred in the early phases of sustainability policy.

Although the principles of integrated coastal or water management were clearly defined and specified in the policy practices, their vagueness and scientific inexactness required management approaches that utilised more exact terms and coherent knowledge bases, such as ecosystem-based management and adaptive co-management. Circular economy, a term associated with integrated management, is more clearly oriented to the model of ecosystems, where three functions that complement each other in a circular process are systematically differentiated: production, consumption and reduction. The intention of the idea of a circular economy is intuitively clear—that a reduction function should be built into the social organisation of economic systems to reduce the tremendous loss of resources in the form of waste—and in this sense the idea could be applied at every level of the economy, from individual firms and households to national economies and the global economy. Yet the idea is far too simple in its functionalistic reasoning for the organisation of complex social systems such as the economic system, where many more functions and conditions need to be taken into account than in ecosystems. A workable definition of the circular economy requires more components than the basic differentiation of three functions.

A further problem with applying the circular economy model to the social organisation of the economy in human societies is that ecosystems are evolving and transforming naturally as well as through human modification in agriculture and industry: if the basic functions of ecosystems are not maintained over time, the decoupling and disturbance of social and ecological systems becomes a serious threat to societal and economic development. Differentiating the economic organisation of the dominant modes of production in human history—hunting and gathering, agriculture and industry, with further differentiations—provides concepts for different modes of production and types of economy, both ecologically and socially. The ecological consequences of resource use in the industrial and market-based modern economy, with its extreme division of labour and specialisation requiring high levels of exchange of resources, differ significantly from those of the earlier agricultural and hunter-gatherer societies with regard to two ecologically relevant parameters: the quantitative levels of resource use and the levels of specialisation, which were much lower in the agricultural societies in early modernity, before the industrial revolution. It is easier to apply the model of the circular economy to the forms of production and subsistence in earlier agricultural and hunter-gatherer societies. In Europe around the year 1800 about 90% of the population was rural, mainly small agricultural producers, and the peasant and farming economy had to obey the laws and functions of ecosystems—they lacked the knowledge and technologies to manipulate them that were later developed with industrialisation. The circular economy model could more easily be applied to these earlier societies than to industrialised economies, although even early agricultural systems caused serious environmental problems through deforestation and the ruin of fertile soils (through salination, erosion and desertification), and most of them vanished or collapsed in part as a result of ecological overshoot.

(2) Ecosystem-based resource management: The idea of holistic, ecosystem-based management or ecosystem-based approaches gained influence in the 2010s in different fields for terrestrial and maritime resources. It is closely linked with sustainable development approaches and can be seen as a form of transition to sustainability. It is of interest for developing interdisciplinary, ecology-based approaches to sustainable resource management for several reasons, including its system-theoretical foundation, its dealing with complexity and uncertainty in knowledge production and application, and its orientation to ecological principles and management practices (Waltner-Toews et al. 2008).

Ecosystem-based management tries to practice inter- and transdisciplinary knowledge integration by way of process-oriented knowledge integration as part of the management process itself, not by way of adoption of scientific knowledge in the conventional forms of knowledge transfer and technologies. It can be part of a holistic approach where local ecosystems and human activity are analysed in an attempt to follow the web and coupling of social and ecological systems in horizontal (spatial) and vertical (hierarchical) couplings from the local to the global level. With regard to sustainability transition, this implies taking into account various knowledge requirements.

The management of ecosystem processes requires long temporal perspectives by all actors (scientists, political decision-makers, resource users and owners, civil society actors). Moreover, the process, perspective and context of long-term change in the coupled social and ecological systems needs to be observed when making short-term decisions. Some of this management of temporal complexity is possible using the methodology of scenario analysis rather than traditional forms of modelling, trend analysis, prognostics and forecasts; further methods of future-oriented management need to be developed (see Chap. 10). The management of the spatial complexity of interacting social and ecological systems, and the networking and the spatial arrangements of resource management across spatial scales, requires that networking institutions and procedures be built as part of the interdisciplinary knowledge regimes. Such knowledge regimes become part of the transformation of the global economy proceeding from both the local and the global level, developing institutions for earth system governance and forms of ecological networking, and nesting local, national and global dimensions of the modern world economic system.

The management of complexity in the sustainability process can be realised and translated into procedures of decision-making and collective action by using tools and methods of systems analysis described in Waltner-Toews et al. (2008). This approach reintroduces to ecosystem-based approaches the social complexity caused by unequal development and the global North–South divide, framing the situation to develop system descriptions, the analysis of system dynamics, and the synthesis of knowledge from the preceding analyses to describe potential future consequences of present action, possible pathways of future development and change of the systems involved, and potential side effects and unintended consequences of decisions regarding the future development of the resource use under consideration. This process includes the requirements of sustainable resource management and governance discussed in the sustainability discourse: to become aware of the paradigms guiding analysis and action, of the dependence of societal systems on nature and ecosystems, of the specific relationships between social and ecological systems and the forms of their coupling. The coupling implies the conceptualisation of the biophysical aspects of the economy and bringing them back into analyses of the interaction of economic and ecological systems, which is attempted in ecological and institutional economics (Box 4.4).

Box 4.4 Ecosystem Approaches and Emergent Complexity

The differentiation of ordinary and emergent complexity (Funtowicz and Ravetz in: Waltner-Toews et al. 2008: 310ff) is relevant for the sustainability discourse: society and nature are examples of emergent complexity. Ordinary complexity is found in systems with linear development and stability (for example, mechanical systems); emergent complexity is found in systems with non-linear development, self-organisation, uncertainty, emergent behaviour and properties. Emergent systems with human actors show additional capacities of consciousness, intentional action, autonomy, anticipation, symbolic communication and morality. Complex emergent systems develop a hierarchy involving power, domination and control by a dominant species in an ecosystem, or a dominant group or class in society. Modern industrialised agriculture can be seen as an example of an emergent system where human domination and control of agro-ecosystems requires more and more complicated technologies and more and more input of energy and materials to maintain systems which are ecologically unstable. Markets as they are conceptualised in conventional economic theory, that is, as systems striving towards equilibrium, appear as examples of ordinary mechanistic complexity; however, attempts to regulate markets by way of politics produces elements of emergent complexity.

(continued)

Box 4.4 (continued)

Based on the description of emergent complexity by Funtowicz and Ravetz, complemented by their theory of "postnormal science" (Funtowicz and Ravetz 1993), the concept of emergent complexity can be used to discuss the methodological and practical problems of natural resource management under conditions of global environmental change and in the transition to global sustainability. Post-normal science describes a scientific practice that appears with new forms of knowledge production in inter- and transdisciplinary research: facts are uncertain, values are disputed, stakes are high and decisions are urgent, as the authors argue. The global economy can be seen as an emergent complex system; the methodology of orthodox economics, as any other disciplinary and specialised science, is not able to deal with the emergent complexity of the system epistemologically and methodologically: it tends make complexity simple, excluding and suppressing certain options, reducing, simplifying and fragmenting systemic complexity through specialisation, where new problems are addressed through further subdisciplinary specialisation, not by interdisciplinary knowledge integration. A reductionist methodology, formalised model constructions and simplified forms of economic rationality based on selective assumptions about human behaviour cannot deal with the complexity of the modern economic system. Analysis of the modern economy must deal with uncertainty, different forms of behaviour and concepts from different theories and interdisciplinary forms of analysis, and different worldviews; "no particular partial view can-capture the whole. It is therefore necessary and legitimate for the dialogue … to include persons representing all different interests, which may also include concerns for children, nonhuman species, and ethical values. This point reflects the growing practice in the resolution of global environmental issues, where earlier attempts to privilege one perspective corresponding to a hegemonism of Western culture, have proved to be inadequate at all levels including the practical, economic, political, and ethical" (Funtowics and Ravetz in Waltner-Toews et al. 2008: 316).

Source: mentioned in the text

Beyond the forms of coupling in spatially specific and limited social and ecological systems (Bruckmeier 2016: 183ff), a conceptual model for the interaction between societal and ecological systems can be applied at all levels, from local to global systems. This model includes a threefold relationship: (1) ecological systems provide the context for societal systems; (2) societal systems can modify the structures and processes of ecological systems; and (3) societal systems can modify the context for the self-organising processes of ecological systems (Kay and Boyle in Waltner-Toews et al. 2008: 66). Societal systems include functionally specific subsystems such as the modern economy, political systems, and the science and education systems. Human action and agency is not only a question of individual actors and their rationality and interests that direct and specify natural resource use, as is assumed in orthodox economics and other approaches that follow an epistemology of methodological individualism. The forms of collective action and the systemic nature and emergent forms of societal systems direct, channel and constrain individual and collective behaviour.

(3) Adaptive co-management: Adaptive management and co-management (for a detailed description see Bruckmeier 2016: 246ff) are ecological approaches to resource management that developed from simpler approaches aiming at resilience, collective learning and reducing uncertainty through management experiments (Allen et al. 2011: 1340). Adaptive co-management involves more explicit forms of cooperation and power-sharing in the resources management process, based on the idea that those with claims to the resource should share rights and responsibilities (Plummer 2009). Adaptive governance renews these ideas, addressing the broader social and ecological contexts of resource use in a similar way to ecosystem-based management, including rules and conditions for collective action and coordination (Folke et al. 2005). The similarity to the ecosystem approach can be seen in the use of knowledge from ecology, economics and political science, but adaptive co-management comes closer to the application of knowledge and the practice of environmental governance (Armitage et al. 2007). Adaptive management and co-management differ from earlier approaches in their use of inter- and transdisciplinary knowledge with regard to local social and ecological systems. The new approaches are not tied to a single economic criterion (maximisation of yields), are not guided by single theories, and do not work with assumptions of linearity, single causes or exact knowledge and quantitative data; instead, they aim to continually improve management practice based on collective learning (Bruckmeier 2016: 278).

As in ecosystem-based approaches, adaptive co-management aims to manage the complexity of resource systems with regard to their long-term maintenance. Given this complexity, there is no single, simple and right approach to resource management, and no single efficiency criterion to apply. With such approaches change also the roles of scientists and experts to include as well production of knowledge for problem solving as participation in governance processes; the scientists do not only provide knowledge and advice but share information and deliberate with other actors about options and trade-offs. The guiding idea of the new approaches is that decision-making under conditions of complexity must be broadly participatory (Kay in Waltner-Toews et al. 2008: 80). Finally, in the integrated resource management approaches the social complexity and social situations of the actors and resource users are neglected, reduced to cooperation and participation in the research and governance processes. It should not be forgotten, however, that for the actors in the sustainability process, the aim of sustainable development is the same as for all humans: to improve their conditions of life and well-being, not mainly through higher income but through improved conditions of life and work and work–life balance. The ecological indicators described in the Chap. 3 (Table 3.1) said little about these social improvements. Moreover, the conventional social indicators to measure human, social and cultural development that are used in the policy and governance processes do not say much about activities to improve social well-being in the sustainability process (Table 4.1), except poverty reduction in the countries of the Global South, which is part of the Sustainable Development Goals of the UN. The indicators of the social conditions of life are rather detailed; there is no lack of information about the—mainly bad—conditions of life and well-being for the majority of the population in most countries. In the sustainability process it will become more and more important to find ways of improving quality of life and well-being through new forms of development that are not "locked into growth" and leading to unequal distribution.

(4) Circular economy as organising principle of integrated and ecosystem-based approaches:

A circular economy can be defined as an economic system that aims at the reduction of waste and the continued reuse, remaking and recycling of resources. The idea derives from an analogy to ecosystems, comparing resource flows in the economy to the circular flows in ecosystems, where all waste is decomposed and absorbed in the system. A circular economy differs from a "linear economy", where the products end up as waste and the materials are not reused. The first formulation of the idea of a circular economy in the economic discourse was that of Boulding, based on principles of an economy for the "coming spaceship earth" (1966). He used the terms "open" and "closed" economy to describe the differences between circular and linear economy. The circular economy is not a scientifically and theoretically elaborate concept but rather is a practical term applied to natural resource management and environmental policies. The production and disposal of waste in the modern economy has resulted in increasing environmental problems and damage, including from both household and industrial waste. The problems of plastic waste and plastic pollution in the oceans are a good example. The ideas of a circular economy were developed in the policy context in a 1976 report for the European Commission titled "The Potential for Substituting Manpower for Energy" by

Table 4.1 Social indicators—human well-being and social quality of life

Name	Definition	Interpretation	Sources
HDI (Human Development Index) of the UN	The HDI measures three dimensions of human well-being: life expectancy, education and income per capita	Published annually since 1990; instead of economic growth and income as development indicators the HDI aims to give a simple index for human capacities that should define quality of life	United Nations Development Programme 2019
Social indicators: UN	Country-specific indicators about population, health, housing, education, work	The basic indicators for all UN countries; not very detailed; not updated in recent years (system under change)	United Nations, Demographic and Social Statistics
Well-being: OECD	Country-specific indicators for OECD members, more detailed than the social indicators of the UN, including environmental indicators	Indicators about human well-being/quality of life: current well-being and resources for future well-being (natural, human, social and economic capital)	OECD 2020 (last report)
Social indicators: EU	Country-specific indicators for EU countries, more detailed than the UN and OECD indicators	Indicators of human well-being/quality of life; includes indicators for social protection and social inclusion, health, pensions and investing in children	EU 2019

Sources: mentioned in the table

Stahel and Reday. A 2012 report for the Ellen MacArthur Foundation titled "Towards a Circular Economy and Business Rationale for Accelerated Transition" further developed and specified the ideas of a circular economy.

The concept of circular economy can easily be connected with the concepts of integrated and ecosystem-based management and can be applied to sustainable development. Whether the rules for integrated and ecosystem-based management are sufficient to transform the economic systems to sustainable systems can be doubted. The advantage of the idea of a circular economy is its flexibility and adaptability. It can be applied to many economic sectors and at different levels, from the household and firm to the national economy (the practices to maintain such systems or to manage recycling systems are more complicated). The ideas of a circular economy are heuristic principles for the continual search for new technical and practical solutions for the reuse of waste and the saving of energy and material resources in economic processes of production and consumption. The scientific debate about the circular economy is discussed in many economic, ecological and engineering journals. A discussion of the circular economy as a new sustainability paradigm was initiated by Geissdoerfer et al. (2017); the connection between sustainable growth and the circular economy is discussed by Lin (2020).

4.3 Institutions of Resource Management: Property Rights in Natural Resource Use

The social transformation of natural resource use and management through property rights: Property rights and ownership of natural resources are not intensively discussed in economics, but they do appear in the debates in environmental and ecological economics (Bromley 1991, 1992, 2014; Hanna et al. 1996; Munsinghe 2013) and in the sociology of property rights (Carruthers and Ariovich 2004). While the sociological debate on property rights highlights problems of economic inequality, the ecological debate takes up the problems of unintended consequences of resource use and the need to regulate property regimes through institutions of resource management.

4.3.1 Property Rights: The Commons, Common Pool Resources and Privatisation

Important institutions for the use of natural resources are property rights, which can be roughly differentiated in four groups: state property, private property, commons and "open access" or no property rights specified (see Chap. 3). State and private property are seen as modern property rights effective in socialist and capitalist societies. Commons in the traditional sense refers to local common property of a specific group or resource use; for example, in agricultural societies this includes the common land of a local community or village such as the forest or pastures or other resources, especially water. They differ in form of institutional regulation and are not necessarily formally regulated through legal rules. Open access resources can be freely used without restrictions. For reasons of conceptual clarity, "commons" as a cultural and institutional form of property should be differentiated from "common pool resources" for the ecological state of a resource (for example, the fish in a lake). In modern economics the debate is reduced to two property-related forms, public and private goods. Intensive research on the conditions for success and the possible failure and negative side effects of property rights has occurred in recent decades in the fields of human ecology, cultural and ecological anthropology, environmental and ecological economics, and others, with differing and controversial results.

The "tragedy of the commons": Long before Hardin (1968) formulated his ideas about the tragedy of the commons (see Box 4.5), the problem was discussed as that of economic and ecological advantages and disadvantages of different property forms. The assumption that privatisation and commodification can protect the land and prevent overuse better than the commons can remains controversial. Hardin (1968) saw the commons as an imperfect form of collective property that supports their overuse and ruin (mainly land); therefore he argued, more speculatively than empirically, for a transformation to modern forms of state or private property to prevent overuse. Hardin's interpretation of commons as necessarily involving overuse has been criticized (he understood commons more in the sense of open and unlimited access than as a specific common form of property of local resource users). Ostrom (2007: 242) discussed Hardin's concept and a similar one proposed by Gordon (1954) as examples of critical approaches that brought new ideas to orthodox economics and political science such as the new institutional economics, behavioural game theory and public choice theory. She saw the significance of Hardin's hypothesis of a tragedy of commons through overuse due to the users' short-term maximisation of benefits not in the strength of his reasoning but in its stimulation of empirical and theoretical work which showed that different results—overuse or protection of the resource—are possible in different forms of property regimes.

Box 4.5 A Classical Controversy About the "Tragedy of the Commons"

The expression "tragedy of the commons" dates back to the essays of Lloyd (1833) and Hardin (1968) to describe a situation where a natural resource or property in common use was overused by local users. This was described as not behaving in the common interest to maintain the resource base, for example, common land, in a good state: the individual interest in a short-term benefit through overuse appears as misuse of the commons (the example of a medieval commons used by peasants as grazing land was applied by both authors to show the consequence of overgrazing). Hardin broadened the discussion of commons, illustrating it with the herding example, but his use of the term commons was ill-defined: he interpreted it not as a specific form of local common property but, as he later clarified, as "unregulated commons" (Hardin 1994, 1998), which meant in fact "open access", in comparison to property rights such as local common property, state property or private

(*continued*)

Box 4.5 (continued)

property. In the current research on commons or common pool resources the concept is used for many forms of collective or shared use of resources such as fish stocks, lakes and rivers, or the global commons of the oceans and the atmosphere. The concept of commons can also be used for collective infrastructure and transport systems such as roads. As the concept is broadened it becomes inexact and overlaps with that of public goods used in economics.

The discussion about the use and overuse of commons is linked to other problems of natural resource use and population growth. The classic dilemma discussed as the tragedy of the commons is whether commons tend to be overused and ruined, whereas resources in state or private property are not (Hardin saw state and private property as modern property forms that help to avoid overuse). The question in the commons debate can be generalised: is there a form of property that is optimal for maintaining resource use over a long period without overusing the resource base? When the question became one of empirical research, it turned out that overuse happens in different property forms; overuse depends in the last analysis on the rules regulating the use of the resource and users' compliance with them. Tragedies of overuse were found in some forms of commons, but many commons were well managed and maintained by local user communities (the research is summarised by Ostrom 1990). Further research found "tragedies of enclosures" of privately owned resources (Martinez-Alier): private property regimes do not necessarily better protect the resources. In the sustainability discourse the commons dilemma needs to be connected with economic growth (usually implying private property) and environmental protection in more systematic analyses and empirical research, which Hardin did not take into account: this happened in the intensive research on commons in recent decades.

The research on the commons is documented in a series of publications that continue to demonstrate the influence of Ostrom and her theoretical and empirical work: McCay and Acheson (1987), Feeny et al. (1990), Ostrom (1990), McCay (1995), Becker and Ostrom (1995), Ruttan (1998), Ostrom (1999), Ostrom et al. (2002), Adams et al. (2003), Agrawal (2003), Dietz et al. (2003), Low et al. (2003), Acheson (2006), van Laerhoven and Ostrom (2007), Armitage (2008), Poteete et al. (2010), Nkhata et al. (2012), Anderies and Janssen (2013), Berge (2015), Lara (2015), Mathias (2017), Partelow et al. (2020), van Laerhoven et al. (2020).

Sources: mentioned in the text

Management of the commons and of common pool resources: Ostrom et al. (2002: 456f) provide a synthesis of the commons research, concluding that institutional regimes for commons and common pool resource management differ and cannot be subsumed under one model. (1) The metaphor of the tragedy of the commons, reasoning that commons result in overuse and ruin of resource stocks, is valid under special conditions only: when the resource users cannot communicate with another and build mutual trust. (2) Three basic conditions for the functioning of commons regimes found in the research include the importance of the resource for the users, the autonomy to devise and change rules and the necessity to communicate and bargain with other users. (3) No particular institutional design fits all commons and common pool resources. (4) For resource users, other necessities beyond sustainable resource use, such as livelihoods and well-being, are also important. (5) Indirect effects of resources, for example costs of monitoring and rule enforcement, are important for the functioning of commons regimes. *Global commons* is a term used to describe international and global resource domains which can be understood as commons of humankind, including typical examples such as the oceans, the Antarctic area, the atmosphere and outer space. The commons concept here becomes inexact, showing the cognitive limits of the commons discourse, where a more differentiated terminology is required.

Research on the commons has increased in recent decades and gained importance in the sustainability discourse. Its importance is connected with natural resources that cannot be entirely privatised, if at all (such as fish), and the global commons. The empirical research brought partial clarifications on the conditions under which common property regimes work. The research on commons and common pool resources has broadened and deepened (it can be followed in the "International Journal of the Commons"). It has become part of other research fields such as environmental governance and sustainability, global environmental change and social-ecological systems. New forms of commons beyond the traditional research on natural resources include the global commons, knowledge commons and technological commons.

Property rights—the privatisation of natural resources: Ecological and economic research about the "tragedy of enclosures" (Martinez-Alier) is connected with the system questions of social and ecological limits to growth. It cannot be dealt with sufficiently with the logic of ecological modernisation, where the problems of overuse and limits to growth are reduced to the integration of economic and ecological rationality in the economic production and consumption processes. The further requirements to be discussed in the sustainability discourse and process include the forms and consequences of the use of public and private goods, of individual and collective resource use, and of complementary institutions for natural resource management that are not included in property rights systems: regulatory functions of the state and governments, allocative and distributive functions beyond the market, the rebuilding of local, community-based economies and the reduction of global exchange and value chains.

With the current debate and development of the commodification of nature, natural resources and ecosystems comes a new debate about intellectual property rights and the environment: TRIPS (Agreement on Trade-Related Aspects of Intellectual Property Rights). The new form of privatisation of natural resources is not mainly about land or water, but about privatisation and commercialisation of genetic resources from plants, animals and humans (patenting genetic resources so that they can be traded, bought and sold). This process began with the Convention on Biological Diversity resulting from the United Nations Conference on Environment and Development (UNCED) in 1992, which had three goals: the conservation of biological diversity, the sustainable use of its components and the fair and equitable sharing of benefits arising from genetic resources. One consequence of TRIPS was new forms of ecological distribution conflicts.

4.3.2 The Economics of Ecosystems and Biodiversity

The project "The Economics of Ecosystems and Biodiversity" (TEEB), initiated by Germany and the European Commission and later hosted by the United Nations Environment Programme (UNEP), can be seen as a renewal of earlier attempts to economise or commodify nature. TEEB has faced controversies about the role, significance and application of economic knowledge for purposes of environmental protection and sustainability. Since its beginning it has been criticised as reducing the complexity of ecological dynamics through economic ideas and idealised models, broadening and deepening the commercialisation of nature (MacDonald and Corson 2012: 159). The main criticism is that the economic valuation of ecosystem services, even when it is intended to provide information for assessing the value of ecosystems or to make the value of nature visible, leads to the marketisation, monetarisation and commercialisation of ecosystem services by the financial and banking systems. O'Neill (1997; see also O'Neill 1998, 2007) also argues against attempts to price nature according to the neoliberal policies that guide economic globalisation, which leads to the introduction of market mechanisms in the spheres of nature and life, which previously were protected from markets, resulting in more and more commodification of nature and natural resources for the sake of profit (see also Spash 2011).

TEEB (Kumar 2011) started from a critical review of the post-war history of economic growth as unsustainable, the limits of neoclassical economics, its market-dependent view and the limits of economic valuation. It was a response to the argument for the need for profound changes in society and economy that came with the sustainability discourse, arguing instead for a new approach to protect ecosystem services and biodiversity. The project assumes that protecting nature requires the calculation of costs and benefits, and that markets are the only institution that can serve as the economic instrument for biodiversity protection. Sukhdev, the project leader, argued for a financial and banking sector approach to "capture values" of biodiversity, thus supporting the trading of biodiversity and ecosystem assets. The market-oriented thinking and the profit motive are obvious in this project; the arguments that this creates new forms of extended environmental protection are less convincing. The predominance of economic knowledge in TEEB, the reduction of nature to natural capital and the practice of calculating the monetary value of ecosystems are examples of economic knowledge employed in the interest of the private and financial economy. To achieve its aims, TEEB needs to deal with the ecological criteria of ecosystem assessments, but it does this in the distorting form of the monetary valuation of nature which follows an economic, not ecological logic of valuation (using the conceptual framework of ecosystem services for the Millennium Ecosystem Assessment 2005). The proponents of the project have adopted some critiques of the market and growth-based economy as a distorting and dysfunctional system, in a value-based rhetoric of "economics for a changing society": awareness building about nature and natural resource use should be encouraged through the project.

The critique of economic growth and market-dependent environmental regulations by the TEEB project seems to be simply a rhetorical reference to the critical discussion about limits to growth. In practice the project renews and broadens growth-based strategies, camouflaged with the terminology of green and sustainable growth. The renewal of growth in a situation where growth becomes more critically seen for its negative environmental consequences remains controversial in the sustainability discourse. The economic reasoning behind green and sustainable growth clashes with the various forms of critical ecological and institutional economics arguing for the development of degrowth and similar strategies (Spash 2011, 2018; see Chap. 6). Environmental movements and organisations have responded critically to TEEB, but some among them have defended growth and commodification as means of protecting nature.

4.4 Discussion and Conclusions

The core problem of sustainable development, the use and sharing of natural resources, cannot be solved with knowledge from economics alone but requires interdisciplinary knowledge. A more complicated issue is determining which knowledge from which disciplines and research areas is required in which parts of the political sustainability process. Such questions about knowledge use require analyses of the processes of change and transformation that influence sustainable development—social, political and ecological changes connected with changes in the market- and growth-based economy (for the discussion of growth and sustainability see: Victor and Dolter 2017). The discussion of property rights and their environmental consequences is controversial in economics and in ecology, which makes the sustainability process more complicated. In order to resolve the controversies, further research and knowledge beyond property rights is useful, including research on access to natural resources (Sen 1999), the differentiation and development of integrated and holistic approaches to natural resource management, and the development of a social-ecological science of complexity that began with Ostrom's research.

The development of social-ecological research encountered difficulties with the concepts of economic and ecological resources; these are abstract and imply many types of resources that require theoretically based classifications to deal with different interpretations and to become applicable in sustainability governance. One difficulty in the ecological classification of natural resources is how to deal with complex systems as resources—ecosystems, landscapes, the earth system and the biosphere. In the classification of economic resources theoretical concepts for different resource types are rare; economically relevant resources include more than the classical differentiation of production factors as land, labour, capital and their optimal combination. Terms such as economic, social, human and natural capital need to be transformed epistemologically and methodologically into scientific concepts. Beyond the classification of resources, a main cognitive and methodological difficulty is how to combine economic and ecological criteria and resource concepts in interdisciplinary sustainability analyses. These difficulties become evident in the discourses of environmental economics and ecological economics (see Chaps. 5 and 6).

Summary of Chap. 4: Natural Resource Management
Natural resource management, in sector-specific forms (agriculture, fishery, forestry, industry) or in integrated forms of natural resource management (for example ecosystem-based management), requires balancing multiple economic and ecological criteria to achieve sustainability. The main questions to be discussed in natural resource management are those of the sustainability discourse and sustainability transformation of the economy: how to use and combine various economic concepts and ideas (market economy and a new form based on solidarity, sharing and redistribution) and to avoid confusion in the use of the multifaceted concept of economy in economics, ecology and the interdisciplinary sustainability discourse. Further problems include the combination and balancing of ecological and economic principles in the sector-specific management of natural resources; the forms of property rights and their consequences for resource use (especially common or private property: "tragedy of the commons" or "tragedy of enclosures"); and the development of new forms of property rights for the commodification of nature and natural resources (intellectual property rights regarding genetic resources and economic use of ecosystems and biodiversity). The problems of knowledge use and natural resource management are connected with the core question of sustainable development: the future of economic growth.

Appendices: Further Information and Material

Learning Exercise 1. Definitions of Economics and Economy and Their Practical Utility

Discuss as a group the descriptions and definitions of economics as a scientific discipline and the views of the economy (using Box 4.1). You can also watch short videos on "economics" and "economy" on the internet to prepare for the discussion.

The following questions should be answered: what are the consequences of the pluri-paradigmatic state of economics (that there exist many views, paradigms, approaches and theories in economics and in subdisciplinary subjects such as environmental economics)? (How) can one work with different theories simultaneously?

Thinking about definitions of economics and economy: what are the purposes of general definitions of economics—what can they say about the practise of teaching and research in economics? What are the limits of abstract definitions of economics? Answer the same questions for definitions of the economy as a process or system.

Learning Exercise 2: The Tragedy of the Commons

Discuss, individually or as a group, based on the information given in Box 4.3 of this chapter, the question:

What is the significance of the discourse of the tragedy of the commons for the current problems of natural resource use at local, national or global scale? Further questions that come up in the search for an answer should also be discussed: is the concept of the commons still relevant today (when most of the resources are exchanged, bought and sold on markets and under private property regimes)? Should the concept of the commons be used in more specific and historically concrete variants (as a form of property rights), or in broader, generalised and more inexact forms (as all forms of collective or shared use of resources)? Can you specify under what conditions common property of resources works better for maintaining the resource use in the long run, and under what conditions private property forms work better?

If you want to deepen the discussion, read the article by Hardin (1968) *and his later short updates* (Hardin 1994, 1998).

Learning Exercise 3: The Circular Economy as a Component of Sustainability Transformation

Discuss as a group the advantages and limits of a circular economy as it is described in the chapter. To prepare for the discussion you can also watch one of the videos about circular economy on the internet or read the executive summary of the 2012 report "Towards the Circular Economy: Economic and Business Rationale for an Accelerated Transition" from the Ellen MacArthur Foundation (which can be found on the internet).

The following questions should be discussed: is the circular economy a coherent concept and strategy, or a collection of various ideas and principles? Is the circular economy sufficient for sustainable development? What else in terms of economic, social or ecological criteria is required to achieve sustainability? Is the circular economy a new form of the market economy, or part of the transformation to another form of post-industrial and post-growth economy?

Learning Exercise 4: Bioenergy—A Controversial Case in the Building of Sustainable Energy Systems

The energy sector is key to the transformation of the economy to sustainability, which began in many countries with the use of renewable energy sources to curtail the severe environmental damage caused by the use of fossil resources such as coal, oil or gas. Particularly difficult problems arise from the use of bioenergy in the building of sustainable energy systems.

Bioenergy includes resources such as wood, waste and agricultural residues such as straw, forests and grasslands, and feedstocks in agriculture; its production causes CO_2 emissions, which are a main cause of climate change and global warming. The question of whether bioenergy use is environmentally friendly is not easily answered; much depends on the conversion technologies and their commercial use. For example, biofuels or biodiesel for cars cause further problems when the systems of transport are not changed further; better in terms of sustainable energy use seems to be electrification based on renewable sources and sustainably produced biomass (which is not the case with all biomass production). Discussion of the prospects of renewable energy sources and bioenergy from agricultural land involves value-based and ethical questions and decisions, including the following:

Should agricultural land be used for production of bioenergy or for food production? How much bioenergy can be produced sustainably, and by what means, through sustainable forms of land use? How can sustainable development be operationalised to determine the sustainability of bioenergy production (the book suggests criteria in terms of climate protection, biodiversity maintenance, soil protection, and socio-economic criteria such as food security and health risks through energy use)? Critical questions include the consequences of indirect land use change caused by biofuel production and the time it will take for CO_2 emissions to be significantly reduced, while in the meantime global warming continues.

Suggested tasks and questions for further reflection and discussion (individually or in a working group): Read about bioenergy production and its environmental effects on the internet; discuss how bioenergy production and use can be made sustainable. Discuss the ethical problems with bioenergy production from agricultural land instead of food production under conditions of hunger and malnutrition. Use the discussion of bioenergy production to find out more about the roles and forms of technologies in achieving sustainability.

Learning Exercise 5: Critical Discussion of the TEEB Project of Economics of Ecosystem Services

Discuss (individually or as a group) the information in Sect. 4.3.2 of this chapter on the problems with the commodification and privatisation of living resources, genetic resources and ecosystems, using the example of TEEB.

TEEB raises critical ethical questions about the limits of commodification and privatisation of natural resources, and about the continuation of economic growth through the exploitation of the natural resources in terrestrial and marine ecosystems. Your discussion should address the following questions: for which natural resources (physical or biological resources) should commercial use and monetarisation, buying and selling, be allowed, and for which resources should they be prohibited (based on ethical and ecological arguments)? What are the (wanted and unwanted) consequences when complex resources (ecosystems, biodiversity) are commodified and traded? Can ecosystems and biodiversity, as TEEB is arguing, be protected through privatisation and commodification? Who benefits most from TEEB, and who is disadvantaged?

These questions are not easily answered, but preliminary answers regarding the advantages and disadvantages of the commercialisation of nature and natural resources can be found through the discussion. If you want to deepen the discussion, you can easily find information about TEEB on the internet.

Further Readings Suggested: Deepening, Thematically Specialised

Policy-related—private and public natural resource management: Morrison-Saunders, A., Pope, J., & Bond, A. (Eds.). (2015). *Handbook of Sustainability Assessment*. Cheltenham, UK and Northampton, MA: Edward Elgar

Science-related—resource economics, environmental management: Leal Filho, W. (Ed.). (2018). *Handbook of Sustainability Science and Research*. Cham, Switzerland: Springer Nature

Scientific journals (international peer reviewed) with research relevant for the themes of this chapter: "Agricultural Economics", "Agriculture, Ecosystems and Environment", "Annual Review of Resource Economics", "Coastal Management", "Conservation Letters", "Ecological Indicators", "Ecosystem Services", "Environmental Policy and Management", "Environmental Management", "Food Policy", "Forest Policy and Economics", "Journal of Bioeconomics", "Journal of

Economics and Sustainable Development", "Journal of Environmental Economics and Management", "Journal of Industrial Ecology", "Marine Policy", "Marine Resource Economics", "Natural Resources Journal", "Resource and Energy Economics", "Review of Environmental Economics and Policy", "Urban Studies"

References

Acheson, J. M. (2006). Institutional Failure in Resource Management. *Annual Review of Anthropology, 35*, 117–134.

Adams, W. M., Brockington, D., Dyson, J., & Vira, B. (2003). Managing Tragedies: Understanding Conflict over Common Pool Resources. *Science, 302*, 1915–1919.

Agrawal, A. (2003). Sustainable Governance of Common-Pool Resources: Contexts, Methods, and Politics. *Annual Review of Anthropology, 32*, 243–262.

Albert, H. (1967). Modell-Platonismus: Der neoklassische Stil des ökonomischen Denkens in kritischer Betrachtung. In H. Maus & F. Fürstenberg (Eds.), *Marktsoziologie und Entscheidungslogik*. Neuwied, Berlin.

Allen, C. R., Fontaine, J. J., Pope, K. L., & Garmestani, A. S. (2011). Adaptive Management for a Turbulent Future. *Journal of Environmental Management, 92*, 1339–1345.

Anderies, J. M., & Janssen, M. A. (2013). *Sustaining the Commons*. Center for the Study of Institutional Diversity, Arizona State University. (ebook). Retrieved from http://sustainingthecommons.asu.edu.

Anderson, D. A. (2019). *Environmental Economics and Natural Resource Management*. New York: Routledge.

Armitage, D. (2008). Governance and the Commons in a Multi-Level World. *International Journal of the Commons, 2*(1), 7–32.

Armitage, D., Berkes, F., & Doubleday, N. (Eds.). (2007). *Adaptive Co-Management, Collaboration, Collective Learning, and Multi-Level Governance*. Vancouver and Toronto: UBC Press.

Barnett, V. (1998). *Kondratiev and the Dynamics of Economic Development*. London: Macmillan.

Becker, E., & Jahn, T. (Eds.). (1999). *Sustainability and the Social Sciences: A Cross-Disciplinary Approach to Integrating Environmental Considerations into Theoretical Reorientation*. London and New York: Zed Books.

Becker, C. D., & Ostrom, E. (1995). Human Ecology and Resource Sustainability: The Importance of Institutional Diversity. *Annual Review of Ecology and Systematics, 26*, 113–133.

Berge, E. (2015). On the Commons of Developed Industrialized Countries. *International Journal of the Commons, 9*(2), 469–485.

Biggs, R., Peterson, G. D., & Rocha, J. (2018). The Regime Shifts Database: A Framework for Analyzing Regime Shifts in Social-Ecological Systems. *Ecology & Society, 23*(3), 9.

Billé, R. (2008). Integrated Coastal Zone Management: Four Entrenched Illusions. *SAPIENS, 1*, 2. Retrieved from http://journals.openedition.org/sapiens/198.

Binswanger, H. C., & Chakraborty, R. N. (2000). *The Economics of Resource Management*. Paper Commissioned by the European Commission, Directorate General Enviropnment.

University of St. Gallen, Institute for Economy and the Environment. Retrieved March 20, 2020, from https://ec.europe.eu/environment/enveco/waste/pdf/binswanger.pdf.

Boulding, K. E. (1966). The Economics of the Coming Spaceship Earth. Environmental Quality in a Growing Economy. In H. Jarrett (Ed.), *Essays from the Sixth RFF Forum* (pp. 3–14). Baltimore: John Hopkins University Press.

Brockington, D. (2012). A Radically Conservative Vision? The Challenge of UNEP's Towards a Green Economy. *Development and Change, 43*(1), 409–422.

Bromley, D. (1991). *Environment and Economy: Property Rights and Public Policy*. Oxford: Blackwell.

Bromley, D. (Ed.). (1992). *Making the Commons Work: Theory, Practice, and Policy*. San Francisco: ICS Press.

Bromley, D. (1999). *Sustaining Development: Environmental Resources in Developing Countries*. Cheltenham, UK: Edward Elgar.

Bromley, D. (Ed.). (2014). *Institutions and the Environment*. Cheltenham: Edward Elgar.

Bruckmeier, K. (2016). *Social-Ecological Transformation: Reconnecting Society and Nature*. London: Palgrave Macmillan.

Bruckmeier, K. (2019). *Global Environmental Governance: Social-Ecological Perspectives*. Cham, Switzerland: Palgrave Macmillan and Springer Nature.

Caldwell, B.J. (1994). *Beyond Positivism: Economic Methodology in the Twentieth Century*. Revised Edition. London and New York: Routledge.

Carruthers, B. C., & Ariovich, L. (2004). The Sociology of Property Rights. *Annual Review of Sociology, 30*, 23–46.

Carson, R. (1962). *Silent Spring*. Boston: Houghton Mifflin.

Catton, W. J. R., & Dunlap, R. (1978). Theories, Paradigms and the Primacy of the HEP-NEP Distinction. *The American Sociologist, 12*, 256–259.

Cicin-Sain, B., & Knecht, R. (1998). *Integrated Coastal and Ocean Management: Concepts and Practices*. Washington, DC: Island Press.

Colander, D. (2000). The Death of Neoclassical Economics. *Journal of the History of Economic Thought, 22*(2), 127–143.

Conroy, M. J., & Peterson, J. T. (2014). *Decision Making in Natural Resource Management*. New York: Wiley-Blackwell.

Cooney, R. (2004). *The Precautionary Principle in Biodiversity Conservation and Natural Resource Management*. IUCN, Gland, Switzerland: IUCN Publications Services Unit.

Corson, C., & MacDonald, K.I. (2012). Enclosing the Global Commons: The Convention on Biological Diversity and Green Grabbing. *Journal of Peasant Studies, 39*, 2, 263–283.

Costanza, R., Cumberland, J., Daly, H., Goodland, R., & Norgaard, R. (2010 [1997]). *An Introduction to Ecological Economics*. Enyclopedia of Earth.

Dale, V. H., & Haeuber, R. A. (Eds.). (2001). *Applying Ecological Principles to Land Management*. New York: Springer.

Daly, H. (1994). Steady State Economics. In C. Merchant (Ed.), *Key Concepts in Critical Theory: Ecology*. Amherst and New York: Humanity Books.

Debeir, J.-C., Deléage, J.-P., & Hémery, D. (1991 [1986]). *In the Servitude of Power*. London and Atlantic Highlands, NJ: Zed Books.

De Marchi, N., Blaug, M. (Eds.), (1991). *Appraising economic theories: Studies in the methodology of research programmes*. Aldershot, Hants, UK and Brookfield, VT: Edward Elgar.

Devezas, T. (2006). *Kondratieff Waves, Warfare and World Security*. Amsterdam: IOS-Press.

Dietz, T., Ostrom, E., & Stern, P. C. (2003). The Struggle to Govern the Commons. *Science, 302*, 1907–1912.

EU. (2019). *Social Indicator Dataset*. Brussels. Retrieved March 14, 2020, from https://ec.europa.eu/social/main.jsp?catId=818&langId=en.

Feeny, D., Berkes, F., McCay, B. J., & Acheson, J. M. (1990). The Tragedy of the Commons: Twenty-Two Years Later. *Human Ecology, 18*(1), 1–19.

Fischer-Kowalski, M., Haberl, H. (Eds.). (2007). Socioecological Transitions and Global Change. *Trajectories of Social Metabolism and Land Use*. Cheltenham, UK and Northampton, MA: Edward Elgar.

Folke, C., Hahn, T., Olsson, P., & Norberg, P. (2005). Adaptive Governance of Social-Ecological Systems. *Annual Review of Environment and Resources, 30*, 441–473.

Funtowicz, S., & Ravetz, J. (1993). Science for the Post-Normal Age. *Futures, 25*(7), 739–755.

Geissdoerfer, M., Savaget, P., Bocken, N., & Hultink, E. (2017). The Circular Economy – A New Sustainability Paradigm? *Journal of Cleaner Production, 143*(1), 757–768.

Georgescu-Roegen, N. (1975). Energy and Economic Myths. *Southern Economic Journal, 41*(3), 347–381.

Goodland, R., Daly, H., Serafy, S. E., & von Droste, B. (Eds.). (1991). *Environmentally Sustainable Economic Development: Building on Brundtland*. Paris: UNESCO.

Gordon, H. S. (1954). The Economic Theory of a Common Property Resource: The Fishery. *Journal of Political Economy, 62*, 124–142.

Grinin, L., Korotayev, A., & Tausch, A. (2016). *Economic Cycles, Crises, and the Global Periphery*. Heidelberg, New York and London: Springer International Publishing.

Hall, C. A. S., & Klitgaard, K. A. (2012). *Energy and the Wealth of Nations: Understanding the Biophysical Economy*. New York et al.: Springer.

Hanna, S., Folke, C., & Mäler, K.-G. (Eds.). (1996). *Rights to Nature*. Washington, DC: Island Press.

Hardin, G. (1968). The Tragedy of the Commons. *Science, 162*(3859), 1243–1248.

Hardin, G. (1991). Paramount Positions in Ecological Economics. In R. Costanza (Ed.), *Ecological Economics* (pp. 47–57). New York: Columbia University Press.

Hardin, G. (1994). The Tragedy of the Unmanaged Commons. *Trends in Ecology & Evolution, 9*(5), 199.

Hardin, G. (1998). Extensions of "The Tragedy of the Commons". *Science, 280*(5364), 682–683.

Harrison, M. (1975). Chayanov and the Economics of the Russian Peasantry. *Journal of Peasant Studies, 2*(4), 389-417.

Harrison, M. (1977). The Peasant Mode of Production in the Work of A.V. Chayanov. *Journal of Peasant Studies, 4*(4), 323–336.

Hawken, P., Lovins, A., & Lovins, H. (1999). *Natural Capitalism: Creating the Next Industrial Revolution*. New York: Little, Brown & Company.

Heynen, N., McCarthy, J., Prudham, S., & Robbins, P. (2007). *Neoliberal Environments: False Promises and Unnatural Consequences*. New York: Routledge.

Jones, B. D. (1999). Bounded Rationality. *Annual Review of Political Science, 2*, 297–321.

Jorgensen, T. H., Remmen, A., & Mellado, M. D. (2006). Integrated Management Systems – Three Levels of Integration. *Journal of Cleaner Production, 14*(8), 713–722.

Kumar, P. (Ed.), (2011). *The Economics of Ecosystems and Biodiversity: Ecological and Economic Foundations*. London: Earthscan from Routledge.
Lara, A. (2015). Rationality and Complexity in the Work of E. Ostrom. *International Journal of the Commons, 9*(2), 573–594.
Leal Filho, W. (Ed.). (2018). *Handbook of Sustainability Science and Research*. Cham, Switzerland: Springer Nature.
Lee, K. (1989). *Social Philosophy and Ecological Scarcity*. New York: Routledge.
Lin, B. C.-a. (2020). Sustainable Growth: A Circular Economy Perspective. *Journal of Economic Issues, 54*(2), 465–471.
Linné, C. v. (2010 [1749]). Oeconomia naturae. Whitefish, Montana: Kessinger Publishing (Kessinger Legacy Reprints).
Lloyd, W. F. (1833). *Two Lectures on the Checks to Population*. UK: Oxford University. Oxford: Collingwood, Printer to the University.
Low, B., Ostrom, E., Simon, C., & Wilson, J. (2003). Redundancy and Diversity: Do They Influence Optimal Management? In F. Berjket, J. Golding, & C. Folke (Eds.), *Navigating Social-Ecological Systems: Building Resilience for Complexity and Change*. Cambridge et al.: Cambridge University Press.
Luhmann, N. (2012–2013). *Theory of Society*. 2 Vols. Stanford: Stanford University Press.
Martinez-Alier, J., & Muradian, R. (Eds.). (2015). *The Handbook of Ecological Economics*. Cheltenham, UK and Northampton, MA: Edward Elgar.
Mathias, J.-D. (2017). Multi-Level Policies and Adaptive Networks – A Conceptual Modeling Study for Maintaining a Polycentric Governance System. *International Journal of the Commons, 11*(1), 220–247.
McCay, B. J. (1995). Common and Private Concerns. *Advances in Human Ecology, 4*, 89–116.
McCay, B. J., & Acheson, J. M. (Eds.). (1987). *The Question of the Commons: The Culture and Ecology of Communal Resources*. Tucson: The University of Arizona Press.
Miller, A. (1999). *Environmental Problem Solving: Psychosocial Barriers to Adaptive Change*. New York: Springer.
Moore, J. (2015). *Capitalism in the Web of Life*. Eugene, OR: Cascade Books.
Morgan, J. (Ed.). (2016). *What Is Neoclassical Economics? Debating the Origins, Meaning and Significance*. New York: Routledge.
Morrison-Saunders, A., Pope, J., & Bond, A. (Eds.). (2015). *Handbook of Sustainability Assessment*. Cheltenham, UK and Northampton, MA: Edward Elgar.
Munsinghe, M. (2013). Property Rights and Ecological-Social Interactions. *Encyclopedia of Earth*. Retrieved January 30, 2020, from https://editors.eol.org/eoearth/wiki/Property_rights_and_ecological-social_interactions.
National Research Council. (2002). *The Drama of the Commons. Committee on the Human Dimensions of Global Change* (E. Ostrom, T. Dietz, N. Dolsak, P. C. Stern, S. Stovich, & E. U. Weber, Eds.). Washington, DC: National Academy Press.
Nefiodow, L. A., & Nefiodow, S. (2017 [2014]). *The Sixth Kondratieff: The New Long Wave in the Global Economy*. Sankt Aufgustin. Retrieved from www.nefiodow.de.
Newell, A., & Simon, H. A. (1972). *Human Problem Solving*. Cambridge, MA: Harvard University Press.

Nkhata, B. A., Breen, C., & Mosimane, A. (2012). Engaging Common Property Theory: Implications for Benefit Sharing Research in Developing Countries. *International Journal of the Commons, 6*(1), 52–69.

O'Neil, J. (2007). *Markets, Deliberation and Environment*. London: Routledge.

O'Neill, J. (1997). Managing Without Prices: On the Monetary Valuation of Biodiversity. *Ambio, 56*, 546–550.

O'Neill, J. (1998). *The Market: Ethics, Knowledge and Politics*. London: Routledge.

Odum, E. (1997a). *Ecology: A Bridge Between Science and Society*. Sunderland, MA: Sinauer Associetes Inc. Publishers.

Odum, E. (1997b). Commentary: Source Reduction, Input Management and Dual Capitalism. *Journal of Cleaner Production, 5*(1), 1–2.

Odum, E. (2012 [1998]). *Ecological Vignettes: Ecological Approaches to Dealing with Human Predicaments*. London: Routledge.

OECD. (2020). *How's Life 2020? Measuring Wellbeing*. Paris.

Ostrom, E. (1990). *Governing the Commons: The Evolution of Institutions for Collective Action*. Cambridge et al.: Cambridge University Press.

Ostrom, E. (1999). Coping with Tragedies of the Commons. *Annual Review of Political Sciences, 2*, 493–535.

Ostrom, E. (2007). Challenges and Growth: The Development of an Interdisciplinary Field of Institutional Analysis. *Journal of Institutional Economics, 3*(3), 239–264.

Ostrom, E., Dietz, T., Dolsak, N., Stern, P. C., Stonich, S., & Weber, E. U. (Eds.). (2002). *The Drama of the Commons*. Washington, DC: National Academy Press.

Parsons, T., & Smelser, N. (1956). *Economy and Society*. Glencoe, IL: The Free Press.

Partelow, S., Abson, D. J., Schlüter, A., Fernández-Giménez, M., von Wehrden, H., & Collier, N. (2020). Privatizing the Commons: New Approaches Need Broader Evaluative Criteria for Sustainability. *International Journal of the Commons, 13*(1), 747–776.

Pauly, D. (2019). *Vanishing Fish: Shifting Baselines and the Future of Global Fisheries*. Vancouver and Berkeley, CA: Greystone Books.

Pauly, D., Christensen, V., Dalsgaard, J., Froese, R., & Torres, F. (1998). Fishing Down Marine Food Webs. *Science, 279*, 860–863.

Plummer, R. (2009). The Adaptive Co-Management Process: An Initial Synthesis of Representative Models and Influential Variables. *Ecology and Society, 14*(2), 24. Retrieved from http://www.ecologyandsociety.org/vol14/iss2/art24/.

Polasky, S., Kling, C. L., Levin, S. A., Carpenter, S. A., Daily, G. C., Ehrlich, P. R., et al. (2019). Role of Economics in Analyzing the Environment and Sustainable Development. *PNAS, 116*(12), 5233–5238.

Poteete, A. R., Jansson, M. A., & Ostrom, E. (2010). *Working Together: Collective Action, the Commons, and Multiple Methods in Practice*. Princeton, NJ: Princeton University Press.

Rees, W. (1996). Carrying capacity revisited: Area-based indicators of sustainbility. *Population and Environment, 17*: 195–215.

Rice, J. (2007). Ecological Unequal Exchange: Consumption, Equity and Unsustainable Structural Relationships Within the Global Economy. *International Journal of Comparative Sociology, 48*(1), 43–72.

Ricklefs, R. E. (1993 [1973]). *The Economy of Nature: A Textbook in Basic Ecology* (3rd ed.). New York: W.H. Freeman.

Rosenberg, A. (1992). *Economics: Mathematical Politics or Science of Diminishing Returns*. Chicago: University of Chicago.
Rosenberg, A. (1994). The Cognitive Status of Economic Theory. In R. Backhouse (Ed.), *Nature of Economic Method* (pp. 216–235). London: Routledge.
Ruttan, L. M. (1998). Closing the Commons: Cooperation for Gain or Restraint? *Human Ecology, 26*(1), 43–66.
Scoones, I. (1999). New Ecology and the Social Sciences: What Prospects for Fruitful Engagement? *Annual Review of Anthropology, 28*, 479–507.
Sen, A. (1999). *Commodities and Capabilities* (2nd ed.). Delhi and New York: Oxford University Press.
Simon, H. A. (1979). Rational Decision-Making in Business Organizations. *American Economic Review, 69*, 495–501.
Simon, H. A. (1995). Rationality in Political Behaviour. *Political Psychology, 16*, 45–61.
Spash, C. L. (2011). Terrible Economics, Ecosystems and Banking. *Environmental Values, 20*(2), 141–145.
Spash, C. L. (2018 [2017]). *Routledge Handbook of Ecological Economics: Nature and Society*. Abingdon and New York: Routledge.
Swilling, M. (2016). Preparing for Global Transition: Implications of the Work of the International Resource Panel. In H. G. Brauch et al. (Eds.), *Handbook on Transition and Sustainable Peace* (pp. 391–418). Cham: Springer.
Trepl, L. (1994 [1987]). Geschichte der Ökologie vom 17. Jahrhundert bis zur Gegenwart. Weinheim: Beltz Athenäum.
United Nations Development Programme. (2019). *Human Development Report 2019*. New York.
van Laerhoven, F., & Ostrom, E. (2007). Traditions and Trends in the Study of the Commons. *International Journal of the Commons, 1*(1), 3–28.
van Laerhoven, F., Schoon, M., & Villamayor-Tomas, S. (2020). Celebrating the 30th Anniversary of Ostrom's Governing the Commons: Traditions and Trends in the Study of Commons, Revisited. *International Journal of the Commons, 14*(1), 208–224.
Victor, P. A., & Dolter, B. (Eds.). (2017). *Handbook on Growth and Sustainability*. Cheltenham, UK and Northampton, MA: Edward Elgar.
Wallerstein, I. (1979). *The Capitalist World Economy*. Cambridge: Cambridge University Press.
Waltner-Toews, D., Kay, J. J., & Lister, N. M. E. (Eds.). (2008). *The Ecosystem Approach: Complexity, Uncertainty, and Managing for Sustainability*. New York: Columbia University Press.
Worster, D. (1979 [1977]). *Nature's Economy: The Roots of Ecology*. Garden City, NY: Anchor Books.

5 Environmental Economics: Orthodox Perspectives

The research themes of environmental economics overlap with those of a series of other economic subdisciplines such as resource economics, sector-specific economics such as agricultural economics, development economics and other fields of specialisation. With the development of environmental and sustainability research, the economic discourse is increasingly confronted with questions of interdisciplinary cooperation, research and knowledge exchange. Environmental economics is described in this chapter with regard to the knowledge and ideas from economic research relevant for sustainable development, not in the dogmatic form of traditions of academic teaching. For an overview of the history of environmental economics, its themes, approaches and methodologies, the textbooks and handbooks reviewed at the end of the chapter can be used as additional sources.

Since the 1970s, with the intensification of national and international environmental discourses and politics, environmental economics has developed from a small subdiscipline of economics and from research on natural resource use to an influential subject by adopting new environmental themes, concepts and approaches through interdisciplinary knowledge exchange. It is concerned with the development and differentiation of the economic growth debate, the integration of ecological knowledge in environmental economics to deal with externalities, complementary approaches, and a critique of economic growth that intensifies through the sustainability discourse and the transition to sustainable forms of the economy.

The core themes and controversies in environmental economics relevant for the sustainability discourse are visualised in Fig. 5.1.

K. Bruckmeier, *Economics and Sustainability*,
https://doi.org/10.1007/978-3-030-56627-2_5

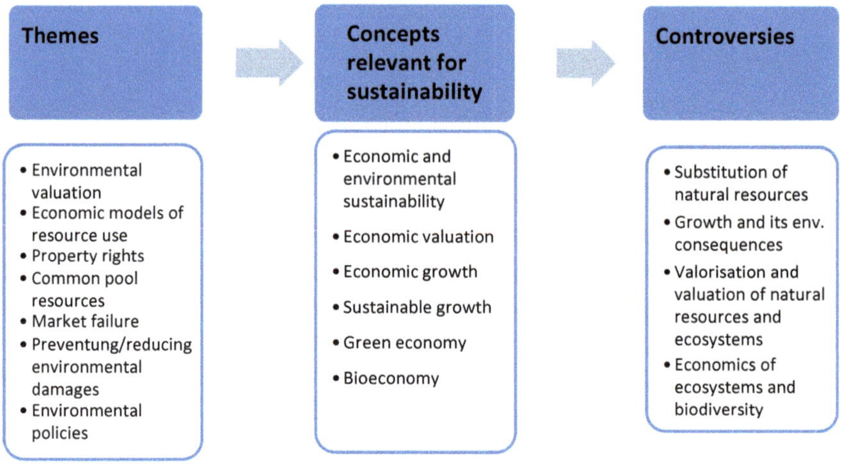

Fig. 5.1 Environmental economics: themes and controversies

5.1 History and Development of Environmental Economics

A main reason for the rapid development and thematic differentiation of environmental economics in recent decades has been the need to seek *alternatives to market-based valuation of natural resources*; many natural resources cannot be appropriated privately and traded on markets (Dasgupta and Mäler 2004: 1). Exchange and valuation are possible for some natural resources, especially land, which can be divided up and appropriated and has been privatised to a large extent in the transition to the modern economy; it is more difficult to regulate access to and appropriation of natural resources that are located on the land (water and groundwater), growing on it (wild and agricultural plants) or living on it (wildlife), or that are located below it in the soil (mineral resources). In the case of water-based resources the transformation from commons to private property is more difficult, for example the privatisation of fishing rights in semi-private forms and imperfect forms of quotas for catch in form of tradeable fishing quotas. Appropriation and use of resources without defined property rights (open access) continues to occur in modern economies where privatised resources are dominant, for example the so-called global commons such as the oceans or the air and the atmosphere, for which privatisation is difficult or impossible because of their physical nature and size. Cultural and ethical norms and legal rules may prevent the privatisation and commodification of certain natural, physical and living resources, for example human bodies and genetic resources. These rules are changing with the increasing appropriation and economic use of genetic resources of plants, animals and humans as a result of globalisation and deregulation of markets. Many ethical rules of natural resource use are not observed in "low ethics countries" with insufficient regulation of research, business practices, working conditions, health protection and social or environmental protection. The United Nations' Human Development Index provides information on the differences between countries in terms of ethical rules and practices.

For water, imperfect forms of private property exist: a part of coastal waters adjacent to privately owned land can, for example, be privately owned, and the creation of greater and greater exclusive economic zones in the coastal waters of a country defines quasi-property rights for limited groups of users. Production, exchange and trade of food products happen in the modern economy on the basis of property rights—common, state or private property are used to describe the land and water that are necessary for production.

Environmental economics as a subdiscipline has developed in response to the global environmental crisis since the second half of the twentieth century as part of the effort to prevent or limit environmental destruction. Economic and political discourses are closely linked in this process, and the economic concept of resources is critically discussed in all forms of environmental regulation. With the valorisation of nature and the commodification of ecosystems and living resources, the term economic resources has been extended to include resources that were previously exempt from appropriation and commodification. Many observers consider the commodification of living resources and ecosystems undesirable and risky.

5.1.1 The Changing Thematic Scope of Environmental Economics

Different traditions and approaches such as neoclassical environmental economics, natural resource economics, ecological economics, economics of sustainability and environmental policies to regulate natural resource use have influenced environmental economics. Not only did it have to find new economic methods for the valuation of resources, it also had to face thc problem of interdisciplinary integration and knowledge sharing, especially for global change processes of climate change and biodiversity loss. The integration of economic knowledge with other types of knowledge has been limited, but has been practised more in critical or heterodox forms of economics: institutional, development, ecological and interdisciplinary economics. Environmental economics as it is described in three popular textbooks (Tietenberg and Lewis 2018; Harris and Roach 2018; Hackett and Dissanayake 2019) focuses on four main themes as presented in Box 5.1.

Box 5.1 Thematic Scope of Environmental Economics

1. *The economic approach*: The production of economic knowledge in environmental economics works with (a) methods and models of economic analysis and valuation (circular flow models for resources, goods, services, money and capital, environmental accounting) and (b) concepts and theories (for example, principles of environmental economics, externalities, market failure, common property resources, public goods). The methods, concepts and normative assumptions guide and frame research about resource use and pollution in production and consumption, about allocation and distribution, markets and trade, property and access rights for resources. The differences between environmental economics (understood as applying the principles of orthodox or neoclassical economics to the analysis of environmental problems and natural resource use) and ecological economics (understood as interdisciplinary broadening of economics to include knowledge from other disciplines and constructing the economy as embedded in ecosystems) are controversial.
2. *Natural resource economics*: According to the production theory of classical economics, land was seen as a production factor beside labour and capital. Environmental economics research focuses on exhaustible and renewable, physical and biological, material and energy resources, and common pool resources. Research themes include the economic, social and environmental consequences of natural resource use in the form of pollution, toxic substances, climate change, biodiversity loss and land use change. The changing state of the natural resource base and the availability and distribution of resources through the institutions in the market economy are part of the discussion between economics and ecology.
3. *Knowledge application and practices of resource use*: Policy-related knowledge includes the economics of pollution control, decision-making metrics, cost–benefit analyses and economic assessment methods, economic and other policy instruments and incentives, regulation, ecosystem management, adaptation and "greening of the economy" through new principles of resource use and environmental accounting. In the application of economic knowledge for environmental policies, the internalisation

(*continued*)

Box 5.1 (continued)

of externalities is a core problem discussed in terms of policy instruments (for example environmental regulation using cost–benefit analysis, combatting pollution through taxes, tariffs and tradeable quotas, improving thc functioning of property rights regimes).

4. *The future*: How can environmental justice and a sustainable, post-industrial and post-capitalist economy and society be achieved by way of economic change or transformation,? This transformation is not achieved with the knowledge form empirical economic research; it requires theoretical knowledge, interdisciplinary knowledge integration, and cooperation of many actors; the possibility of continued economic growth becomes a critical question.

Sources: Tietenberg and Lewis 2018; Harris and Roach 2018; Hackett and Dissanayake 2019

Of the three textbooks reviewed in Box 5.1, Tietenberg and Lewis (2018) is the oldest and most conventional, discussing sustainable development in a simplified form without including the current state of research and discussion. Harris and Roach (2018) integrate environmental and natural resource economics and discuss environmental economics from a more pluralistic perspective, where conventional environmental economics and ecological economics are seen not as conflicting and competing but as complementary. The authors assume that a pluralistic analysis is required to understand the complex relations between the modern economy and the environment and to link the economic analysis with policy analyses. Hackett and Dissanayake (2019) also approach environmental economics from an interdisciplinary perspective, discussing sustainability in depth in several chapters.

Of the four topical areas of environmental economics, the first three represent the conventional knowledge from economics, the principles and concepts, the toolbox and the core themes of research, and the application of economic knowledge in practice and for decision-making. Only the discussion of the future, with sustainable development as a core theme, is more innovative and less structured based on available economic knowledge, following instead the public and policy discourses about the environment and resource use. This theme is the weakest in the textbooks, as the knowledge reviewed lags behind the interdisciplinary sustainability discourse and the requirements of interdisciplinary knowledge integration. Meanwhile, texts and textbooks about the economics of sustainability (Goldin and Winters 1995; Bartelmus 2013, 2018) are more advanced in addressing interdisciplinary knowledge for environmental valuation. In terms of integrating economic knowledge in the broader inter- and transdisciplinary sustainability discourse, there is still a long way to go from economics to an interdisciplinary environmental or sustainability science (Polasky et al. 2015).

The theme of environmental valuation has become dominant in environmental economics with the intensification of the environmental policy discourse (Pearce et al. 1989; Pearce 1992; Hodge 1995). Valuation, according to the conventional view, is difficult when no markets and prices for the environment exist. Attributing monetary value to the environment or natural resources is seen as necessary in order to make them comparable, to judge their importance and to find solutions to scarcity and other problems of resource use (Hodge 1995: 58). The methodology for environmental valuation is broadened and pluralised, using methods of both economic valuation (for example surveys, expert panels) and non-economic research (see Appendix, Learning Exercise 1). Environmental valuation is being developed through the critique and change of disciplinary knowledge practices towards interdisciplinary practices (Box 5.2).

Box 5.2 The Knowledge Practice of Specialisation in Environmental Economics

The development of environmental economics is documented in many handbooks: Bromley 1995; van den Bergh 1999; Markandya et al. 2002; Mäler and Vincent 2003; Gallagher 2008; Aronsson and Löfgren 2010; Bennett 2011; Halvorsen and Layton 2015; Fouquet 2019; Soytas and Sari 2019; Chichilnisky and Rezai 2020.

The deepening specialisation of research is visible in the trend towards more thematically specialised handbooks on energy economics, economics of climate change, or valuation and accounting, orientated to the application of economic knowledge in environmental policy and governance and economic decision-making. The dominant approach in the textbooks and handbooks is that of specialised disciplinary research and limited interdisciplinary knowledge exchange. Environmental economics as a subdiscipline is developing through the application and refinement of concepts, models and methods from orthodox economics for the analysis and solution of environmental and resource use problems. Economic knowledge is transferred into practice through two channels for decision-making in (a) environmental policies and (b) the economy, that is, in enterprises. Sustainable development and other forms of transformation of modern economies are reduced to the

(*continued*)

Box 5.1 (continued)
adaptation of economic institutions and processes to a changing natural environment with the knowledge and methods of conventional or orthodox economics: the legitimacy and functioning of the global economy is not questioned, and a critical discussion of the transformation of the economic system or of economic growth is avoided.

The problems and knowledge practices in environmental economics as a specialised economic subdiscipline are critically discussed by Bowles and Carlin (2020). With regard to the teaching of economics, they conclude that new problems such as climate change, economic disparities, the future of work and the global financial crisis challenge the content of introductory courses in economics; some methods for studying these problems exist in economics, for example dealing with limited information, instability and path dependence, agency dilemmas and principal–agent models (Eisenhardt 1989). In the future, economics needs to become interdisciplinary, integrating micro- and macroeconomics and broadening the study of institutions and their effectiveness, including economic and political institutions. The critical questions are those of changes in the economic system: the deregulation of markets, their changing structures through globalisation and their adaptive limits, which appear in dysfunctional growth with negative environmental effects and in concentration of capital and monopolies. The ecological dysfunctionality is becoming a bigger problems than the social dysfunctionalities of inequality and unequal distribution of resources.

Sources: mentioned in the text

The two overarching themes in the sustainability discourse are *economic growth and transition to sustainability*. Environmental economics is challenged by the critical debates about limits to growth: it cannot develop in isolation from the complex and global problems that are dealt with in the interdisciplinary discourses of sustainability in science and politics. The traditional views of academic science, disciplinary specialisation, and the division of labour between science, politics and social practices of knowledge application are outdated. Regarding the core themes, the views of economic growth and sustainability transformation have clearly changed since the turn of the century under the influence of the broader sustainability discourse.

Pearce (2002) gives an example of redefining sustainable development in orthodox economic terms by way of discussing economic growth as an indicator of welfare or wellbeing. The question of whether growth measured as gross domestic product (GDP) growth at national levels is an indicator of wellbeing and quality of life has become increasingly controversial in recent debates, resulting in the differentiation of the terms of growth, wellbeing and sustainable development in the economic and ecological discourse (see Box 5.3).

Box 5.3 Economic Growth and Wellbeing: Pearce 2002

Pearce discusses sustainable development in economic terms, with the notion of an economic system where well-being per capita increases over time. To reflect the conditions of the creation and distribution of well-being, traditional growth theory should be reinterpreted in terms of sustainable development:

> Defining sustainable development in terms of sustained per capita increases in well-being needs to be distinguished from stating the conditions for the achievement of such increases. Like traditional growth theory, the modern theory of sustainable development is based on notions of capital assets as the means of generating well-being. Along with developments in growth theory, e.g., endogenous growth theory, capital is decomposed into human-made capital, human capital (knowledge, skills), natural or environmental capital, and social capital. The condition for sustainability (or potential sustainability) is that the sum of these assets, i.e., total wealth, should increase in per capita terms over time. Intergenerational rules of behaviour follow immediately, e.g., that each generation should bequeath to the next generation a capital endowment no less than the one it has now. This interpretation of sustainable consumption was debated in the 1930s, with Hicks in 1939 setting out the notion that it involved maximum consumption consistent with maintaining capital assets intact … The resulting *Hicksian income* measure can then be interpreted as 'the interest on total wealth.' (Pearce 2002: 62)

The construction and discussion of a "modern theory of sustainable development" by Pearce gives an example of blending the terms of economic growth, creation of well-being, utility, consumption and sustainability—a conceptual reductionism and distortion of the forms of abstraction that show the heritage of orthodox and neoclassical economics. Such a "theory" based on capital assets as a means of generating well-being is a form of economising society and nature; it does not consider other ecological and social-scientific research and critical views of economic dysfunctionalities. Rather than creating a substantiated theory of symbolic and material capital,

(continued)

Box 5.3 (continued)

the metaphoric concepts of capital used by Pearce can be seen as extending the notion of capital in fictitious forms to non-economic resources that do not correspond with the capital functions in the global economy. Thus the economic views and valuations are distorted in ways that are criticised in the economic and ecological degrowth discourses as forms of "colonised" and "disembedded" thinking, as a "growth fetish": the market logic is generalised for all natural resources, ecosystems and nature. The transformation of natural resources into economic resources needs to be critically reassessed in the sustainability discourse and process.

Source: Pearce 2002

Dasgupta (2008) criticised economics for its neglect of nature and tried to open it up to communication with environmental science, thus providing arguments for the development of environmental economics. Economics largely failed in this respect, remaining under the influence of neoclassical thinking, although in the debate about economic valuation of the environment some advances in forms of non-market valuation were achieved. Dasgupta (2014) summarised the discussion about wealth and the valuation of natural capital in environmental economics, repeating his earlier critique: the problem of environmental economics is to overcome the limits of neoclassical economics, which ignores natural capital in the production and creation of wealth. As long as nature continues to be ignored in the established economic accounting of GDP, there will be no real understanding of the wealth of nations. This is an advance in the deconstruction of the blended concepts of income, wealth, welfare and well-being, beginning with the critical question about indicators of economic growth: what do the indicators really measure? Dasgupta expects the development of green national accounting (discussed by Bartelmus (2013, 2018) to be the solution to the valuation problems (Box 5.4).

Box 5.4 Ignorance of the Worth of Natural Capital as a Knowledge Barrier: Dasgupta 2014

> The absence of natural capital in growth and development models has meant that contemporary national accounts continue to be prepared without mention of the environmental resource base. Although the United Nations statistical office

(*continued*)

Box 5.4 (continued)

has constructed satellite accounts that include natural capital, few countries treat them as anything more than the proverbial footnote. … When national accounts are better prepared, health and natural capital will in all probability be found to be the most significant component of the wealth of nations. …

But even if figures for natural resource stocks were available, the deep problem of imputing values to them would remain. Market prices may be hard facts, but shadow prices are soft. The issue isn't merely that the role that natural capital plays in production and consumption possibility is uncertain, but also that people differ in their ethical values. The sensitivity of wealth estimates to shadow prices should become routine exercise in national accounts. (Dasgupta 2014: 28)

That people may never agree on the wealth of nations is, however, no reason for abandoning wealth as the object of interest in sustainability analysis. Our ignorance of the economic worth of natural capital remains the greatest barrier to an understanding of the history of economic development. Until that ignorance is lifted, policy analysis will remain crippled, and sustainability will continue to be a notion we admire but cannot put into operation. (Dasgupta 2014: 29)

Dasgupta discusses critically the difficulties with environmental valuation, but these do not vanish when the economic or monetary value of natural capital can be better assessed; the consequences of the reformulation of nature, ecosystems or natural resources as natural capital are not reflected. Before the appropriation of natural resources by way of privatisation, they did not have the economic functions of accumulating capital, and after appropriation it is still doubtful whether all natural, social and human forms of capital can be subsumed under the logic of reasoning and the notion of capital: they do not have the functions of monetary capital in the modern economy. The more the concept of capital is extended, the more the view of natural resource use in economics is distorted, de-contextualised and dehistoricised: natural resources are used in the economy, but they do not become transformed into economic categories.

Source: Dasgupta 2014

Bartelmus (2013, 2018) expanded the critical reflection about measuring nature and environmental quality in economic terms. In his critical review of the environment and growth discussion in economics, he tries to combine environmental and ecological economics for developing indicators of economy–nature relations to find practical solutions for sustaining prosperity and the quality of the environment. His discussion of the limits of indicators and what they really show brings some clarification of the concepts and assumptions in economic and ecological valuation. Economic indicators and models of growth are deconstructed, showing the assumptions on which they are based. The limitations of economic indicators for measuring environmental quality and human well-being—which cannot be measured with GDP—requires the development of differentiated and combined indicators. The discussion is thus opened up to the construction of new indicators for well-being and welfare beyond GDP growth (see Table 5.1). This is not limited to the question of how and how far monetary and non-monetary forms of valuation can be combined; it also becomes part of the critical ecological debate on limits to growth.

5.1.2 Trends in the Economic Growth Debate

Continued economic growth is assumed to be necessary in mainstream economics, including for solving environmental and resource use problems. However, the consequences of economic growth need to be discussed more critically in environmental economics, including social inequality, distributional injustice and deteriorating environmental quality, which are not easily corrected through new forms of growth. The normative assumptions of the necessity of growth, and its justification under conditions of economic inequality, concentration of power and capital, and deregulated markets, have become blurred in attempts to specify growth through mathematical modelling or empirical research limited to specific conditions and cases. In the economic discourse, growth analysis is absorbed in the construction of a series of mathematically formalised growth models where the social and environmental consequences are not always explicit. Enterprises in the modern economy act under different market conditions and conditions of resource availability which are continually changing.

The discussion of growth in classical political economy has been the focus of critical research that has questioned both the dominant assumption of growth as necessary and advantageous and the possibility of a growing throughput of natural resources in the modern economy. The factors studied in economics that influence growth include the division of labour, specialisation and economies of scale; sur-

Table 5.1 Economic indicators—economic growth and development

Name	Definition	Interpretation	Sources
GDP	Monetary indicator for economic wealth and growth: market value of all final goods and services produced annually in a national economy	GDP can be measured in different ways: according to the production approach (OECD), the income or the expenditure approach; controversial discussion: what is GDP measuring (just wealth, or also development and human well-being)	International standard for measuring GDP is the "SNA 93" (System of National Accounts 1993) Critique of GDP: Stiglitz et al. 2010
GGEI (Global Green Economy Index)	GGEI measures the economic performance of the green economy and perceptions of it in 130 countries, in four dimensions: leadership and climate change, efficiency sectors, markets and investment, environment	The GGEI and the green economy are criticised as reductionist approaches that are not able to deal with the limits to growth and global environmental change; rather, they support the commodification of nature by extending the monetary valuation of natural resources	Dual Citizen LLC: 2010–2018 Global Green Economy Index™ (GGEI)
GPI (Genuine Progress Index) and ISEW (Index of Sustainable Economic Welfare)	The GPI indicator includes the GDP and additional data about the costs and negative effects of economic activities (costs of crime, ozone depletion, environmental pollution and other costs); a forerunner of the GPI was the ISEW (developed by Daly and Cobb), which also included distributional inequality	The GPI is a critical indicator to correct the GDP by accounting for positive and negative results of economic growth; it exists in different variants; GPI and ISEW are not calculated for all countries	GPI: Kubiszewski et al. 2013 ISEW: Lawn 2003
MEFA (material and energy flow accounting) and MFA (material flow accounting)	Accounting for energy flow and material flow is the basis for quantifying the biophysical flows in an economy (at national or global levels), in addition to monetary flows	MEFA and MFA help to assess the environmental sustainability of natural resource use in production and consumption and to quantify hidden flows (materials that are extracted or moved, including waste, but do not enter into the economy)	MEFA: Haberl et al. 2004 MFA: Fischer-Kowalski et al. 2011

Sources: mentioned in the table

plus of labour, capital and trade; extension of markets through foreign trade; combination of capital and labour; technological development and entrepreneurial innovations; investments of enterprises, accumulation forms and regimes; and the ratio of savings to investments. The changing definitions, descriptions and explanations of economic growth in the history of the discipline show that there is no single, coherent and durable accepted theory of economic growth, but rather multiple competing paradigms, theories and models that try to explain the possibilities, the obstacles and limits, and the phases and changes in economic growth (see the review of growth theories by Pietak 2014). The mathematically formalised growth models do not offer empirical knowledge and explanations of growth, but rather a casuistic collection of cases with varying assumptions about conditions and requirements of growth. The normative assumption in economics of a gap between unlimited human needs and limited resources, and the construction of rational choice behaviour connected with it, become points of critique in the debate on limits to growth. The basic assumptions require discussion of the worldviews, paradigms and pre-analytic visions of economic thinking and research. This is often neglected in the routines of economic research and in specialised research on growth, where the creation of knowledge is focused on model-based data collection (see Appendix: discussion of theories and models of economic growth).

In modern economics the trends in growth research are towards (a) *specialisation*, replacing theories of growth with more limited models of growth based on empirical research; (b) *pragmatic orientation*, turning away from theoretical explanations towards studies of the political forms and conditions of influencing and maintaining economic growth; and (c) studying *varying* conditions of growth, including the critical discussion of limits and ultimately the global limits to growth. Factors relevant for the ecological critique of economic growth and its limits include arguments about population growth, human needs, welfare and scarcity. These factors are discussed in orthodox economic theories but in reductionist and selective forms, based on abstractions and generalisations that ignore large parts of social-scientific and ecological knowledge about the effects of growth and its limits, or as hypotheses about the substitution of natural resources with other resources to maintain growth. A significant explanatory factor for economic growth that is often highlighted is technology, which is seen as improving the productivity of labour, for example in the neoclassical exogenous and endogenous growth theories (Sredojevic et al. 2016). This can be intuitively understood in the mathematical formalisation as well as in the neoclassical growth model, where the production function is used to measure the growth and identify the equilibrium of an economy where demand and supply are equal, based on the equation representing the relationship between labour and technology $Y = F(K, AL)$, where Y means the GDP of an economy, K the share of capital, L the amount of unskilled labour and A the determinant level of technology. Practical controversies and difficulties arise in the operationalisation and measurement of the factors determining growth.

The discussion of economic growth that influenced the development of environmental economics during the twentieth century reappears in sustainability discourse and politics in three overlapping fields of research: in the economics of exhaustible resources, in the system analyses of economic growth in natural resource economics and in new growth-critical analyses of the global economy.

(1) Economics of exhaustible resources—the conventional economic research on growth: The classical analyses of Gray (1914) and Hotelling (1931) were long neglected in the economic discourse. The debate about exhaustible resources was taken up in the early 1970s, when the political and public discussion of global limits to growth also began. Hotelling, referring to dwindling exhaustible natural resources such as minerals or forests, discussed the physical limits to growth, criticising the static economic equilibrium theory as inadequate for exhaustible resources where a constant rate of production cannot be maintained indefinitely (Hotelling 1931: 139). The "Hotelling rule" says that the price of an exhaustible resource needs to grow in parallel to the rate of interest, in efficient extraction forms and in a competitive resource economy that tends towards equilibrium. These assumptions show the characteristic form of the discussion of exhaustible resources in economics since then: on the basis of a model of economic production, introducing a dynamic element and periodisation in the model, and comparing prices under monopoly and competitive conditions. The arguments are formulated in formal logical form, in mathematical equations, and they take as a given the functioning of the market economy; more specific assumptions and conditions under which exhaustible resources can be exploited in a market and growth-based economy are successively discussed. Hotelling introduced ecological arguments into economic thinking in a disciplinary- and specialisation-based view of economics, which consequently did not open up economics to ecological knowledge.

The economics of natural resource use received more attention in the 1970s as the debate about fossil energy resources and the limits to growth began. Solow (1972, 1974) initiated new debates on intergenerational equity and exhaustible resources based on new models. He tried to develop an economic theory of such resources, for which he formulated some basic conditions: the importance of a long-term view and of sufficient information about reserves, technology and demand in the far future (being aware of the uncertainty and the problems of knowledge about futures trading). Solow suggested indicative planning and regulation of markets for exhaustible resources as a possible means to direct the economy away from "ultimately inferior paths" (Solow 1974: 13f). The problems with Solow's modelling and its limits are critically reviewed by Klitgaard (2013; see Box 5.5), showing the de-contextualised reasoning and abstraction from relevant historical and social conditions of economic resource use—a prominent critique of orthodox economics in institutional economics (Kapp 1968).

Box 5.5 Institutional Economics: Critique of Neoclassical and Orthodox Economics

Kapp 1968: Regarding demand and supply, "the institutionalist would not be satisfied to derive these fundamental categories from axioms of rational conduct, marginal rates of substitution and static production functions but would push the analysis beyond supply and demand by inquiring into the factors of market power and market domination including high pressure salesmanship and would point out that goods and services also derive their relative importance from the fact that they serve as symbols of rank and status" (Kapp 1968: 4).

Klitgaard 2013: "Rather than a 'decennial cycle' of prosperity and depression, the normal state of monopoly capital was longterm stagnation or slow economic growth. The source of the stagnation was a rising economic surplus that could not be fully absorbed by the spending outlets available: investment, consumption and waste. … Investment creates additional capacity even as it serves as a spending outlet (or absorbs the economic surplus). Spending is short lived, while the investment is long lived, and the problem becomes both perpetual and unsolvable by further incremental investment. Moreover, the system is burdened by excess capacity, which is a chronic condition of monopolistic industrial organization …, and one cannot rely on vibrant investment in a time when much capital remains unutilized. Consequently, the growth trajectory of a capitalist economy is unstable. Mainstream economic growth theory results from a critique of the work of Domar, along with that of Roy Harrod. In 1956, Robert Solow … contended that Harrod and Domar postulated fixed technical conditions of production (although this assumption appears explicitly in neither the original books nor papers of Harrod or Domar.) Solow, claiming that resource substitutability is a 'crucial' assumption, substituted a Cobb-Douglas production function for Harrod and Domar's supposed fixed-production isoquants. Presto! The

(*continued*)

Box 5.5 (continued)
instability of the system disappears, and a fundamental social problem of economic instability is transformed into an easily-solvable technical problem. Yet, despite Solow's prominence and the virtual disappearance of the original work of Harrod and Domar from the teaching of economic growth theory, the vast social problems of stagnation and unemployment persist even in today's economy" (Klitgaard 2013: 284).

Sources: mentioned in the text

Exhaustible resources for which the physical and ecological conditions of use need to be accounted for in economic resource use became the paradigmatic case to discuss limits to growth in natural resource use economics.

(2) Natural resource economics—economic and ecological systems analysis to identify limits to growth: Natural resource economics developed before environmental economics and later became a part of it. The differentiation between natural resource economics and environmental economics is sometimes reduced to that between input (natural resources) and output (environmental pollution) of the economy. Prior to the development of environmental economics, Barnett and Morse (1963) discussed how economic growth could be maintained and global welfare achieved in the context of exhaustible resources. These became key questions in economics that cannot be answered through simple or partial models. Only with the development of computer-based world models, first used for the study of "limits to growth" in the report for the Club of Rome in 1972, did the modelling of long-term global processes and economic growth begin. The discussion remains controversial in economics and in sustainability discourse in the broader context of the relations between the developed countries of the Global North and the developing countries of the Global South. The industrial countries have been able to maintain uneven development and technological progress and to avoid diminishing returns. In practice this was a brutal process of conquest, colonialism, war, unequal development and exchange to maintain their privileged positions in the modern world economic system. Barnett and Morse formulated a critique of the global economy as an

> open question, whether modern economies have also learned to maintain social progress, to continue improving the quality of life, to avoid qualitative diminishing returns. Even more open is the question whether these nations, in cooperation with those that have not yet become industrialized, will be able to assure both quantitative

> and qualitative progress for mankind throughout the world. Thus interpreted, our obligation to future generations poses difficult problems—problems that touch the sensitive core of our value standards, our methods of calculating benefits and costs, and our decision-making procedures. (Barnett and Morse 1963: 250f)

This formulation anticipates the core problem of sustainable development that appeared later in the Brundtland Report. The authors do not have solutions; they discuss the future-related problems of natural resources and quality of life in modern societies only briefly in a concluding chapter. Hicks (1965: 201) fails to offer a solution as well, as his critical reflections on economic growth show: the dynamic problem of positive economics, of the actual behaviour of an economy, with imperfect foresight, becomes unclear when studied in detail; the actual path which the economy (even the model economy) will follow cannot be determined. The question of the links between economic growth and resource use problems has continued to be controversial. Dasgupta and Healy (1979) discuss within the neoclassical paradigm of economics possible limits to growth: they do not exclude the possibility of an economic collapse due to overuse, thus implicitly supporting the reasoning in the "limits to growth" report of 1972. The collection of essays by Kemp and Long (1984) and Heal and Garrett (1993) show the difficulties of developing an economic theory of exhaustible resources.

Sušnik (2016) derives growth limits with regard to water withdrawals at individual and national levels in seven GDP growth scenarios for the year 2100. He finds that the sustainability of resource use varies with the scenarios; strong growth implies excessive use of natural resources, which leaves two possibilities for sustainable resource use: consuming less and reducing economic growth, or finding ways to decouple growth and resource use; such decoupling has not yet happened. The model-based and formal analyses of the use of exhaustible resources under market conditions do not show the social and biophysical reality in their selective assumptions. With these models a definite form of optimal growth, of limiting growth, or a theory of growth that includes economic, social and ecological conditions and limits are not possible; they can only provide preliminary solutions and search for substitutions, technological innovations and new natural resources that can be economically exploited to continue growth (the ideas currently offered in the green economy strategies). The epistemological critique of the limits of modelling and the knowledge produced with the use of formalised economic models that build mainly on logical and mathematical reasoning is well established. Already in the 1960s, at the time of the positivism dispute between critical rationalism (Popper) and critical theory (Adorno) in the German social sciences, Albert, a proponent of critical rationalism, criticised the "model Platonism" of neoclassical economics, arguing that economic models are abstract exercises that play with varying assumptions and data but do not capture the social reality of economic processes with their idealisation and simplification. He argued against exercises in decision logics, favouring instead the development of an empirical science of economics as "sociology of markets" (Albert 1998). This critique is still relevant for the growth-critical debate in the sustainability discourse.

Solutions to the problems of global depletion, extinction and exhaustion of resources based on global environmental change cannot be sufficiently explored and explained through specialised economic research and modelling of market processes. Scholars have sought to overcome the limits of economic modelling by combining economic and ecological modelling, and these can also become part of future interdisciplinary research and knowledge practices for sustainable development; the models are simpler for local resource use practices and more complicated for global analyses of resource use. Simple forms of bio-economic models have long been in use, for example in fisheries, agriculture or forestry, to optimise resource use, while more complex models have also been developed (overview in Kraqt 2012; Flichman and Allen 2013; for fisheries and marine resources: van den Bergh et al. 2006; Anderson and Seijo 2010; Aronsson and Löfgren 2010; for agriculture Flichman 2011; for forestry: Graves et al. 2007; see Journal "Bioeconomics"). The epistemological critique of orthodox and neoclassical economics, including environmental economics, concerns the limits of modelling and the knowledge produced with the use of formalised economic models that build mainly on logical and mathematical reasoning. The development of the economic debate of limits and alternatives to growth requires more than models: non-economic knowledge about the systemic and institutional contexts of economic growth, about the different constituents and functions of social systems and ecosystems, and about the coupling of such systems and their integration as social-ecological systems is essential.

(3) New growth-critical debates in economics—critical theoretical system analyses of economic growth: Lowe (2011) made clear the difficulties involved in assessing and predicting changes in resource availability, long-term growth possibilities and instability of growth paths due to insufficient or problematic data based on simplified assumptions about economic behaviour. The discussion of economists published by Smith (1979) sparked a broader debate about scarcity and growth in which critical economists such as Georgescu-Roegen and Daly participated. Resources for the Future (RFF), a non-profit organisation from the United States, is until today supporting the development of environmental and natural resource economics and organising critical debates about resource scarcity. Zinn (1980) analysed economic growth critically. Both books are examples of the opening up of the economic growth discourse to critical and interdisciplinary analyses and the unfolding of growth- and globalisation-critical reasoning that became part

of the sustainability discourse. Key themes of this debate include (a) the dis-embedding of the economy from the individualised modern market economy, which has destroyed the social quality of life and undermined social cohesion; and (b) the risks of political action and decision-making in the global economy that are "colonised" and dominated by powerful economic actors and their interests (Zinn 1980: 12). Similar reasoning developed later in the globalisation critical discourse about "post-politics" (Swyngedouw and Wilson 2014). Analyses of continued growth and estimations of the possibilities of sustainable growth with exhaustible resources are discussed for single countries (for example, Scholl and Semmler (2000) for the USA). Research on global environmental change and the development of ecological indicators for the limits of growth brought about a more critical discussion of growth of the global economy.

The growth-critical debate intensified after the millennium under the influence of research on global environmental change and sustainable development: it became a public discourse, beyond economics, fostered through social movements such as the global degrowth movement. The intensifying debate about climate change, in combination with the global economic crisis after 2007, brought a breakthrough in the debate about the "folly of growth" ("New Scientist", 2008), renewing the earlier discourse of limits to growth from the Club of Rome report. The limits-to-growth discourse became more specific, focusing on economic growth and its consequences, and more realistic, investigating and discussing different possibilities of transforming the modern economy to a non-growing economy. In these discussions the ideas of sharing, distribution and reciprocity have been rediscovered as alternatives to the growth-based market economy; so far they have mainly been discussed in the heterodox economic discourse of institutional economics, taking up ideas from Polanyi (1944).

5.2 Transformation of the Growth Debate in the Sustainability Discourse

The research on growth in the history of economics includes many different and specific models that interpret, explain and justify growth differently and at different levels of economic processes, in micro-and macroeconomic perspectives and under specific historical situations. Therefore it is impossible to formulate a dominant growth factor or to argue generally for the necessity of economic growth. As growth is only a means to achieve specific economic ends, it requires further arguments to explain or justify it based on its ends and to specify its negative social and ecological effects. The common justification that economic growth translates through incomes into welfare, not only of producers but also of consumers and all social groups in modern society, although unequally, is insufficient with regard to the social and ecological effects. A widespread critique of economic growth as the unequal distribution of wealth involving negative ecological consequences in terms of overuse and depletion of natural resources is now on the agenda of the sustainability process.

Because GDP does not allow for specifying the value of nature or natural capital, new economic indicators have been developed. "Inclusive wealth" (Arrow et al. 2012; Dasgupta 2014; Polasky et al. 2015) is an attempt to develop an economic indicator for measuring whether an economy is approaching sustainability, including social, economic and environmental sustainability. In constructing such an indicator, the difficulties of economic valuation become visible in comparing valuations and estimating the aggregate value of all capital assets, including natural, manufactured, human and social capital, most of which are based on monetary value only to a limited degree or not at all. These variants of the capital concept can be interpreted differently; interpretations vary between analogies, metaphors, comparisons, generalisations of the economic concept of capital, or new theoretical concepts that cannot be explained with economic terms. Much of what is discussed with the new non-economic capital concepts was formerly discussed and explained without these, especially in critical economic and social theories. Instead of struggling with the capital concepts, the sustainability discourse shifted the focus of the economic growth debate away from its economic limitations and towards the ignored social and ecological conditions and limits of growth that are not explained by the new capital concepts.

The primary controversy that arose in the sustainability discourse was about the social and ecological consequences of exponential growth in the global economy. Economic methods and models of growth, insofar as they do not include social and ecological data, play only a limited role in this debate. The debate about environmental growth was long centred around the discussion of methodological questions and models and the clarification of terminology. In the 2010s the debate advanced to one about possibilities of combining and integrating economic, social and ecological indicators and information to measure different aspects of interaction between nature and society. Although the existence of ecological and social limits to growth is less controversial today, no consensus has been achieved on the limits to growth and the possibility of transforming the modern economy into a non-growing system. Economic growth has only lost its status as an economic dogma and a value that cannot be doubted. The discussion of market failures has entered the public and political discourses, advancing from a critique of the imperfection of markets that can be corrected to public policies and complementary measures, to ideas of re-embedding markets or seeking alternatives to them. Meanwhile, new proponents of competition and growth have appeared in the economic discourse (Haucap 2018) to argue for continued growth under the conditions of globalisation and the digital economy.

The interdisciplinary discussion of economic growth and its natural and social limits (Victor and Dolter 2017) makes visible what was ignored in earlier economic research on growth or filtered through the methodology of modelling and the theoretical assumptions and hypotheses of orthodox economics, with anthropological assumptions about the nature of humans in the rational actor model of the "economic man". The processes of abstraction and knowledge filtering are a strategy of reducing social relations between humans to the pursuit of individual economic interests in competition with others, based on the maximisation of income, yields and profits. The market economy appears in abstract form, whereas the social preconditions and the systemic qualities of the modern economic system are ignored.

The social and ecological consequences of growth and the normative justifications of growth require empirical research and theoretical reflection, explanation and critique to demystify popular beliefs that appear in the growth models. These beliefs in modernisation, innovation, technical progress and economic rationality as drivers of socially advantageous economic growth are fading. The controversies over how wealth can be measured, whether growth in terms of individual income or aggregate growth in terms of GDP are indicators of welfare or human well-being, are changing in the sustainability discourse but have not yet disappeared. This began with the clarification of differences between growth and development, or quantitative growth and qualitative growth. Two points are important for the future discussion of sustainability transformation: (a) there is no single "objective" indicator for measuring multifaceted concepts such as wealth or welfare that require several and specified indicators; and (b) GDP growth is a form of quantitative growth that may be interpreted as an indicator of wealth, when this is understood as monetary income; it is not an indicator of welfare and well-being in terms of further and non-economic components of quality of life.

Quality of life depends only to some degree and indirectly on monetary income (measured, for example, in the Human Development Index). The growth of a national economy in terms of GDP appears in the economic discourse as creating welfare in various forms that include profit for enterprises, future investments, workplaces, financing of public institutions and social policy, public education, healthcare and infrastructures. Not all the social improvements made through development can be attributed to economic growth alone. An economic picture of the welfare society already includes components of redistribution through public policies and taxing, public goods that cannot be produced through markets, and social improvements and change in forms of education, healthcare and public services that are not based primarily on monetary or market transactions and that cannot be subjected to the functional logic of markets. Trying to introduce economic reasoning to these activities by way of contingent valuation, "willingness to pay", "shadow prices" and assessment of non-economic social and ecosystem services through economic valuation becomes a misleading exercise in "the commodification of everything".

For the debate about the transition to sustainability and the question of whether economic growth can help to achieve environmental sustainability, the GDP is of interest not as an indicator of wealth but as an indicator of throughput of natural resources in economic growth to show how economic growth is connected with the growth of natural resource use (Daly 2013: 21). The growth of natural resource use measured with ecological indicators, especially in the accounting of material and energy flows in physical terms, is more concrete and informative regarding the biophysical reality of overuse of resources and environmental destruction. Thus far there has been no decoupling of aggregate growth and throughput growth, and whether and how far decoupling is possible remains a controversial issue. The economic globalisation process of recent decades led to a change in the nature of wealth-increasing growth. The deregulation of markets devalued the national redistribution processes that corrected social inequality to some degree and created new and larger forms of inequality, showing the dilemma of the growth-based industrial economies. Countries and national economies compete in the processes of industrialisation and economic growth, and only a part of the global population is able to gain from growth and the appropriation of the resources necessary for growth. Industrialisation in all countries seems impossible because of the global limits of natural resources (which have already been reached for important resources). The discussion of development without growth becomes necessary in the sustainability discourse, including redistribution and sharing of resources between countries and within and between generations. The discussion of social distribution and redistribution of resources introduces further questions that have not been taken into account in the economic growth and welfare constructions: the functions of the state, the concentration of economic power in international corporations, the connections between micro- and macroeconomic growth and the decoupling of economic growth from resource use.

5.2.1 The Blocked Growth Discourse and Its Renewal

In economics and environmental economics the question of whether environmental sustainability can be achieved with the market economy and through economic growth has hit a dead end. The orthodox arguments for growth lose their scientific cogency and are mainly enforced in the sustainability discourse through an alliance of powerful and influential government-dependent institutions and economic elites in controversial programmes of green and blue growth. There is no intensive dialogue between the proponents and opponents of growth about future strategies for sustainable development, and renewal of the growth discourse has come about slowly, supported by empirical studies showing that sustainable growth cannot be achieved by all countries and currently has only been achieved by a few Organisation for Economic Co-operation and Development (OECD) countries (see Chap. 6). Furthermore, such sustainable growth has only been reached as part of the global economy, with the majority of countries continuing to experience unsustainable and environmentally destructive growth that is intensifying in the late industrialising countries. In a globalised economy and under conditions of global environmental change, national-level action cannot be truly effective: the few exceptional countries that have achieved sustainable growth continue to suffer from the global deterioration of the environment, the ecological limits to growth and the crises of the globalised economy (see Chap. 6). The improvement of sustainable development and its governance practices will require the solution of two main problems:

1. *Transformation towards sustainability* cannot be achieved through economic knowledge and economic strategies. The process is too complex to be assessed and regulated with economic knowledge and under assumptions of functioning competitive markets; the many non-economic factors influencing economic processes and distorting market processes and price mechanisms cannot be solved through enforced growth. This can already be seen in the national economies, for which growth in terms of GDP is a misleading indicator: much of the wealth generated in a national economy is not counted at all in the GDP, because it is not paid in money nor created through markets and exchange processes. This wealth includes the huge amount of labour and the creation of physical wealth through reproduction work in households, through subsistence work, through voluntary and non-paid work in many forms, through work and wealth created outside markets in newly developing forms

of sharing economy, solidarity economy and barter, not to mention the informal economy and black-market labour. GDP growth implies ecologically destructive processes of resource use: the costs of repairing ecological damages create only income or profit for the enterprises doing the repair work; for the whole society they are costs for the maintenance of ecosystem functions that undermine the possibility of long-term economic growth the more complex the damages become.

2. *Prevention of environmental damages* through forms of regulation other than in terms of externalities and markets: These damages are only insufficiently dealt with as "externalities"; internalisation of externalities only occurs for some environmental damages and those that can be exactly assessed and translated into economic costs. An assessment of environmental sustainability on the basis of economic criteria and costs is incomplete. Large parts of environmental damages will have to be regulated without economic compensation, restoration or repair, through political regulations and ecological criteria that do not follow economic rules. Especially difficult and controversial is the calculation of externalities that have to be borne by future generations; translating the interests of future generations into monetary terms and taking them into account in economic or political decision-making is only a weak instrument of accounting for the future. Strong or lasting improvements to the environment will only be possible when the economic institutions of the global market economy and its capitalist form are no longer valid and a new economy based on sharing and redistribution is created. To achieve a sustainability transformation and the reduction or prevention of externalities will require forms of valuation of the environment other than economic ones. Restricting the economic valuation of natural resources and the commodification of the environment and ecosystems, and "re-embedding" the economy in ecological and social systems, are processes that create possibilities for further non-economic and non-monetary regulation without environmental damage and social divisions.

Schilling and Chiang (2011) discuss some of the knowledge problems with and the limits of economic attempts to define sustainable development paths on the basis of market processes (see Box 5.6).

Box 5.6 Critique of Market- and Growth-Based Transition to Sustainability

Schilling and Chiang (2011) conclude that the arguments for a market-based sustainability transition are weak: "A policy for sustainable development would seek to minimize non-sustainable externalities in order to facilitate a smooth transition towards a sustainable economic development. But a complex of mutually dependent factors poses obstacles to time the transition accurately and makes it difficult to determine an optimal path towards the employment of renewable resources. As well, the rate of depletion changes over time according to the development of new technologies, the employment of substitutes, new discoveries of natural resources, changes in consumer behavior, and changes of the structural framework of the global economy. Therefore, there is uncertainty about the actual price development of both renewable and non-renewable resources. Finally, governmental intervention is necessary where the market system itself does not take sustainability sufficiently into account" (Schilling and Chiang 2011: 997).

This critique is not yet arguing for a transformation of the economic system, only that the transition is a more complex and complicated process that probably cannot be dealt with on the basis of economic knowledge. The limits of knowledge in the orthodox economic approach and the problems of inter- and transdisciplinary knowledge integration require more critical reflection. Becoming aware of the consequences of economic thinking and research involves criticising the universalising logic of "economising ecology" (Harmin et al. 2017) and learning how sustainability transformation can become possible in other ways.

Cumming and von Cramon-Taubadel (2018) went further in reflecting the limits of orthodox economics, showing with national data and a mathematical model that economic growth alone cannot achieve environmental sustainability. They discuss two different growth regimes in the countries of the Global North and South which move toward one of two different equilibrium points, a more economically developed equilibrium or a less economically developed equilibrium, "because of their different patterns of natural resource use and responses to population growth. At the less economically developed equilibrium point maintained by 'green-loop' feedbacks, human populations depend more directly on ecosystems for income. At the more economically developed equilibrium point maintained by 'red-loop' feedbacks, nonecosystem

(continued)

Box 5.6 (continued)
services(e.g., technology, manufacturing, services) generate the majority of national gross domestic product (GDP), but increasing consumption of natural resources means that environmental impacts are higher and are often exported (via cross-scale feedbacks) to other countries. Feedbacks between income and population growth are pushing countries farther from sustainability. … economic growth alone cannot lead to environmental sustainability …, current trajectories of resource use cannot be sustained without breaking feedback loops in national and international economies" (Cumming and von Cramon-Taubadel 2018: 9533).

Sources: mentioned in the text

The critical discussion of economic growth reintroduces, in inexact and selective form, the analysis and discussion of the systemic nature of the modern world economic system as divided into two structurally heterogeneous components, the Global North (centre) and South (periphery). This implies a more critical analysis of growth and its consequences and of its social and ecological limits. A bifurcation of the growth debate can be expected in the sustainability discourse along two increasingly divergent development paths:

1. The first approach continues to search for pathways of growth that allow for the maintenance of environmental quality in the form of green accounting or sustainable growth (Victor and Dolter 2017; Bartelmus 2018), but the indicators are differentiated because it is impossible to achieve a single and encompassing economic indicator for measuring environmental quality and human well-being in monetarised form.
2. The second approach continues to develop alternative pathways of economic development without growth, phasing out growth and transforming the global economy into a new, post-industrial and post-capitalist form of production. This requires further and more interdisciplinary knowledge creation and integration and theoretical reflection and analysis to develop strategies for transforming the global economic system. The degrowth discourse marks the beginning of the search for alternative pathways (see Chap. 6).

Confronted with these alternatives, the debate on economic growth and environmental valuation is driven by the critique of environmental economics and by the interdisciplinary environmental discourse in which social- and natural-scientific research is used not only to criticise the limits of orthodox economics but also to build a social-ecological theory and knowledge basis for the sustainability transformation.

Although more economists are now discussing growth critically, the dominant view in mainstream economics remains that growth is necessary, and the ecological debate about limits to growth is met with defensive arguments such as appear in the ideas of green or sustainable growth. These ideas assume that economic growth can be decoupled from the growth of natural resource use. This decoupling has mainly taken the weak forms of modification of growth and growth-supporting mechanisms of substitution of finite resources, technological change and innovations, reduction of emissions and pollution, and saving and recycling of natural resources—ideas which can be translated into values and lifestyle changes. This reasoning of decoupling economic growth from growth of natural resource use does not take into account the complexity of economic systems and processes when it comes to decoupling growth and resource use. At the level of a single enterprise it may be simple to achieve decoupling and zero emissions of CO_2 without transforming economic systems. At the levels of national and global economic systems, decoupling and zero growth are much more complicated and seemingly impossible without the social-ecological transformation of the global economy. Without such transformation the chances of sustainable development are limited.

5.2.2 Technological Innovation and the Digital Economy from the Perspective of Sustainability

Technological innovation and its significance for the transition to sustainability: The discussion about transforming the economy through technological change and innovation, especially through the rapidly developing digital economy that is sometimes called the "fourth industrial revolution" (Schwab), moves between economic knowledge in the scientific discourse, descriptions of the practical forms and changes to the economy, and visions of a future society that at times resemble science fiction fantasies. Most of the ideas about technology-driven change are dominated by *technical fix* thinking, expecting the solution to the problems of growth to be found in new environmentally and socially adapted technologies. An alternative approach to these visions of a self-transforming modern economy that advances through evolution-like processes is to experiment with alternatives to the market economy. This approach is inspired and motivated by ideas that are relevant to the sustainability process: new forms of solidarity-based economies, renewal of the commons, the sharing economy and the solidarity economy. In these alternatives to the market economy technologies have another function: to prove their value and validity not through the market, growth and profit, but through their positive social and environmental consequences—keeping in mind that technologies can have unanticipated negative effects.

The ideas and the discussion of technological transformation and its limits introduces questions about whether and how far the changes are endogenous to the system, or whether they require a social-ecological transformation of the economic system that is driven from outside the economy. Such a transformation requires actors other than economic ones, including collective actors and new social and environmental movements that have also initiated and driven environmental policy and governance, or transformative action groups in which different actors and institutions cooperate. Knowledge from economics about the functioning and change of the modern economy is necessary for both endogenous and exogenous change, but the perspective needs to be broadened beyond these two forms. More alternatives are required for the transition to sustainability than growth and market-based pathways of "self-correction" and internalisation of externalities, especially alternatives that do not exclude technological innovations in the digital economy. The digital economy needs to be connected with the search for non-market-based forms of economy. The economic transformation can begin with the building of new economic sectors in the market economy, as happened with the sub-sectors of fair trade, organic farming and renewable energy sources.

The digital economy and its significance for the transition to sustainability: The digital economy, with the internet at its core, has developed rapidly since the turn of the century but has only slowly revealed its transformative potential. It is not only a driving force for economic and technological modernisation, unleashing new growth potential in the market economy. In the future sustainability process new cultures of social use, control and regulation of the internet and the digital economy can be developed to support the transition to sustainability. The sustainability transformation begins in the market economy with the development of new economic sub-sectors and markets, for example fair trade, organic farming or locally produced high-quality food. The ecological and ethical sectors of the economy that are developing rapidly in many countries are not only part of a new green or bio-economy. They are seminal forms of a new economic system beyond the market and the capitalist economy that needs to be developed with several forms of economy (market, sharing, solidarity, redistributive economy). Several new forms of sharing economy have appeared along with the digital economy: they have developed as part of the capitalist sharing economy, but they have introduced new forms of a future economy in which access to resources is more important than ownership. The consolidated definition by Schlagwein et al. (2020: 818) is that of an "IT-facilitated peer-to-peer model for commercial or non-commercial sharing of underutilized goods and service capacity through an intermediary without a transfer of ownership". It is not yet a transformative economy, but non-commercial forms are part of the sharing economy. The digital economy as a knowledge-based economy does not necessarily pave the way to a post-industrial economy, although it can have this function when it helps to abolish polluting industrial technologies. New ideas of "smart" or "sustainable" local, regional or national economic systems, of high-tech economies where human labour is less and less necessary for subsistence and survival, are not yet an economic reality. Reductions in work time and significant changes to ways of life and consumption (practically relevant components of a non-growth society) become possible through the digital economy that can pave the way to a post-growth economy when it is dissociated from the growth mechanism. Currently, in the context of the unequal division of labour and the economic power relations in the global economy, the digital economy is strengthening social and economic divides and creating new ones. The fact that "clean" growth is now a development strategy for the OECD countries, and "dirty" industrial growth is a strategy for countries in the Global South, reinforces the older global divide in the economic world system. The two strategies together stabilise the maladaptive growth-based change in the global economic system. Relocating industrial production to the Global South and importing industrial products from there helps to maintain the privileged position of the OECD countries in the global economy and to extend industrial growth, with its environmental damage, indefinitely. In connection with economic globalisation and the deregulation of markets, the digital economy has created new spaces of wealth in global cities (in contrast to new global hinterlands) and extreme forms of concentration of economic power.

5.3 Complementary Approaches to Environmental Economics

Economic growth cannot be discussed and assessed without analysing the societal and ecological functions and factors that enable growth. The social and ecological contexts of growth as part of the technological modernisation of industry are reflected in the theory of ecological modernisation, which is split between a growth-oriented sociological variant and a non-growth-oriented variant from political science, a political strategy of transforming the modern economy to environment-friendly forms (Box 5.7).

5.3.1 The Theory of Ecological Modernisation of the Economy

Ecological modernisation is an interdisciplinary theory that does not support polluting forms of industrial growth but sees the potential for self-transformation of the industrial economy to less or non-polluting forms, making sustainable growth possible (see Box 5.7). The theory supports the assumption of the Brundtland Report that growth is necessary during the transition to sustainability. However, while the report conceived of growth as primarily necessary for the developing countries of the South, more recent ideas about green and sustainable growth that entered the agenda of sustainable development with the Rio + 20 Conference in 2012 pertain to industrialised countries in the OECD and the European Union (EU) (Hickel and Kallis 2019: 1).

Box 5.7 Ecological Modernisation

Ecological modernisation theory conceives the transition of the modern economy to sustainability as environmental adaptation through the development of a new ecological rationality and as a continuing process of technological modernisation. In this process the environmental movements and their ecological critique of industrial society are seen as integrated in the environmental policy process and cooperating with other actors in reform processes to drive the economy gradually into environment-friendly forms of resource use. Ecological modernisation converges with ideas from environmental economics, including that growth can become environmentally sound through policy reforms, based on technological innovation, dematerialisation, saving of material and energy resources in production and consumption, and accompanying changes in environmental behaviour. A business strategy of sustainable growth for the ecological modernisation of industry is developed through policy incentives for the internalisation of externalities and the institutionalisation of a second, ecological, rationality criterion that should be integrated with economic rationality.

Although ecological modernisation uses ideas that are discussed in economics, it originated not as an economic concept or theory but in research in the political sciences in Europe (Jänicke created the idea of ecological modernisation and saw it as part of political strategies for environmentally friendly and more efficient use of resources that do not require the high

(*continued*)

Box 5.7 (continued)

growth rates of the global economy, based on the assumption that environmental policy enables the industrial economy to stay within the ecological limits of growth and find social acceptance: Jänicke 2007) and in sociology (Mol and Spaargaren: Mol et al. 2009; Huber 2004, who complemented the approach with a theory of monetary reform: Huber 2017). Economists participating in the early discussion of economic modernisation include Huiting (1992) and Simonis (Ayres and Simonis 1992). The concept gradually became part of the broader environmental discourse in science and politics (Hajer 1995; Weizsäcker et al. 1998). Non-European scientists participating in the debate include the sociologist Sonnenfeld (Mol and Sonnenfeld 2000) and the physicists Rosenfeld (Rosenfeld et al. 2001) and Lovins (Hawkens et al. 2000). The multifaceted ecological modernisation theory was reviewed by Buttel (2000), who concluded that its sociological core theory could be strengthened by connecting it with the research and literature in related fields such as civil society and political sociology, thus making explicit and strengthening its core component of a theory of politics and the state. Fisher and Freudenburg (2001) broadened this approach to formulate the theory more precisely and connect it with more differentiated and rigorous empirical research.

Sources: mentioned in the text

Ecological modernisation became a model for conventional, business-oriented strategies for sustainable development. Economic growth is assumed to become environment-friendly or sustainable, and new technologies can help to make possible continued economic growth without exceeding nature's limits. In this view, the social-ecological transformation of society and the economy becomes superfluous: a post-industrial society can rather be reached in evolutionary forms, through structural change, with industry "growing into" another economic system through technological innovation and continuous policy-induced improvements. The practical forms of ecological modernisation include the increasing efficiency of energy and resource use, the reduction of emissions and clean technologies, the substitution of dangerous substances, and environmental and supply chain management. Strategies for changing environmental behaviour, consumption and resource-intensive ways of life are also discussed. Such ideas have reappeared in the recent policy discourse on the green growth strategies supported by the United Nations

(UN), OECD and EU, where a bio-economy and the commodification of living resources are seen as triggering new growth potential.

The concept and theory of ecological modernisation adopted technological, innovation and growth-based modernisation of the economy as its core idea, although it supports environment-friendly, low and sustainable forms of growth within the global ecological limits. The theory evoked criticism that the transition to a sustainable economy within the limits to growth, the successful protection of the environment and the sustainable maintenance of the natural resource base are not possible based on the industrial and capitalist economy but require its transformation (York and Rosa 2003, whose argument was later confirmed by new environmental indicators and analyses of global material and energy flows: see Chap. 6). The transformation capacity of ecological modernisation and of green and sustainable growth is limited. The increasing throughput of natural resources, which has been accelerating globally since the beginning of the twenty-first century, introduces new arguments against economic growth (on the degrowth discourse, see Chap. 6).

The second fundamental criticism of ecological modernisation is that it has become a model and strategy for the affluent countries of the Global North to modernise their industries profitably and to maintain their dominant positions in the modern world system by relocating the polluting industries to the newly industrialising countries of the Global South, from which the industrialised countries in the Global North buy large parts of the industrial production. European countries have not become sustainable through de-industrialisation on their own territories. Rice (2007) shows that isolated and national economic strategies are no longer effective under conditions of global social and environmental change. Further criticisms of ecological modernisation theories and strategies refer to the social and economic inequalities connected with the modern world economic system and the asymmetrical power relations and distribution of resources between industrialised and developing countries, and the resulting problems of economic and environmental justice, which have not been resolved by globalisation and growth-based development (Harvey 1997; Everett and Neu 2000; Rice 2007). Reacting to critiques of ecological modernisation, Mol and Sonnenfeld (2000) and Mol and Spaargaren (2005) refined the theory in attempts to integrate economic and ecological theory (to be achieved through policy reforms). Later they merged ecological modernisation theory into a "sociology of global flows", where the flows of resources, goods and information are seen as so complex that the causes and consequences of growth can no longer be identified. Furthermore, the idea to dematerialise production and consumption was also unrealistic for achieving a sustainable economy, since the saving of energy and materials in economic processes (the "factor 4" and "factor 10" debates on improving efficiency: Weizsäcker et al. 1998) is offset by continuing exponential growth. While the concept of ecological modernisation is ageing and becoming less convincing, its basic ideas have reappeared in the current discussion of green growth and the green economy.

5.3.2 Environmental Policy: Green Growth as a Transition Strategy to Sustainability

The concepts of green growth and blue growth (for maritime resources) were supported in the global sustainability discourse by influential global players including the UN, the United Nations Environment Programme (UNEP), the OECD, the EU, the World Bank, the Food and Agriculture Organization (FAO) of the UN and the International Chamber of Commerce (ICC). Green growth is becoming the dominant policy strategy for sustainable development. Such forms of sustainable growth are presented as alternatives to the previous form of industrial growth and its resultant pollution and overuse of the natural resource base. Green growth is defined in preliminary and inexact forms as growth that does not harm the environment; how this aim is achieved remains an open question. The fact that the strategies are supported by powerful political and economic institutions, and that they further the interests of the OECD countries and their privileged position in the global economic system, does not increase their persuasiveness; the conservative approach of the UNEP's (2011) green economy strategy aims "to gain traction in the (even more) conservative minds of the business and economic leaders" (Brockington 2012: 411). The scientific persuasiveness of the strategy is based on sophisticated macroeconomic modelling, using a global model (based on the "Threshold 21" or "T 21" models where the twenty-first century is seen as a threshold period for humankind) which differs from ordinary growth scenarios (business as usual) and macroeconomic models through the inclusion of natural resources as a factor of production (UNEP 2011). The belief in green or sustainable growth in the UNEP report cannot be substantiated by the reasoning in the report or its global model (which looks good for the environment only because the business-as-usual scenarios do not include in their calculations the state of natural capital). Empirical research that has already begun (see Chap. 6) will show whether a strategy of green growth can be realised for all countries.

Other institutions of the green economy coalition make similar arguments. The OECD (2011) defines green growth as economic growth that ensures natural assets continue to provide the resources and environmental services on which human well-being relies. It is seen as a flexible approach to be achieved through varying policy measures and instruments according to national, regional and local conditions. The critical point, of which the proponents of green growth are aware, is that natural resources often cannot, or can only to a limited degree, be substituted by other resources; thus, strategies of green growth need to anticipate the reduction of growth and limits to growth. The concept of green growth needs to be seen as part of the green economy strategies and their assumptions for a sustainability transition. For the transition to a green economy, the ICC (2012) formulated ten conditions in a "Green Economy Roadmap": open and competitive markets; metrics, accounting and reporting; finance and investment; awareness; life cycle approach; resource efficiency and decoupling; employment; education and skills; governance and partnership; and integrated policy and decision-making. Similar to other policy documents on green growth and the green economy, the principles are vague and do not give clear explanations of how the whole economic system is to be transformed, although indicators are used to measure progress towards a green economy, including a global green economy index. Evidently the green economy remains a market- and growth-based economy, and the transition is one driven by technological innovation, sometimes described using a term from the broader transformation debate as "great technological transformation". How the framework translates in sector-specific measures and which economic sectors are important for the transformation is not evident from the criteria and conditions of the roadmap. Changes beyond the economy are only discussed insofar as they influence economic processes directly, for example education and skills. Altogether the framework does not provide an open and transparent programme. This and other policy documents from the UN, OECD and EU do not provide sufficient arguments to defend the green economy strategy against the criticism by social and environmental movements (etcgroup 2011; Lander 2011) that the green economy could become a strategy for increasing corporate control and further intensifying resource use. Given the weaknesses and the abstract quality of the green growth concept, alternative paths to green growth must be sought. The long-term trends towards reduced and low growth that the OECD expects to see in all national economies during the twenty-first century, first in the OECD countries and later in the developing or newly industrialising countries, does not fit with the picture of the future presented in the roadmap. The prospect of the long-term slowdown of growth in the global economy (as predicted by the OECD) already seems to be accepted by the same institutions that propagate the green economy strategy.

The evidence that green growth opens a pathway towards global sustainability is weak and increasingly being disproved by further economic and ecological data about economic growth and material and energy flows in the global economy in several ecological indicators (see Chap. 6). Hickel and Kallis (2019) conclude their discussion of the possibility of green growth with two arguments against it:

1. The possibility of green growth is not supported by empirical evidence: the necessary conditions to decouple growth from resource use and from CO_2 emissions differ widely between the national economies; under exceptional conditions in some high-income countries decoupling may be possible, but it cannot be expected at the global level; green growth and decoupling of growth from resource use and from global CO_2 emissions is highly unlikely.
2. The argument for green growth under conditions in which empirical evidence is growing that its global spread is impossible seems to be politically motivated, using a defensive reasoning of the following kind: "it is not politically acceptable to question economic growth … no nation would voluntary limit growth in the name of the climate or environment; therefore green growth must be true, since the alternative is disaster" (as Wackernagel and Rees argue) "… 'the politically acceptable is ecologically disastrous while the ecologically necessary is politically impossible'" (Hickel and Kallis 2019: 15).

5.4 Critique of Environmental Economics

The concepts and methods of environmental economics have been criticised in conventional academic economics and according to the influential neoclassical paradigm (Klitgaard 2013). The critique spread through the scientific discussion on the development of the modern economy (Graves 2007; Hahnel and Sheeran 2010; Naess and Price 2016; Roy 2016) and in the policy and governance discourses, where the critique of growth is the focus. Barry summarised the critique of environmental economics, referring to the publication of Pearce et al. (1989) as trying to translate environmental problems into economic ones and dealing with them by correcting market externalities, and politically by internalising negative environmental effects (such as pollution by way of taxing and pricing to address the market failure). The problems and deficits of this way of "economising ecology" and of addressing environmental problems are, according to Barry (2007: 224ff), a consequence of the conceptual and methodological heritage of orthodox economics:

1. *Economic reductionism*: social and environmental problems are reduced to economic ones, constructing nature as an economic resource or as natural capital.

2. *Economising nature*: nature is redefined as natural capital, ignoring the fact that nature in terms of ecosystems is not natural capital (Holland 1997: 127); the preservation of natural capital will not necessarily lead to the preservation of nature, ecosystems, species or landscapes.
3. *Methodological individualism*: this economic reasoning in environmental valuation tends towards reductionism, individualism or atomism, isolating resources from the social and ecological context in which they function, thus preventing integrated and holistic approaches (Dryzek 1987).
4. *Discounting the future*: the economistic view of nature and environmental problems tends to be from a short-term perspective and prioritises technological solutions to environmental problems ("techno-fixes": O'Riordan 1981).
5. *Assuming the necessity of growth*: environmental economics, and orthodox economics more generally, assumes the necessity of continuous economic growth to generate wealth and welfare, ignoring the fact that growth undermines long-term economic and ecological sustainability, supporting socio-economic inequality and rejecting the policy of redistribution (Mulberg 1995).

The critique and its normative implications are consistent with critical social and ecological views of economics and the modern market economy which can be found in many examples, arguing for re-contextualising economic knowledge production in social and ecological perspectives and for re-embedding the economy in social institutions in order to be able to regulate it. This critique distinguishes environmental and orthodox economics from critical approaches such as institutional economics, ecological economics or development economics. The critique remains valid on many points despite the more recent interdisciplinary opening up and broadening of environmental economics. To date sustainable development has not been dealt with adequately from a disciplinary economic perspective: too much knowledge from other social-scientific and ecological research is excluded or ignored. The argument that the orthodox, market-based reconstruction of the economic system is ignoring long-term development and environmental sustainability is similar to the critique in the "limits to growth" debate since 1972.

A critique of environmental economics similar to that summarised by Barry has been formulated in institutional, development, interdisciplinary and ecological economics and from the perspective of ecology. Critical questions include the integration of economic knowledge and ecological knowledge, and how the different knowledge perspectives limit each other. Disciplines and disciplinary knowledge cultures are no longer automatically accepted in environmental and sustainability policies (see Box 5.8).

Box 5.8 Economics and Environmental Policy: Disciplinary Hegemony and Interdisciplinary Cooperation

In the sustainability discourse and policy process, it must be decided which knowledge counts in the controversial debate on economic growth and its social and environmental consequences. In the "limits to growth" debate since the 1972 report of the Club of Rome, the critique of the negative consequences of economic growth has intensified. Much of the growth-critical reasoning in this and later reports was derived from system-analytical and mathematical thinking, built on modelling and logical reasoning. However, a concrete discussion of the social and environmental consequences of growth cannot be based solely on economic models; it must include interdisciplinary concepts, theoretical reflection and empirical environmental research from different disciplines in both the social and natural sciences.

The dominance of economic forms of reasoning and argumentation in environmental policy: Pearce (1992: 8), defends the economisation of politics, reasoning that "politicians and their advisors are engaged in the activity of trading off environment against economic activity …. Defending the environment means presenting the arguments in terms of units that politicians understand … adducing evidence that the environment does matter in economic terms is important, especially as the record of decision-making in the absence of such valuations is hardly encouraging for the environment."

Criticising the dominance of economic knowledge: Barry (2007: 223f) argues that it is difficult to use other forms besides economic argumentation in public policy, where environmental problems are often translated into economic problems and courses of action on the basis of the economic costs or benefits: "… in the classic case of environmental protection versus development, it is very often the case that environmental campaigners have to couch their case in economic terms and language as well as have an economic reason for environmental preservation. … To base one's case for environmental protection on the intrinsic value of the environmental space or landscape in question (as opposed to some economic instrumental value it may possess) is to adopt a strategy that would be difficult to persuade or influence the environmental policy-making. … This is not to deny the importance of economic considerations, but simply to note how an economic approach to and understanding of social-environmental problems can (and does) 'crowd out' non-economic forms of environmental valuation and argumentation. The privileged position occupied by economics in environmental policy-

(*continued*)

Box 5.8 (continued)

making has the effect of drowning out other 'voices', other forms of reasoning, valuing and thinking about the environment."

Disciplinary hegemonies—how to deal with the limits of orthodox economics: The circular flow of exchange between firms and households, according to the classical economics textbook by Samuelson, "is the basic conceptual starting point for economics. Like all highly stylized diagrams, it's obviously an over-simplification, but, even considering that, is there something missing here? Where do all the inputs (raw materials and energy) from production come from? Where do all the waste outputs go? Unfortunately, this isn't an isolated instance where economics has a little 'blind spot'. It's actually quite pervasive in economics to believe that the physical world is irrelevant to the economy. … Samuelson … considered man-made capital and natural capital (or resources) to be so substitutable that it was redundant to mention them as two separate entities." Herman Daly criticised the failure of mainstream economics to address the physical context of the economic process: "it is 'as if biology tried to understand animals only in terms of their circulatory system, with no recognition of their digestive tract'" (Stallworth et al. 1997: 3).

New interdisciplinary knowledge cultures: "Controlled broadening" of the disciplinary research agenda in economics to include environmental themes has not yet allowed for an interdisciplinary approach. Interdisciplinarity requires cooperation across disciplinary boundaries, bi-directional knowledge exchange, knowledge integration, and acceptance of different perspectives and valuation criteria for nature and natural resources, not dissolving the differences in one final economic valuation system that guides environmental policy and governance.

Sources: mentioned in the text

For the scientific sustainability discourse and for the policy and governance practices of sustainable development, the most important theme is economic growth and its environmental consequences. The growth critique influences the search for new, sustainable strategies of growth and the alternative of "sustainable degrowth" that unfolds in the critical discourse of ecological economics.

5.5 Discussion and Conclusions

Economic growth has progressed from an academic theme to a core theme in the sustainability discourse and process. In order for the modern economic system to be transformed into a sustainable system, the vague idea of sustainable development needs to be reformulated and specified as a social-ecological transformation or change in the mode of production. *How can sustainable development be restructured as a transformation process, and where does the knowledge for this transformation come from?*

The knowledge does not come from ordinary economic research about the functioning of the modern economy, its institutions and the markets. Even when the dynamics of economic development include structural change and adaptation to new economic, social and environmental conditions, they do not bring about transformation. The process of systemic change cannot be driven by an economic system that is organised to maintain the systemic identity, in the case of the modern economic world system an identity characterised by contradictory forms and unequal development. The systemic identity of the global market system based on exponential economic growth is called into question in the ecological and sustainability discourse with the question that has been asked in economics since the classical political economy of the early nineteenth century: *what will happen with the economy when industrial growth has come to an end?*

Instead of dealing with the big question, the economic discourse was reduced to small questions of valuation of resources and price finding on the markets, first reducing the whole economy to the market, then reducing the market to an information-theoretical concept of prices, where the knowledge relevant for the agents in economic practice became information on prices. Even after Arrow warned that "agents know too little when they only know prices" (Arrow 1973: 139f), orthodox economics, with its formal conception of the economy, has not found its way back to the bigger questions, including the limits of the system, its ruling principles, and its inherent dysfunctionality, social inequality and exclusion. For the maintenance of the economic system, the only additional knowledge that was sought was that from political science about political regulation and its instruments. It is often repeated in the interdisciplinary discourse about ecology, the economy and society (Scheffer et al. 2002: 212) that the development of modern society depends on the interaction and complementary functions of the economic and political systems. But these systems depend in the final analysis on the individual and collective actions of persons and groups that either practise environment-friendly behaviour and ways of life or reject them. With that comes a complexity in social integration and change that has often been ignored in economic and social-scientific thinking. The action of persons cannot be described in the simplified forms of the rationality of "economic man" or the "oversocialised conception of man" (Wrong 1961, criticising the sociological theory of Parsons) where social and economic actors appear as puppets on a string rather than as autonomous actors.

The critique of economic thinking in the interdisciplinary discourse of environment and sustainability raises questions about the social and ethical limits to growth as well as the ecological limits. The valorisation and commodification of nature and ecosystems introduced new ethical problems to modern capitalism, with the key questions concerning the privatisation of common land to subject it to market conditions and the commodification of human labour in the form of creating a labour class dependent on the selling of the capacity for work on the labour market. Subsequently ever more difficult ethical questions have arisen over the commercialisation of ecosystem services, living resources including human bodies, and genetic resources. In all these cases economic and other scientific knowledge is no longer sufficient: ethical values and criteria need to be discussed, which involves criticism of the processes of economic valuation and valorisation of natural resources (Anderson 1993; Radin 1996; Jaeggi 2001). This discussion includes the questions asked by Anderson (what is the proper scope of the market) and by Radin (what is the proper relation between market and non-market aspects in the modern economy and society; see Jaeggi 2001: 401). In the ethical critique of modern capitalism, Anderson draws attention to the tensions in the affirmative reasoning of the proponents of free markets: the liberal values of freedom and autonomy that are used to defend markets are threatened by ongoing commodification, which undermines the preconditions for realising these values. The social and ethical conditions of freedom and autonomy need to be discussed critically, including social conditions such as humankind's dependence on a wide variety of social spheres and institutions (not just markets). The ethical arguments for the limiting and social re-embedding of markets are necessary to strengthen the knowledge-based critique of economics in the sustainability discourse (Anderson 1993: 141).

Regarding strategies of sustainable growth, not much is known from the analysis of the market mechanism; no experience is available to rebuild the market economy into an economy that does not destroy the environment and does not overuse the natural resource base but stays within the global limits to growth. Regarding strategies of sustainable degrowth (Chap. 6), a similar "veil of ignorance" exists. In both cases the building of the future sustainable society and economy begins in a situation where not much is known about the process; more knowledge needs to be acquired on the path towards sustainability. The prospects for a future economy of sustainable growth are uncertain. The few exceptional cases where it has worked cannot not be generalised for all countries or for the global economy. In economic practice and in national and international politics, doubts about continued growth and the commodification of ecosystems and natural resources are growing. The abstract concepts of qualitative, green or sustainable growth do not provide ideas or knowledge on how to transform the modern economy into a sustainable form.

As long as the search for possibilities of continued growth is the main focus, the principles of sustainable development, sharing and the redistribution of economic wealth between rich and poor countries, which will require a discussion of alternatives to growth, will be neglected. Sharing and redistribution as principles of solidarity are not yet practised in sustainable forms in the digital and sharing economy. The Economics of Ecosystems and Biodiversity (TEEB) discusses the idea of sharing the benefits and solving the problems of redistribution only in terms of the market logic of growth and profits; how fairness and sharing can be achieved under such conditions remains unclear. Redistribution happens in the practice of international development in the form of charity and disaster aid, which cannot be seen as a strategy for redistribution and sharing of resources to achieve sustainability. Meanwhile, socially and ecologically destructive forms of "primitive" industrial growth continue to characterise the newly industrialising countries in the Global South. Attempts to achieve sustainable growth have yet to reduce, let alone stop, the global overuse of resources and destruction of the environment.

The controversy over sustainable growth versus degrowth is deadlocked, as the analyses of Hickel and Callis and Wackernagel and Rees (see Sect. 5.3.2) show. Attempts to ignore the scientific, public and political discussion about degrowth, as Daly (2013) experienced when invited to such a discussion, cannot continue indefinitely. Strategies of ignoring and blocking have been used since the beginning of the new environmentalism and environmental policies in the past half-century, and since the global environmental and limits-to-growth discourse in 1972. Big environmental problems such as climate change, pollution of the oceans and phasing out the use of fossil energy sources entered the policy agendas only after they had long been blocked and delayed by powerful economic and political actors. The themes of the sustainability discourse have received similar treatment, such as sharing and redistribution of resources between the countries in the North and South, reforms of international institutions and the initiation of a social-ecological transformation of the economy. The inclusion of ideas such as social-ecological transformation on the agenda of the sustainability process represented an important breakthrough in the 2010s. Now the transformation needs to be defended against attempts to water it down, as in the strategy of the green economy.

To renew the sustainability process, it makes sense to experiment for some time with strategies of both sustainable growth and degrowth. This makes sense for several reasons, including the lack of knowledge on how to achieve a sustainable economy based on a significant reduction of resource use without it leading to impoverishment and catastrophe. It is clear from recent research (see Cumming and von Cramon-Taubadel 2018) that sustainability requires more than the greening of growth; further changes are required, including the search for non-market- and growth-bound economic value creation, production and resource use. In the continuing processes of political, economic and ecological transformation it will be necessary to focus on two connected problems:

1. *building a new world order that is effective for earth system governance*, to escape the dead end of neoliberal policies and economic globalisation through a new form of globalisation—this process includes the phasing-out of economic growth and the transition to new non-growing economic systems;
2. *continuing to build more effective global environmental governance in the form of climate and sustainability governance*: this is the process of creating the elements of a new sustainable society and enabling the transformation and building of such a society.

Summary of Chap. 5 Environmental Economics

The discussion about a transition to a new, sustainable economy in the discourse of environmental economics is stuck in the growth debate, hindering the discussion of alternative strategies for a transformation to sustainability. Without opening up a broader, interdisciplinary debate in which economic knowledge is critically assessed with regard to its limits and embedded in ecological knowledge about limits to growth and consequences of global environmental change, there is little hope of progress in the building of a knowledge stock for sustainability transformation, or for applying the continually improved knowledge to the policy and governance processes framing and guiding the sustainability process. In the short term, there is little chance that the two clashing paradigms of sustainable growth and degrowth can be mediated and connected in an overarching ecological rationality of knowledge integration and application in politically organised governance processes at national and international levels. The attempts to maintain growth and defend it with new ideas and constructions such as that of green or sustainable growth will continue, although the empirical knowledge from

(*continued*)

(continued)
ecological research is increasingly demonstrating that the throughput growth through use of natural resources has exceeded ecological limits.

This situation should not to be seen as a collapse of the sustainability process, but rather as a crisis that can help to renew it. There is not yet enough knowledge available for the structuring and governing of the long-term transformation process, but it is clear that industrial, resource-intensive and polluting growth cannot be the way to sustainability nor to lasting prosperity. It is not yet known how to make growth sustainable, nor how to limit growth and develop a non-growing economy. Further knowledge creation, experience and collective learning can create new pathways to sustainability.

Appendices: Further Information and Material

1. Questions and Individual Exercises

Learning Exercise 1: Discussion and Critique of Economic Growth Models

(1) Read the summary article by Roy (2016; get to know the journal "Consilience. The Journal of Sustainable Development", a scientific journal produced by undergraduate and graduate students); (2) view one of the videos on economic growth available on the internet. Discuss the text, the video and the following summary of economic growth theories with the questions: Are there possible ways to combine the contradictory strategies of growth (the green economy and ecological modernisation) and non-growth (sustainable degrowth, social-ecological transformation) in the sustainability process? If this does not seem possible, how could the sustainability process advance with the irreconcilable ideas of growth and non-growth?

A variety of factors, functions and consequences of growth have been highlighted at various times in the study of economics. As a result, there is no single coherent picture or general theory of economic growth in modern economies, but rather historically and geographically specific and contingent forms of growth.

1. *Theories in classical political economy*: division of labour, economies of scale; Malthusian concepts of limits to growth—the growing gap between population growth and growth of subsistence or food production, generalised in the Neo-Malthusian models of limits to growth through resource degradation and exhaustibility of resources;

2. *Keynesian economics*: macroeconomic discussion of demand-side factors, of aggregate demand or total spending in the economy and possibilities of stimulating growth in crisis situations, including deficit spending (practically through policies of stimulating investments through combinations of reduction of interest rates and governmental investment in infrastructure); contrasted by Keynes to the aggregate supply in focus in classical economics (Say's law: "supply creates its own demand").
3. *Neoclassical growth theories (Solow et al.)*: the theories include exogenous growth, explained through supply-side factors (productivity of labour, size of workforce, etc.) and macroeconomic theories of exogenous growth ("turnpike theories"). The growth debate in orthodox economics does not allow for the formulation of one generally valid and best model; factors limiting growth need to be discussed more specifically.
4. *Endogenous growth theories*: growth through human capital (especially knowledge) and technological innovation (since Schumpeter), focusing on factors that are also dominant in defending growth under conditions of environmental pollution and overuse of the ecological resource base.

Different conditions and temporal limits of growth were already evident in the early growth debate in economics (Dorfman et al. 1958: 331). Summarising the main factors discussed in economics to explain growth, Cumming and von Cramon-Taubadel (2018: 9533) mention neoclassical economics with labour, capital and technology, and the new growth theory of Romer adding knowledge and its institutional embedding; but they all neglect environmental factors that limit growth, in spite of the extensive discussion of exhaustible resources. The growth discourse shows historically changing forms and conditions of economic growth. One conclusion about possibilities of continued growth is formulated by Romer (1990) in a mathematical model for long-term growth that incorporates the idea of learning: finding ever better possibilities to maintain growth and better ways to use the limited resources available obviously requires some form of "trial and error" joint learning.

Learning Exercise 2: Environmental Valuation Methods

Environmental valuation in economics involves the attribution of economic or monetary value to natural resources, public goods and environmental media such as air, water and soils and changes in their quality due to pollution or environmental change.

The valuation debate has been argued from two different perspectives: Post-Keynesian environmental economics and orthodox economics, especially the neoclassical principles and models that neglect uncertainty, endogenous preferences, path dependence, a foundational role for money, the distribution of income, complex system dynamics and economic power relations (Perry 2013). The methodological problems with environmental valuation begin with the different value concepts used in economics (market and non-market values, use and non-use values, consumptive and non-consumptive values, etc.) and the subjective nature of values, which makes valuation of environmental goods contingent, sometimes arbitrary, incomparable, difficult to aggregate and changing over time. Valuation is also used in the practices of natural resource management and environmental policies to assess the effectiveness, efficiency and consequences of application of different environmental policy instruments aiming to improve environmental conditions and reduce energy and material throughput.

Discuss as a group several or all of the following valuation methods:

Valuation can be done with different methods, for which their applicability and limits, possibilities of combination and integration are controversially discussed. To follow this discussion and make your own assessment of the valuation methods, first read (in one of the textbooks or handbooks on environmental economics referred to in this chapter: see below) the descriptions of the following main methods of valuation: contingent valuation, choice experiments, travel costs, hedonic price methods, willingness to pay/accept, payment cards, Delphi survey, meta-analysis.

Try to find out, by learning about each method, which natural resources it can be applied to and what its advantages and disadvantages are in estimating and calculating the monetary value of natural resources.

To deepen the discussion of valuation methods, first read Chaps. 5 and 6, and discuss thereafter valuation in connection with the problems discussed in the sustainability process, especially the possibilities of continued economic growth and conservation of nature and natural resources. Next, discuss whether the problems of environmental damage and overuse of natural resources can be reduced or solved through double valuation of natural resource use (with economic criteria or indicators and with ecological indicators). Can various economic, ecological and social criteria of valuation be combined?

2. Further Reading Suggested: Deepening, Thematically Specialised

The standard economic terminology and the methods for analyses of growth, resource use and environmental valuation have not been discussed in detail in this chapter. Basic information about the concepts and methods can be found in the following textbooks and handbooks of environmental economics. Most methods and models in economic research, valuation and assessment are not constructed for use in the process of sustainability transformation or for the utilisation of ecological data and knowledge.

Textbooks—environmental economics:

Three recent textbooks that are used in many courses on environmental economics present the subject from an interdisciplinary perspective, although they still follow the disciplinary tradition of economics, where environmental economics is understood as a specialised, subdisciplinary subject. The books include chapters on sustainability, but not all include detailed information. They are suitable for learning the principles and methods of economic sustainability analyses for non-economics students dealing with environmental, natural resource management and sustainability.

Tietenberg, T., & Lewis, L. (2018). *Environmental & Natural Resource Economics (11th ed.). New York, NY and Abingdon, Oxon: Routledge.* This broad introduction to environmental economics provides detailed information on the themes, theories and methods of environmental economics. The authors understand environmental economics as developing from the orthodox, neoclassical tradition of economics but follow a pragmatic approach in which they also use the more critical discourse of ecological economics. They see the relations between the two fields as complementary, in spite of significant differences in the methodologies and the values influencing the knowledge production. The discussion of sustainability is limited and somewhat selective, based on the definition of the Brundtland Report, including the description and discussion of different scenarios of future growth and some economic and ecological indicators to measure sustainability.

Harris, J. M., & Roach, J. (2018). *Environmental and Natural Resource Economics: A Contemporary Approach (4th ed.). New York, NY and Abingdon, Oxon: Routledge.* The authors, integrating environmental and natural resource economics, open up environmental economics to a more pluralistic perspective, where conventional environmental economics and ecological economics are seen as complementary. The authors assume that a pluralistic analysis is required to understand the complex relations between the modern economy and the environment, also connecting the economic analysis with policy analyses. The pluralism is exercised in the form of presenting core concepts and approaches of standard environmental economics (externalities, public goods, allocation of resources in temporal perspective, valuation of natural resources and cost–benefit analysis) and of ecological economics (macroeconomic scale of economic systems, entropy, green national accounting), with a variety of thematic examples of analyses such as international trade, water and forest resources, fisheries and agriculture, energy and climate change, taking into account the design of policies for sustainable development and "greening" of the economy.

Hackett, S., & Dissanayake, S. T. N. (2019). *Environmental and Natural Resources Economics. Theory, Policy, and the Sustainable Society (5th ed.). New York, NY and Abingdon, Oxon: Routledge.* Several chapters in this textbook provide detailed descriptions of environmental policy and sustainable development. Hackett and Dissanayake describe the main principles of economics, the economic methods for environmental policy analysis, and the principles of natural resource economics illustrated through fisheries and ecosystem services, and they present the economics of sustainability in case studies on sustainable economic development, production and management of common pool resources. The chapters on methodology for environmental economics, environmental policy and sustainability are useful for the learning exercises in this and other chapters.

Handbook:

Victor, P. A., & Dolter, B. (Eds.). (2017). *Handbook on Growth and Sustainability*. Cheltenham, UK and Northampton, MA: Edward Elgar.

Scientific journals (international, peer reviewed) relevant for the research in environmental economics: "Journal of Bioeconomics", "Current Research in Environmental Sustainability", "Current Opinion in Environmental Sustainability", "Environmental Economics", "Environmental and Resource Economics", "Environmental Sustainability", "Environment, Development, Sustainability", "Forest Policy and Economics", "International Journal of Environment and Sustainability", "International Journal of Sustainable Economy", "Journal of Economics and Sustainable Development", "Journal of Environmental Economics and Policy", "Journal of Environmental Sustainability",", "Resource and Energy Economics", "Business Strategy and the Environment", "Journal of Cleaner Production", "Institutional Economics"

References

Albert, H. (1998). *Marktsoziologie und Entscheidungslogik: zur Kritik der reinen Ökonomik.* Tübingen: Mohr Siebeck.

Anderson, E. (1993). *Value in Ethics and Economics*. Cambridge, MA: Harvard University Press.

Anderson, L.G., Seijo, J.C. (2010). *Bioeconomics of Fisheries Management*. Ames, Iowa: Wiley-Blackwell.

Aronsson, T., & Löfgren, K.-G. (Eds.). (2010). *Handbook on Environmental Accounting*. Cheltenham, UK: Edward Elgar.

Arrow, K. J. (1973). Information and Economic Behavior. In *The Economics of Information* (1984). Cambridge, MA.

Arrow, K. J., Dasgupta, P., Goulder, L. H., Mumford, K. J., & Oleson, K. (2012). Sustainability and the Measurement of Wealth. *Environment and Development Economics, 17*(3), 317–353.

Ayres, R. U., & Simonis, U. E. (Eds.). (1992). *Industrial Metabolism*. Tokyo: United Nations University Press.

Barry, J. (2007 [1999]). *Environment and Social Theory*. London and New York: Routledge.

Bartelmus, P. (2013). *Sustainability Economics: An Introduction*. London and New York: Routledge.

Bartelmus, P. (2018). *Sustaining Prosperity, Nature and Wellbeing – What Do the Indicators Tell Us?* Abingdon, UK and New York: Routledge.

Barnett, H. J., & Morse, C. (1963). Scarcity and Growth. The Economy of Natural Resource Availability. Baltimore: Johns Hopkins University Press.

Bennett, J. (Ed.). (2011). *The International Handbook on Non-Market Environmental Valuation*. Cheltenham, UK: Edward Elgar.

Bowles, S., & Carlin, W. (2020). What Students Learn in Economics 101: Time for a Change. *Journal of Economic Literature*, 58,1, 176–214.

Brockington, D. (2012). A Radically Conservative Vision? The Challenge of UNEP´s Towards a Green Economy. *Development and Change*. https://doi.org/10.1111/j.1467-7660.2011.01750.x

Bromley, D. W. (Ed.). (1995). *The Handbook of Environmental Economics*. Oxford, UK and Cambridge, MA: Blackwell.

Buttel, F. (2000). Ecological Modernization as Social Theory. *Geoforum, 31*(1), 57–65.

Chichilnisky, G., & Rezai, A. (Eds.). (2020). *Handbook on the Economics of Climate Change*. Cheltenham, UK: Edward Elgar.

Cumming, G. S., & von Cramon-Taubadel, S. (2018). Linking Economic Growth Pathways and Economic Sustainability by Understanding Development as Alternate Social-Ecological Regimes. *PNAS, 115*(38), 9533–9538.

Daly, H. (2013). A Further Critique of Growth Economics. *Ecological Economics, 88*, 20–24.

Dasgupta, P. (2008). Nature in Economics. *Environmental and Resource Economics, 39*, 1–7.

Dasgupta, P. (2014). Measuring the Wealth of Nations. *Annual Review of Resource Economics, 6*, 17–31.

Dasgupta, P. S., & Healy, G. H. (1979). *Economic Theory and Exhaustible Resources*. Cambridge, UK: Cambridge University Press.

Dasgupta, P. S., & Mäler, K.-G. (2004). *Environmental and Resource Economics: Some Recent Trends*. Working Paper 7-04. Kathmandu: South Asian Network for Development and Environmental Economics (SANDEE).

Dorfman, R., Samuelson, P. A., & Solow, R. M. (1958). *Efficient Programs of Capital Accumulation. Linear Programming and Economic Analysis*. New York: McGraw Hill.

Dryzek, J. (1987). *Rational Ecology: Environment and Political Economy*. Oxford: Blackwell.

Eisenhardt, K. (1989). Agency Theory: An Assessment and Review. *The Academy of Management Review, 14*(1), 57–74.

etcgroup. (2011). *Who Will Control the Green Economy?* Ottawa. Retrieved from www.etc-group.org.

Everett, J., & Neu, D. (2000). Ecological Modernization and the Limits of Environmental Accounting? *Accounting Forum, 24*(1), 5–29.

Fischer-Kowalski, M., Krausmann, F., Giljum, S., Lutter, S., Mayer, A., Moriguichi, Y., et al. (2011). Methodology and Indicators of Economy-wide Material Flow Accounting. *Journal of Industrial Ecology*. https://doi.org/10.1111/j.1530-9290.2011.00366.x.

Fisher, D. R., & Freudenburg, W. R. (2001). Ecological Modernization and Its Critics: Assessing the Past and Looking Toward the Future. *Society and Natural Resources, 14*, 701–709.

Flichman, G. (Ed.). (2011). *Bio-Economic Models Applied to Agricultural Systems*. Dordrecht, Heidelberg, London, New York: Springer.

Flichman, G., & Allen, T. (2013). Bio-Economic Modelling: State-of-the-Art and Key Priorities. *CGIAR*. Retrieved from www.pim.cgiar.org/files/2013/12/FlichmanThomas_BioEconModeling_Final.pdf.

Fouquet, R. (Ed.). (2019). *Handbook of Green Growth*. Cheltenham, UK: Edward Elgar.

Gallagher, K. P. (Ed.). (2008). *Handbook on Trade and the Environment*. Cheltenham, UK: Edward Elgar.

Goldin, I. & Winters, A. (1995). *The economics of sustainable development*. Cambridge and New York: Cambridge University Press.

Graves, P. E. (2007). *Environmental Economics: A Critique of Benefit-Cost Analysis*. Lanham, MD: Rowman & Littlefield.

Graves, A. R., Burgess, P. J., Palma, J. H. N., Herzog, F., Moreno, G., Bertomeu, M., et al. (2007). Development and Application of Bio-Economic Modelling to Compare Silvoarable, Arable and Forestry Systems in Three European Countries. *Ecological Engineering, 29*, 434–449.

Gray, L.C. (1914). Rent under the Assumption of Exhaustibility. *Quarterly Journal of Economics, 28*, 466–489.

Haberl, H., Fischer-Kowalski, M., Krausmann, F., Weisz, H., & Winiwarter, V. (2004). Progress Towards Sustainability? What the Conceptual Framework of Material and Energy Flow Accounting (MEFA) Can Offer. *Land Use Policy, 21*(3), 199–213.

Hackett, S., & Dissanayake, S. T. N. (2019). *Environmental and Natural Resources Economics. Theory, Policy, and the Sustainable Society* (5th ed.). New York, NY and Abingdon, UK: Routledge.

Hahnel, R., & Sheeran, K. (2010). *A Critique of Environmental Economics: A Heterodox Approach*. London: Routledge.

Hajer, M. A. (1995). *The Politics of Environmental Discourse: Ecological Modernization and the Policy Process*. Oxford: Oxford University Press.

Halvorsen, R., & Layton, D. F. (Eds.). (2015). *Handbook of Environmental and Resource Economics*. Cheltenham, UK: Edward Elgar.

Harmin, M., Barrett, M. J., & Hoessler, C. (2017). Stretching the Boundaries of Transformative Sustainability Learning: On the Importance of Decolonizing Ways of Knowing and Relations with the More-Than-Human. *Environmental Education Research, 23*(10), 1489–1500.

Harris, J. M., & Roach, J. (2018). *Environmental and Natural Resource Economics: A Contemporary Approach* (4th ed.). New York, NY and Abingdon, UK: Routledge.

Harvey, D. (1997). *Justice, Nature and the Geography of Difference*. Malden, MA and Oxford: Blackwell.

Haucap, J. (2018). *Macht, Markt und Wettbewerb: Was steuert die Datenökonomie?* Berlin: Nicolai Publishing & Intelligence.

Hawkens, P., Lovins, A., & Hunter Lovins, L. (2000). *Natural Capitalism: Creating the Next Industrial Revolution*. Boston et al.: Little, Brown and Company.

Heal, G., & Garrett, P. (1993). *The Economics of Exhaustible Resources*. Cheltenham, UK: Edward Elgar.

Hickel, J., & Kallis, G. (2019). Is Green Growth Possible? *New Political Economy*, 1–18. https://doi.org/10.1080/13563467.2019.1598964.

Hicks, J. (1965). *Capital and Growth*. New York and Oxford: Oxford University Press.

Hodge, I. (1995). *Environmental Economics*. London: Palgrave Macmillan.

Holland, A. (1997). Substitutability: Or, Why Strong Sustainability Is Weak and Absurdly Strong Sustainability Is not Absurd. In J. Foster (Ed.), *Valuing Nature? Economics, Ethics and Environment*. London: Routledge.

Hotelling, H. (1931). The Economics of Exhaustible Resources. *Journal of Political Economy, 39*(2), 137–175.

Huber, J. (2004). *New Technologies and Environmental Innovation*. Cheltenham: Edward Elgar.

Huber, J. (2017). *Sovereign Money. Beyond Reserve Banking*. London: Palgrave Macmillan.

Huiting, R. (1992). The Economic Functions of the Environment. In P. Ekins & M. Max-Neef (Eds.), *Real-Life Economics: Understanding Wealth Creation* (pp. 61–69). London: Routledge.

ICC. (2012). *Green Economy Roadmap*. International Chamber of Commerce. Document 213-18/8 – June 2012.

Jaeggi, R. (2001). The Market's Price. *Constellations, 8*(3), 400–412.

Jänicke, M. (2007). *Megatrend Umweltinnovation. Zur ökologischen Modernisierung von Wirtschaft und Staat*. München: Oekom Verlag.

Kapp, W. (1968). In Defense of Institutional Economics. *The Swedish Journal of Economics, 70*(1), 1–18.

Kemp, M. C., & Long, M. V. (Eds.). (1984). *Essays in the Theory of Exhaustible Resources*. Amsterdam: North Holland.

Klitgaard, K., (2013). Heterodox political economy and the degrowth perspective. *Sustainability, 5*, 276–297.

Kraqt, M. E. (2012). Bioeconomic Modelling: Integrating Economic and Environmental Systems? In R. Seppelt, A. A. Voinov, S. Lange, & D. Bankamp (Eds.), *International Environmental Modelling and Software Society (iEMSs) 2012 International Congress on Environmental Modelling and Software, Managing Resources of a Limited Planet*. Leipzig, Germany: Sixth Biennial Meeting. Retrieved from http://www.iemss.org/society/index.php/iemss-2012-proceedings.

Kubiszewski, I., Costanza, R., Franco, C., Lawn, P., Talberth, J., Jackson, T., et al. (2013). Beyond GDP: Measuring and Achieving Global Economic Progress. *Ecological Economics, 93*, 57–68.

Lander, E. (2011). The Green Economy: The Wolf in Sheep's Clothing. *Transnational Institute*. Retrieved from www.tni.org.

Lawn, P. A. (2003). A Theoretical Foundation to Support the Index of Sustainable Economic Welfare (ISEW), Genuine Progress Indicator (GPI), and Other Related Indexes. *Ecological Economics, 44*(1), 105–118.

Lowe, A. (2011 [1976]). *The Path of Economic Growth*. Cambridge et al.: Cambridge University Press.

Mäler, K.-G., & Vincent, J. R. (Eds.). (2003). *Handbook of Environmental Economics: Environmental Degradation and Institutional Responses* (Vol. 1 (vol 2–4 2005 ff)). Amsterdam et al.: Elsevier.

Markandya, A., Harou, P., Bellu, L. G., & Cistulli, V. (Eds.). (2002). *Environmental Economics for Sustainable Growth*. Cheltenham, UK: Edward Elgar.

Mol, A. P. J. & Spargaaren, G. (2005). From Additions and Withdrawals to Environmental Flows: Reframing Debates in the Environmental Social Sciences. *Organization and Environment, 1*, 91–107.

Mol, A. P. J., & Sonnenfeld, D. A. (2000). *Ecological Modernisation Around the World: Perspectives and Critical Debates*. London: Frank Cass.

Mol, A. P. J., Sonnenfeld, D. A., & Spaargaren, G. (2009). *The Ecological Modernisation Reader: Environmental Reform in Theory and Practice*. New York: Routledge.

Mulberg, J. (1995). *Social Limits to Economic Theory*. London: Routledge.

Naess, P., & Price, L. (Eds.). (2016). *Crisis System: A Critical Realist and Environmental Critique of Economics and the Environment*. New York, NY and Abingdon, UK: Routledge.

OECD (2011). Towards green growth. *A summary for policy makes*. Paris. www.oecd.org/greengrowth/480012345.pdf.

O'Riordan, T. (1981). *Environmentalism* (2nd ed.). London: Pio.

Pearce, D. (1992). Green Economics. *Environmental Values, 1*(1), 3–13.

Pearce, D. (2002). An Intellectual History of Environmental Economics. *Annual Review of Energy and the Environment, 25*, 57–81.

Pearce, D., Markandya, A., & Barbier, E. B. (1989). *Blueprint for a Green Economy*. London: Earthscan.

Perry, N. (2013). Environmental Economics and Policy. In G. C. Harcourt & P. Kriesler (Eds.), *The Oxford Handbook of Post-Keynesian Economics* (Critiques and Methodology) (Vol. 2). Oxford: Oxford University Press.

Pietak, L. (2014). Review of Theories and Models of Growth. *Comparative Economic Research, 17*(1), 45–60.

Polanyi, K. (1944). *The Great Transformation*. New York: Rinehart.

Polasky, S., Bryant, B., Hawthorne, P., Johnson, J., Keeler, B., & Pennington, D. (2015). Inclusive Wealth as a Metric of Sustainable Development. *Annual Review of Environment and Resources, 40*, 445–466.

Radin, M. J. (1996). *Contested Commodities*. Cambridge, MA: Harvard University Press.

Rice, J. (2007). Ecological Unequal Exchange: Consumption, Equity and Unsustainable Structural Relationships within the Global Economy. *International Journal of Comparative Sociology, 48*(1), 43–72.

Romer, P. M. (1990). Endogenous Technological Change. *Journal of Political Economy, 98*, 71–102.

Rosenfeld, A. H., Kaarsberg, T. M., & Romm, J. J. (2001). Efficiency of Energy Use. In J. Zumerchik (Ed.), *The Macmillan Encyclopedia of Energy*. Macmillan.

Roy, S. (2016). A Critique on Current Paradigms of Economic 'Growth' and 'Development' in the Context of Environment and Sustainability Issues. *Consilience: The Journal of Sustainable Development, 16*(1), 74–90.

Scheffer, M., Westley, F., Brock, W. A., & Holmgren, M. (2002). Dynamic Interaction of Societies and Ecosystems – Linking Theories from Ecology, Economics, and Sociology. In L. H. Gundersen & C. S. Holling (Eds.), *Panarchy: Understanding Transformation in Human and Natural Systems* (pp. 195–239). Washington et al.: Island Press.

Schilling, M., & Chiang, L. (2011). The Effect of Natural Resources on a Sustainable Development Policy: The Approach of Non-Sustainable Externalities. *Energy Policy, 39*, 990–998.

Schlagwein, D., Schoder, D., & Spindeldreher, K. (2020). Consolidated, Systemic Conceptualization, and Definition of the "Sharing Economy". *Journal of the Association for Information Science and Technology, 71*, 817–838.

Scholl, A., & Semmler, W. (2000). *Sustainable Economic Growth and Exhaustible Resources: A Model and Estimation for the US*. Working Paper 1. University of Bielefeld, Center for Empirical Macroeconomics.

Smith, V. K. (Ed.). (1979). *Scarcity and Growth Reconsidered*. Baltimore and London: Johns Hopkins University Press.

Solow, R. M. (1972). Intergenerational Equity and Exhaustible Resource. *Review of Economic Studies, 41*(5), 29–45.

Solow, R. M. (1974). The Economics of Resources or the Resources of Economics. *The American Economic Review, 64*(2), 1–14.

Soytas, U., & Sari, R. (Eds.). (2019). *Routledge Handbook of Energy Economics*. New York, NY and Abingdon, UK: Routledge.

Sredojevic, S., Cvetanovic, S., & Boskovic, G. (2016). Technological Changes in Economic Growth Theory: Neoclassical, Endogenous and Evolutionary Approach. *Economic Themes, 54*(2), 177–194.

Stallworth, H., Harris, L., & Wise, G. F. (1997). *The Economics of Sustainability. US Environmental Protection Agency*. Watershed Academy Web. Distance Learning Modules on Watershed Management. Retrieved from http://www.epa.gov/watertrain.

Stiglitz, J., Fitoussi, J.-P., & Sen, A. (2010). *Mismeasuring our Lives: Why GDP Doesn't Add Up*. New York: New Press.

Sušnik, J. (2016). *Economic Growth and Resource Use: Exploring the Links. DNC Policy Brief*. Tokyo: United Nations University, Institute for Integrated Management of Material Fluxes and of Resources (UNU-FLORES).

Swyngedouw, E. & Wilson, J. (Eds.) (2014). *The Post-Political and its Discontents: Spaces of De-politicization, Specters of Re-Politicization*. Edinburgh: Edinburgh University Press.

Tietenberg, T. H., & Lewis, L. (2018 [1994]). *Environmental and Natural Resource Economics*. New York NY and Abingdon, UK: Routledge.

UNEP. (2011). *Towards a Green Economy: Pathways to Sustainable Development and Poverty Eradication*. Geneva: United Nations Environment Programme.

van den Bergh, C. J. M., Hoekstra, J., Imeson, R., Nunes, P. A. L. D., & de Blaeij, A. T. (Eds.). (2006). *Bioeconmic Modelling and Valuation of Exploited Marine Ecosystems*. Dordrecht, The Netherlands: Springer.

van den Bergh, C. J. M. (Ed.). (1999). *Handbook of Environmental and Resource Economics*. Cheltenham, UK: Edward Elgar.

Victor, P. A., & Dolter, B. (Eds.). (2017). *Handbook on Growth and Sustainability*. Cheltenham, UK and Northampton, MA: Edward Elgar.

Weizsäcker, E. U., Lovins, A., & Hunter Lovins, L. (1998 [1997]). *Factor Four: Doubling Wealth – Halving Resource Use*. A Report to the Club of Rome. London: Earthscan.

Wrong, D.H. (1961). The Oversocialized Conception of Man in Modern Sociology. *American Sociological Review, 26*(2), 183–193.

York, R., & Rosa, E. A. (2003). Key Challenges to Ecological Modernization: Intitutional Efficacy, Case Study Evidence, Units of Analysis, and the Pace of Eco-Efficiency. *Organization & Environment, 16*(3), 273–288.

Zinn, K.G. (1980). *Die Selbstzerstörung der Wachstumsgesellschaft: Politisches Handeln im ökonomischen System*. Reinbek: Rowohlt.

6 Ecological Economics: Critical Perspectives

Ecological economics is described and discussed in this chapter from an interdisciplinary, social-ecological perspective, asking which ideas, concepts and knowledge from ecological economics are relevant to sustainable development? For a systematic overview of the intellectual history of ecological economics, its themes, approaches and methodologies, the textbooks and handbooks reviewed at the end of the chapter are useful as complementary sources. Key themes of ecological economics that are relevant to the future discourse and transition to sustainability include the critique of economic growth and the assessment of green or sustainable growth that develop in the sustainability discourse and in the policy process. With its search for alternatives to growth, guided by ideas about degrowth and transformation to a post-growth society, ecological economics influences the sustainability process; its critique of markets and technologies as the main instruments for achieving sustainability helps to clarify the limits of economic research and knowledge on sustainable development. With the development of new economic and non-economic instruments for sustainability governance, ecological economics participates in the development of interdisciplinary instruments, for example ecological indicators to measure the progress of sustainable development. It becomes part of an interdisciplinary knowledge base for understanding changing social and biophysical realities.

The multiple sources and variants of ecological economics are shown in Fig. 6.1.

K. Bruckmeier, *Economics and Sustainability*,
https://doi.org/10.1007/978-3-030-56627-2_6

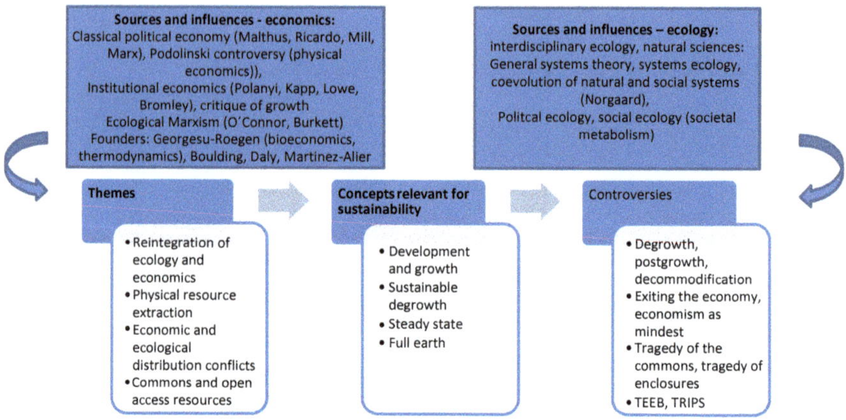

Fig. 6.1 Ecological economics: sources and variants

6.1 Ecological Economics in a Historical Perspective

Ecological economics is part of a variety of critical and interdisciplinary approaches in economics with which it shares ideas, concepts and critical assessments of the state and development of the modern market economy. The critique of the modern, growth-based capitalist economy is much older than ecological economics; economists in the classical political economy, especially Ricardo and Mill, critiqued the limits of the economic growth that began with industrialisation. Twentieth-century critical approaches renewed this critique of growth with research on unequal development and its social and ecological consequences: for example, institutional economics (Kapp, Bromley) on power relations in economic and political processes; development economics on the economic dependency of developing countries on developed industrial countries; feminist economics (Mellor, O'Hara) on the gender divide and the vast amount of unpaid household and reproduction work that is part of uneven development within every country; and heterodox economics, which includes a series of critical approaches in economics (Jo et al. 2019). Ecological economics communicates with these critical economic approaches but differs in various theoretical perspectives: economics of the steady state (Daly), sustainable degrowth (Martinez-Alier et al.), social-ecological economics (Spash, Fischer-Kowalski et al.), co-evolutionary ecological economics (Norgaard) and Marxist ecological economics (Altvater, Burkett, O'Connor). It developed from a critique of conventional environmental economics that originated from the orthodox traditions of economics: critique of the ignorance of ecological limits to growth and of misleading abstractions and fallacies regarding the adaptation of social systems to ecosystems.

Daly and Farley (2004: 30), borrowing from Whitehead, use for the critique of misleading abstractions in neoclassical economics the term "fallacy of misplaced concreteness", meaning, in the case of economics, misplaced concreteness by ignoring the broader contexts of society, culture and nature in which an economy functions. Ignoring these contexts results in misleading information or data, for example idealised models that provide incorrect interpretations of economic agents based on assumptions and generalisations. In its present form, ecological economics has developed since the 1980s (Daly 1980; Martinez-Alier 1987) in Europe and North America; it shares many of the basic ideas of the sustainability discourse, for example the interaction and co-evolution of economic and ecological systems (Norgaard 1994); the view of the economy as a subsystem of society and of the ecological earth system; the rejection of the conventional economic idea that natural capital can be substituted by other forms of capital; and normative views in which environmental justice, intergenerational solidarity and equity gain importance (Daly and Farley 2004; Faber 2008; Söderbaum 2008).

6.1.1 Ecological Discussion in the History of Economics

The interaction between economy and nature has been discussed since classical economics. Critical ecological ideas developed in the classical political economy of the nineteenth century, in the critique of political economy by Marx and in the subsequent development of a critical theory of modern capitalism. During the twentieth century institutional economics took up environmental problems, though not as its main theme. Critical institutional economics re-established an interdisciplinary approach to the analysis of the social context and social institutions on which the modern economy depends. The steps of unfolding the ecological themes in economics include the following:

1. *Classical and critical political economy in ecological perspectives*: from Malthus, Mill, Ricardo and Marx, to the renewal of critical political economy in the twentieth century, in world system theory and ecological Marxism (Altvater, Foster), and in the "limits to growth" discourse, the interaction between economic and ecological systems became part of the system analysis of modern capitalism. In classical political economy, ecological analysis developed through the critique of the Malthusian theory of population growth and eco-

nomic growth, discussing the conditions of economic growth and its continuity. This can be briefly summarised in the main arguments of the discussants as follows (Barnett and Morse 1963: 64ff): Malthus argued with a naturalistic reductionism of political economy that modern society tends to ruin its natural resource base through overuse and through rapid population growth. Ricardo modified the "natural scarcity" described by Malthus into a "social scarcity", generated by social practices of resource use but partially solved through new technologies and resource use practices. Mill forwarded the idea that accompanied classical political economy since Ferguson and Smith as the economy of the stationary state: economic growth and industrialisation are exceptional forms of development that cannot last forever. He foresaw economic and ecological catastrophe after the end of industrial growth. Marx described various environmentally and socially destructive components of modern capitalism: "previous accumulation" and expropriation of agricultural producers, the alienation of humans from nature and the damage to nature caused by industrial production, including the "ecological rift" (interruption of nutrient cycles through modern industrialised agriculture).

That industrial growth can only last for a limited time was not doubted by the classical political economists, even if they interpreted the conditions differently and did not always highlight the idea of limits. Later, in ecological Marxism the concept of the ecological rift was generalised for industrial production and the whole economy. Marx criticised Malthus for his naturalist reductionism in reducing social factors to natural or biological ones. But he did not reject the Malthusian reasoning in toto, accepting it in the hypothetical scenario in which global population growth exceeds the capacity of the earth as an ecosystem, which was, for Marx, plausible in the distant future. This plausible future, however, became a reality at the end of the twentieth century. In world system theory the theoretical analysis of economy–nature relations is called world ecology, a complementary component of the theory of modern capitalism (Moore 2000, 2015).

2. *Institutional economics* has throughout the twentieth century focused on the sociocultural evolution of institutions and the relations between economic and other social institutions on which the functioning of modern economy depends (Ménard and Shirley 2005; Ostrom 2005; Bromley 2006; Kapp 2011). Two theoretical analyses relevant to the sustainability discourse are "The Great Transformation" of Polanyi (1944) and "The Social Costs of Private Enterprise" of Kapp (1950). Polanyi reconstructed the detachment of the growth-based market economy from its social and ecological conditions and the limits of economic growth as a course for the destruction of the economy. In Kapp's analysis the costs of environmental destruction caused by industrial production fall back

on society in form of ecological restoration. *New variants of institutional economics* (Hall and Klitgaard 2012) are broadening the ecological analysis, discussing global limits, especially for fossil energy resources. This form of analysis can be integrated with ecological economics and interdisciplinary research in social and political ecology.

3. *Development economics* is a subdiscipline that offers different approaches to the conditions of growth, development and welfare in the poorer countries of the Global South to support efforts to combat hunger and poverty and improve employment conditions, health and education (Easterly 2002; Meier 2005; Sachs 2005), to reduce the vulnerability of people and states or institutions (Bromley and Anderson 2012), or to develop institutions for solving environmental problems (Bromley 2014). The orthodox variants of development economics see growth as an opportunity to improve well-being in developing countries, as they do with industrialised countries. Critical development economics sees growth in industrial economies and the appropriation of the majority share of the natural resources of developing countries through unequal exchange as causes for the blocked development of the Global South. Achieving welfare and well-being in developing countries requires analysis of the systemic nature of unequal development in the global economy and changes to the institutional, structural and transnational components of the modern economy in the transition to sustainability. *Dependency theory*, which emerged in critical development economics and critical theory of society, developed especially in Latin America on the basis of the Prebisch-Singer hypothesis: in the modern world system, the periphery (developing countries) produces mainly primary goods for export to the centre (industrialised countries), which exports processed industrial goods. The unequal division of labour and power relations in the world system causes unequal exchange and terms of trade: developing countries must export more to obtain a certain value of industrial exports; the benefit from growth, international trade and technological development is thus reserved to rich industrialised countries. Following dependency theory, new approaches in development economics emerged, discussing questions of poverty, unequal development and growth (Sen 1992; Banerjee and Duflo 2011).
4. *Growth-critical analyses* are part of development economics, for example the Harrod-Domar model (see Chap. 5) and other Keynesian and post-Keynesian growth models that questioned the possibility and necessity of balanced growth. Economic cycles and periodical crises show that supply and demand, production and consumption do not continually tend to an equilibrium; economic growth depends, beyond labour and capital, on many exogenous factors that cannot be fully controlled through economic institutions and require political and other forms of regulation. When the debate on economic growth becomes

part of the sustainability discourse, further conceptual differentiations are required: between factors of growth and factors of development; between growth, steady state and degrowth; and between changes in the unequal distribution of resources through market-based instruments and changes through redistribution processes that fall under the principles of intragenerational and intergenerational solidarity articulated in the sustainability discourse. The global redistribution and sharing of resources requires a social-ecological transformation of the global economy and will require the building of international institutions for redistribution in the further processes of global governance and sustainability transition.

5. *Ecological Marxism* connects ecological economics with the critical political economy that has developed since the nineteenth century. What Marxist theory can provide for ecological economics and sustainability is described in the new forms of ecological Marxism by O'Connor (1988, 1998; the second contradiction of capitalism), Foster (2000, 2009; Foster and McChesney 2012; Foster and Burkett 2016: the endless crisis of capitalism and the ecological revolution); Burkett (2006: integration of Marxist analyses in ecological economics with regard to nature and economic value, nature as capital, entropy and economic systems, sustainable development); and Altvater (1992; Promethean revolutions, market failure and social-ecological development paths). Burkett interpreted the relations between Marxism and ecological economics as complementary: Marxist class analysis can help answer questions of ecological economics which in turn can enrich Marxist materialist analysis. In Altvater's work appears a more comprehensive view of the multiple transformations of the economy and the accumulation and social-ecological regimes required to shift towards sustainability.

The variants of ecological Marxism include (as do all theories of environmental destruction, its prevention and solution) analyses of the changing environmental problems and potential solutions of modern capitalism. The fusion of ecological analyses and Marxist approaches leads to an inevitable question: "How will capitalism end?" (Streeck 2016)—the peculiarity of Marxist theory. The answer to this question can no longer be that of classical Marxism: the working class will become the revolutionary subject and overcome capitalism. The peculiar challenge of the present conditions of globalisation and global environmental change is to find a new social subject and new forms of agency to achieve social-ecological transformation; a subject that is, with the limited knowledge available so far, not a social class but a multi-subject comprising different social groups cooperating in changing forms of transformation action groups.

Beyond the questions of class analysis and transformation, the discourse of ecological Marxism recapitulates in the economic and social analyses of the environmental crisis of modern society the big system-related questions that need to be dealt with in any social-ecological transformation towards sustainability: questions about mode of production; social metabolism; material and symbolic reproduction of society and reproduction of nature and life; interaction between nature, society and humans; and connecting the theory and practice of sustainability transformation. Most of these themes are neglected or marginalised in economics and other specialised social sciences but are key themes in the interdisciplinary ecological discourse and the sustainability discourse.

The multifaceted ecological economics shares with the other critical approaches the view of the economy as a system that is part of the whole of modern society, functionally connected with its other social, political, cultural systems and institutions and embedded in nature or ecosystems. These guiding ideas connect with the interdisciplinary integration of ecological knowledge in economic thinking and research. The critical analysis paradigmatic of the global economy does not develop in ecological economics into a single and closed theory. The subdiscipline remains pluralistic and adopts concepts and models from different economic approaches, including from orthodox environmental economics (Costanza et al. 2010; Martinez-Alier and Muradian 2015; Spash 2018). The thematic scope of ecological economics is outlined by Martinez-Alier (1990) with the core themes of energy, environment and society. Daly (1980, 1994, 1997, 2008) developed ecological economics from the key themes of an economy "beyond growth" and "steady state", highlighting the ethical aspects of economy. Daly and Mayumi (2001) see the origin of ecological economics in the bioeconomy of Georgescu-Roegen (1975) with its focus on energy use. The leading authors of the subject represent different perspectives and traditions that are adopted from other critical approaches in economics, ecology and environmental research: Martinez-Alier integrated themes from political ecology and social movement research. Mayumi (2001; Giampietro et al. 2011, 2012) integrated knowledge from energy analysis and complex systems theory. Hardin worked with ecological and biological themes. Vatn (2005, 2016) adopted themes from institutional and evolutionary economics.

6.1.2 The New Ecological Economics

The intellectual history and development of ecological economics has been described by Ropke (2004, 2005). A key theme of ecological economics, as with other critical approaches in economics, is the historical dis-embedding detachment of the economy from its social and ecological contexts; the analyses aim to reconnect society and nature with knowledge about modern society, the capitalist mode of production and ecological knowledge. The differences between environmental and ecological economics are discussed in both subjects, without a final clarification of their scope and limits (Box 6.1).

Box 6.1 Differences Between Environmental and Ecological Economics: Pearce

The environmental economist Pearce (2002: 75ff) highlights the following differences:

1. Ecological economists regard environmental problems as far more serious than environmental economists; they emphasise far more the limits to growth and natural resource use in terms of carrying capacity of the earth; they do not give technological innovation and change for improving or solving environmental problems the same attention as environmental economists. However, Pearce does not go deeper in describing the difference and avoids using the term "technical fix" for the description of environmental economics with its orientation to innovation and technology-based growth.
2. Ecological economists do not accept the assumption that production factors and functions can substitute each other, which is relevant for the assessment of sustainability. The assumption of limits to growth assumes final limits and reduced levels of consumption per capita under conditions of sustainability. This is consistent with the view in ecological economics that overconsumption of natural resources accounts for much of the global environmental degradation and requires lifestyle changes.
3. Ecological economists reject the smoothness of production functions from neoclassical economics, arguing: with the given discontinuities in ecological systems already limited variations in economic variables may result in ecological collapse.
4. Ecological economists are critical of practices that discount the future costs and benefits and monetary assessment of environmental damages.
5. Ecological economists see the known differences between economic efficiency and social welfare functions as reason to revise thinking in welfare economics, emphasising a specific set of social welfare functions that stress the rights of future generations.

Source: Pearce 2002

Pearce summarises the important differences between environmental and ecological economics somewhat inexactly, in a standardised terminology from conventional environmental economics. This can be seen as an example of "economistic" and "colonised" thinking based on a conventional view of science as producing authoritative knowledge for policy and decision-makers, which is critically discussed by ecological economists and in the degrowth movement. Pearce (2002: 78) confirms this "economism" with his formulations that economics journals and the language of economics, formerly often seen as obscure, are now understood by decision-makers, and this will remain so, indicating a success for mainstream economics in influencing political practice. An important difference between orthodox and heterodox economics, relevant to the sustainability process, is visible in the use of economic terms and indicators and in different interpretations of sustainability. For ecological economists it is difficult to accept gross domestic product (GDP) and additional national indicators such as the genuine savings (GS) indicator as measures of well-being, as Pearce does. GDP is seen as a bad indicator, GS as better, but the bigger problem as: these indicators do not account for components of well-being that cannot be monetarised (or only partially). For example, the costs for healthcare can be measured in GDP terms insofar as they include paid goods and services; more difficult is measuring the individual activities of healthcare, or the value of health as a component of human well-being.

The nature and scope of ecological economics is continually discussed. The early description by Daly (Box 6.2) differentiates later, with the unfolding of the discourse in a plurality of approaches that can be seen in the recent handbooks by Martinez-Alier and Muradian (2015), and Spash (2018).

Box 6.2 Ecological Economics: The Description of Daly and Farley

The ecological economists Daly and Farley (2004: 3ff) describe ecological economics as part of economics, a discipline they define conventionally as the study of the allocation of scarce resources among alternative, competing ends. They add as a critical comment that efficient allocation of resources is in ecological economics not an end in itself. Resource allocation is re-interpreted in an ecological perspective as limited through the carrying capacity of the earth as an ecosystem. The limits of natural resources that Boulding describes with the metaphor of "economy of the spaceship earth" Daly and Farley describe with the picture of a ship and its cargo. Their textbook is

(continued)

Box 6.2 (continued)

structures similarly to textbooks of environmental economics, written in a disciplinary tradition and in the differentiation of microeconomics and macroeconomics. They discuss whether the knowledge, concepts, methods, models and theories of economics can be used for the analysis of environmental or natural resource use problems. The ecological knowledge they introduce in economics is complementary to conventional economics or corrective of its deficits. With regard to sustainability, reasoning is not systematically developed, knowledge insufficiently described; the complex process of sustainability transformation requires a broad integration of interdisciplinary and transdisciplinary knowledge that cannot be adequately discussed within a disciplinary economic perspective, particularly when it is described as a "transdiscipline" (Daly and Farley 2004: 6f). The limits of this description of ecological economics can be seen in the critique of neoclassical economics and the growth-based market economy that is highlighted by the authors. Further knowledge from non-economic and interdisciplinary research is not sufficiently discussed by the authors to enable a systematic description of the interdisciplinary knowledge base for the sustainability process.

Martinez-Alier (1995: 22), like Daly one of the pioneers of ecological economics, formulated this critique in a stronger form: the inability of orthodox economics to cope with environmental issues has brought about ecological economics, the study of the long-term compatibility of the human economy and ecosystems. The main question for further discussion in ecological economics becomes whether and how far the reintegration of the modern economy in societal and ecological systems can be achieved with scientific knowledge from economics: *which other knowledge is necessary to make economic knowledge function in the broader discourse and process of sustainable development?*

Source: Daly and Farley 2004; Martinez-Alier 1995

The interdisciplinary and pluralistic knowledge practice of ecological economics fosters internal controversies about economic core concepts and their use in environmental research, including the term "natural capital". This concept is rejected by some ecological economists with arguments against weak sustainability and the substitution of natural with human-made capital: destroyed nature or depleted fish stocks cannot be substituted. In this interdisciplinary perspective economic knowledge appears to be constrained by ecological knowledge, not the other way round. The flows of material and energy in coupled social-ecological systems and in the biologically necessary consumption of energy, food and water by humans cannot be seen as "revealed preferences" in the conventional view of consumption in economics.

The descriptions of environmental and ecological economics discussed so far are those of individual economists. Their views are influential as leading proponents of environmental and ecological economics; however, uncertainty about theoretical, methodological and paradigmatic differences between the two approaches remain. The uncertainty is caused by repeated attempts to clarify by way of surveys among economists either the knowledge profiles of the two subjects (Spash and Ryan 2010) or the arguments of proponents and opponents of growth (Drews and van den Bergh 2017), complemented by surveys among the public in general. These surveys show significant differences in the description of environmental and ecological economics and reasoning for or against economic growth, in subjective perceptions and contingent assessments.

The ideas discussed in ecological economics help frame the sustainability discourse theoretically and use non-economic knowledge about the problems the sustainability process faces. Different aspects of sustainable development are addressed in ecological economics:

1. *Energetic aspects in the thermodynamic view of the economy*: Georgescu-Roegen (1975) developed the idea that the economy is a system dependent on the thermodynamic laws of physics (especially the second law of dissipation of energy). This energy-focused view of economy can be seen as a specific variant of the discussion of limits to growth and resource use. The discussion is of a theoretical nature and controversial because of the interpretations the author makes of thermodynamics (Schwartzman 2012): the earth system is not a closed system but energetically seen an open system that is exchanging energy with the cosmos and is powered by the sun.
2. *The "economy of spaceship earth"* (Boulding 1966; Spash 2013): Boulding's reframing of economics (the economy as a system with limited resources) is a theoretical model of ecological economics. The physical view of the economy as a resource-consuming process with ecological limits draws attention to externalities such as environmental destruction that were ignored during the economic growth of industrialisation. Externalities can be effective over large distances and accumulate over a long time, as the emissions of greenhouse gases show. "*Spaceship Earth similarly invoked the first law of thermodynamics to point out that whatever was taken out of natural resource sectors must reappear in equal weight as waste, which will likely affect the environment when disposed*

of…. As economies expand in the economic sense, so they are likely to expand in terms of physical resource extraction and hence in terms of physical waste emissions to the environment. Worse, economic activity chemically transforms materials and energy into waste gases" (Pearce 2002: 60).

3. *Economic aspects of sustainable development*: concepts adopted in ecological economics such as "full earth", "steady state", "spaceship earth", and "limits to growth" open up the view of economics as a system that is part of larger systems and is limited in time and space. With the concept of steady state and recent discussion of degrowth ecological economics is connected to part of the critical discourse about growth and its limits that has emerged since classical political economy. With global economic and environmental change becoming effective, the discussion about growth becomes more systematic and knowledge from different disciplines is sought. Ecological economics does not only connect with the discourse of limits to growth but extends the debate to that of the future economy and the transition to sustainability. This long-term perspective is less developed in orthodox environmental economics, where the changes required for sustainable development tend to be constructed as short-term changes internal to an economy that can, with the help of sustainability policy and governance, be condensed into programmes for goal attainment, as with the sustainable development goals of the United Nations (UN) in the "Agenda 2030".
4. *The interdisciplinary opening of ecological economics*: new ideas of economically and ecologically unequal exchange in the global economy (Rice 2007) and of economic and ecological distribution conflicts and their resolution (Martinez-Alier 1995) are discussed in ecological economics, adopting knowledge from critical theories and analyses of the modern society and economic world system. With that the process of sustainable development can be discussed in the perspectives highlighted in the Brundtland Report and neglected in the following sustainability discourse: sustainability as the building of a solidary economy that enables resource sharing between the Global North and South and between present and future generations.

6.1.3 Creating Knowledge Through Interdisciplinary Communication and Knowledge Exchange

Ecological economics shares a series of critical analyses, concepts and theoretical forms of reasoning with other critical approaches and theories in economics, other social sciences and interdisciplinary and natural-scientific ecology. Among these are institutional economics, development economics, feminist economics, political-economic approaches, ecological Marxism, world system theory, world ecology, human ecology, social ecology (including the theories of societal metabolism, physical economy and material and energy flow accounting) and political ecology (Bryant 2015). A brief description of these approaches, how they connect with ecological economics and their relevance to the sustainability discourse follows.

1. Institutional economics: in institutional economics environmental analyses of the economy developed before ecological economics (Kapp 1950, 1968). Bromley, who studied property rights (Bromley 1991, 1992), brought institutional economics into contact with environmental and ecological economics (Bromley 1995, in the handbook of environmental economics; 2006, with a critical analysis of rational choice theory and models; 2014, with the analysis of the relations between institutions and the environment) and development economics (Bromley 1999, 2014). How to reformulate economics in a critical and interdisciplinary perspective of sustainable development is described by Klitgaard (2013; see Appendix) in a series of requirements to broaden, change and complement economics for a theory of the transformation of modern economy and society into a sustainable society. He provides an encompassing description of the main requirements and difficulties in creating transformative agency for sustainability—more than that, what economics as a disciplinary science can offer in terms of knowledge, concepts, theories, models and methods.

Klitgaard shows the multiple and difficult changes involved in the long transition to sustainability. But a theory of sustainability needs to structure these changes more systematically in a coherent analysis and explanation. Such a theory, where developing through interdisciplinary knowledge integration, is not a single theory but rather an inter-theory combined of a series of specific concepts and explanations from different theories, focusing on nature-society interaction (Bruckmeier 2018). Such a theory cannot avoid the taboo themes of orthodox economics: the "frontiers of commodification" (Streeck 2016: 61ff). Markets have an inherent tendency to expand the trading of goods to all spheres of life and to nature. The discussion of these frontiers intensifies with the debate of natural limits to growth to one about social limits, especially for the "fictitious commodities" (for which the laws of supply and demand apply only partially or not at all) discussed by Polanyi, labour, money and land or nature. A theory of sustainability needs to analyse and explain ecological limits to growth, but also the socially destructive tendency of the modern market economy towards unlimited commodification and market-based regulation. Furthermore, with the development of an interdisciplinary theory of nature-society relations that provides knowledge for the sustainability discourse, it becomes necessary to rethink and change theory-practice relations for a science of sustainability (see Chap. 8): the standards of disciplinary knowledge application in epistemic models like knowledge transfer, science communication and piecemeal engineering no longer work for the complex systems and processes involved in the transition to sustainability. One of the few attempts to develop new theory-practice relations has emerged from sustainability science (see Chap. 3).

2. *Feminist economics*: ecofeminist authors have reviewed and discussed, in a manner similar to ecological economists, the approaches and theories of economics, questioning the construction of conventional economic concepts, theories and methods with regard to their ostensible objectivity (O'Hara 1999, 2010, 2016). Langley and Mellor (2002: 49) start from the proposition that the global market economy established through neoliberal economic thinking and practice is socially and ecologically unsustainable: this economy cannot generate conditions of socio-economic and ecological well-being for the global population. The authors identify and describe potential sites of "transformative space" in politics and in the economy itself. The core of the feminist critique of orthodox economics is the differentiation between markets and households as separate spheres, where markets appear as exclusively individualistic, self-interested and competitive institutions, and households as exclusively altruistic, harmonious and cooperative institutions. Households are omitted from the economic analysis and from a modern economy that focuses on market practices and reduces households to private consumers (Gardiner 1997: 336). The market economy as an exclusive, value-creating economic sphere is not autonomous; it depends on other economic practices in which the work of women dominates, especially in the reproductive and unpaid work in households. Furthermore, the market economy depends on ecosystems which do unpaid "labour", that of ecosystem services. The embedding of economic practices in nature and society cannot be ignored without distorting the economic processes (Mellor 1997). As in other critical approaches of ecological, institutional and development economics and critical political economy, in feminist economics the value-based reasoning and the power- and interest-based distortions in economic research and economic practice are made visible, with arguments against orthodox economics strengthened through new forms of economic research.

Feminist economics connects with the broader discussion of genderised structures and conflicts in the economy and in the transition to sustainability. Gender inequality, long investigated in academic and critical science, has also reached mainstream economics with the equality discourse (Sen 1992, 1995). In the sustainability discourse, gender equality is highlighted in UN documents and sustainable development goals, but fragmented and neglected in economic processes. In the *Environmental Education and Communication for a Sustainable World* handbook (GreenCom and Academy for Educational Development 2000) gender relations are critically reviewed for gender mainstreaming in international development aid and cooperation. This is not enough to integrate knowledge about and regulation of gender relations in the economy. Franz-Bahlsen (2014) reviewed gender norms in sustainability communications. Gender studies emerged in the 1970s out of the critique of patriarchal authority in feminist studies to ever more complex constructions of gender and diversity, where the interplay of gender, ethnicity, class, age, religion and education is analysed. However, gender equality in the economy, in politics and in education is in most countries still far less advanced, perpetuating old forms of inequality. Assessing gender relations in the sustainability process shows that traditional gender norms and genderised social systems are still dominant: more a form and cause of unsustainability, inhibiting the progress of sustainable development. With all other changes implied in sustainability transition, gender norms must also change. The deep-rooted inequality and injustice in culturally constructed and socially reproduced gender relations needs to be critically reflected and discussed in the sustainability discourse (Franz-Bahlsen 2014: 1987). As the sustainable transition progresses it will become necessary to transform the increasing awareness of gender relations into political strategies for gender equalisation.

The monolithic form of a market economy in the modern world system that is criticised in feminist and ecological economics, in political economy and ecology, and in other critical social-scientific and ecological approaches, requires in the sustainability process a transformation through a series of connected changes. The first step in creating transformative agency can be a revaluation of the multiple forms of economic practices of work, production, distribution and consumption that have developed over the course of human history. Among these the economy of provisioning is a core concept in the discourse of feminist economics to specify life-sustaining forms of human needs, material and immaterial, providing food, shelter and maintaining emotion-based social relations. With the building of an economic system with a plurality of components, the commercial and commodity-based market economy should be regulated and complemented through other economic practices in a new economy with several principles of organisation, rebalancing the household economy (in the Aristotelian sense, distinct from money use) and the market economy. The material forms of production, exchange and consumption cannot only be organised in the monetarised forms of the market economy and require several principles and regulatory forms. More recent attempts to disaggregate the economy in different practices can be found in the works of Braudel and Polanyi, where different forms, layers and levels of abstraction of economic practices can be found (Braudel: capitalism, market economy, material life; Polanyi: exchange, redistribution, reciprocity as circular economic processes).

3. Critical political-economic approaches and ecological Marxism: critical political economy developed in new variants throughout the twentieth century, no longer only in orthodox Marxism, more in critical theories of the developing and dependent economies of the Global South, in connection with Latin American dependency theory (Cardoso and Faletto 1979), critical anthropology (Wolf 1982), environmental theory (Schnaiberg 1980: "treadmill of production"), ecological Marxism (Foster, Burkett, Altvater; see Burkett 2006) or world system theory (Wallerstein 1974, 1980). The common elements of these theories include a system analysis of the changing forms of modern capitalism, its accumulation regimes (the French regulation school), its socio-ecological regimes (new social ecology), the globalisation process and its taming (Held and Koenig-Archibugi 2003) and the new forms of global environmental change (Freedman). The sustainability discourse is not a way to integrate knowledge from these variants, but a way of seeking out new knowledge where it can be found: which components and concepts from these theories are relevant to reformulate environmental, economic and social sustainability in a critical perspective of social-ecological transformation. The sustainability discourse requires that critical theories advance; such theories can be found in the social sciences and economics (e.g. ideas for an economic theory of sustainability are drafted by Klitgaard 2013; see learning exercises described at the end of this chapter).

Ecological Marxism developed from the system analysis of modern capitalism with the components of energy and resource flows between the economy and nature elaborated by Marx in a theory of the ecological rift. This analysis of modern capitalist agriculture, based on the agricultural chemistry of Liebig being developed at that time, showed the interruption of nutrient cycles in modern agriculture as a form of ecological disturbance. In the newer ecological Marxism (Foster) this analysis developed into a critical theory of natural resource use under the conditions of late capitalism. The reconstruction of economics in physical terms, as suggested by Podolinsky (1880) and discussed by Foster and Burkett (2004) in an early version of ecological economics, was in the Marxist economic discourse controversial, providing an example of the dilemma that reappears repeatedly when natural and social-scientific knowledge needs to be integrated in interdisciplinary economic approaches, as it does in ecological economics: with interdisciplinary concept transfer and knowledge integration appear also new forms of reductionism, neglecting either the social or the ecological dynamics of the modern economy.

From world system theory (Wallerstein 1974, 1980) the theory of world ecology influences sustainability theory; it creates under the term "capitalocene" a critical counterpart to the diffuse and simplified economic analysis in anthropocene theory, taking into account the structural change of the global economy through the *globalisation and deregulation of markets*. A theory of world ecology needs to develop through an interdisciplinary social-ecological analysis of global environmental change, which is presently at the beginning.

4. Interdisciplinary ecology—human ecology: human ecology (the "treadmill of production": Schnaiberg 1980), social ecology (the theory and concepts of societal metabolism, colonisation of nature, physical economy and material and energy flow accounting: Fischer-Kowalski and Haberl 2007) and political ecology (connecting to the critical theory and political economy discourse: Bryant 2015) are new interdisciplinary approaches in the environmental and sustainability discourse. Knowledge dilemmas with culturally specific knowledge practices, their explanation and syntheses are not always dissolved in interdisciplinary analyses. In particular, political ecology (Bryant 2015) is fragmented into "small narratives", following the cultural turn in science and reconstructing the different languages of perception and valuation: the small narratives are more fragmenting the analysis of complex systems than generating possibilities for interdisciplinary knowledge integration.

The material and energy flows between ecosystems and social systems has become a main theme of ecological economics as well as of other economic and ecological approaches. Questions of economically and ecologically sustainable energy systems, of energy efficiency and of energy return on energy investment are part of the broader interdisciplinary sustainability research. The transformation of the energy systems of industrial society into ecologically sustainable systems using renewable energy sources has already begun, supported by environmental policy programmes, in European and other countries. What is already evident is that this project of transforming the energy sector is complicated and conflicting and as a process goes well beyond the time horizon of prior sustainability policies. It is in the policy and governance processes that the complexity and challenges of sustainability transformation become finally visible, where different policies and forms of change interact: globalisation, sustainable development, global environmental governance, climate policy and other fields of environmental policy.

The approaches mentioned above are not all integrated in ecological economics. Rather, they offer examples of how interdisciplinary analyses of problems regarding environment, economy and sustainability develop and change with the building of interdisciplinary knowledge cultures (see Chap. 3). For the further development of ecological economics with regard to sustainability, the theme of degrowth becomes the most important with which to renew the growth-critical discourse and to develop new ideas and strategies for the sustainability transformation.

6.2 Degrowth: The Concept and the Controversy About Growth

Degrowth aims at a fundamental change of the modern economic system. This form of growth critique had not previously been discussed until the present degrowth debate, not even as part of the limits to growth debate inaugurated by the Club of Rome in 1972 (see Box 6.1). Degrowth implies, in coherence with the ideas of the 1987 Brundtland Report, but not clearly discussed there, a reduction of production and consumption in the affluent, industrialised countries of the Global North, and with that also a reduction of GDP growth in the sense that the throughput of resources in economic production and consumption is reduced to stay within nature's limits. The aims of social, economic and environmental sustainability should be achieved instead through growth through a transformation of the global economic system into a multifaceted economy, where other forms and institutions become more important than markets: redistribution, sharing and caring and the development of de-commodified, non-market or money-dependent forms of economic production, consumption and value creation (Box 6.3).

Box 6.3 Note on the Concept of Degrowth

Degrowth is a multifaceted idea that fragments across different approaches that are not connected by an underlying theory or system analysis of the global economy but by a set of critical arguments against the socially, economically and ecologically destructive consequences of exponential growth of the global economy (Latouche 2009). Although ideas and motives from different variants of critical theory of modern capitalism, including political economy and other subjects, are taken up, the new discourse is not their practical application, which remains rather eclectic and incoherent. Rather, the new discourse takes up different ideas for which it is not easy to formulate common principles: degrowth is a set of ideas in flux, much like sustainable development, which is criticised by Latouche (2009), on the basis of etymological and semantic reasoning, as an absurd notion.

Degrowth is not simply zero growth of the economy or impoverishment. The formulation that comes closest to degrowth from the discourses on limits to growth and sustainability is "development without growth", meaning: improving welfare, well-being and quality of life without economic growth,

(*continued*)

Box 6.3 (continued)
through qualitative improvement of conditions of life. According to Latouche it is not meant as negative growth or a shrinking economy; rather, it should maintain and improve the quality of life achieved in modern industrial society. Practically this should be achieved by reducing the regular working day to free up more time for individual needs—an old idea traced to the critique of political economy by Marx, which has appeared in different political movements and utopias through history, with modified arguments, but has gained new significance in recent decades in the environmental movements and is now finally adopted in the "concrete utopia" (Latouche) of degrowth.

Degrowth can be specified in different lines of reasoning, one of which is the ecologically based "agenda for sustainable degrowth and relocalising the economy" under the programmatic title "Avoiding Collapse" (Rees 2014). The basic argument here is: "on a planet already in overshoot, there is *no* possibility of raising even the *present* world population to developed country material standards sustainably with known technologies and available resources" (Rees 2014: 3). This argument needs to be connected with a more encompassing theory of sustainability in the sense of Klitgaard (2013) that goes beyond planning theory and its short-term perspectives. Beyond the guiding ideas of sustainable degrowth further ideas can be found in the rapidly expanding degrowth literature.

Sources: mentioned in the text

6.2.1 Development of the Idea of Degrowth

The degrowth discourse began as a new social movement that spread at the beginning of the twenty-first century from France to Spain and Italy and then to other, mainly European, countries. Since 2006 Weiss and Cattaneo (2017) have observed a slow but steady growth of scientific publications on degrowth and a transformation of the term into a multidisciplinary scientific paradigm, which has until now been limited in terms of influence in the scientific and political sustainability discourse. The pioneers and protagonists of this movement were scientists and economists such as Latouche, Martinez-Alier and Kallis (2011, 2017, 2018; D'Alisa et al. 2015). Degrowth cannot only be improved by becoming a scientific paradigm and part of academic and disciplinary science. It needs to be developed through new strategies and projects for sustainability transformation under the perspectives of limits to growth and no-growth, and through joint learning from experiments and experiences with degrowth projects.

Ideas of degrowth are compatible with Polanyi's theory to "domesticate" the markets through social control, the control and regulation of fictitious commodities, and complementing markets through non-market institutions of redistribution (such as governmental forms of tax and subsidies) and reciprocity (such as socio-economic activities that include voluntary and unpaid work and domestic, communal and subsistence work). Through the reintroduction and enlargement of non-market sectors in the formal economy the dominant market and growth mechanism will be reduced, commodification limited and more space created for the development of fairer and solidarity-based economic practices and institutions.

When such complementary economic practices, seen in orthodox economics as non-economic, should be developed, it is necessary to criticise the construction of the modern economy in orthodox economics as ignoring the social and ecological consequences of growth. Economic practices that work without growth cannot simply be ignored or suppressed as outdated, premodern, inefficient or irrational. Reproductive and subsistence work and forms of property other than private are also necessary in a modern economy; not all resources can be appropriated in form of private property and exchanged through markets. There is already a discussion in economics on how to deal with public goods in a modern market economy; for other forms of property and resource use, especially that of commons, the research of Ostrom (2005), and in human ecology that of McCay (McCay and Acheson 1987), helped to revaluate and strengthen marginalised resource use practices. New forms of resource management, for example user-based or ecosystem-based management, adaptive management and governance, and management of the global ecological commons in earth system governance will be increasingly important for building resource management regimes that guide the search for a post-industrial and sustainable economy and society of the future. The few things that can be said about a new "composite economy", where the forms of domestic work, food production, industrial production and service work need to be rebalanced, include the necessity of a global transition strategy that allows for a plurality of transformation paths, does not prevent solutions and does not block other development paths than that of the market-based economy.

6.2.2 Integration of the Degrowth Concept in Ecological Economics

The project of transformation of the modern economy requires a long process of building new economic institutions for which no knowledge exists. Not much can be learned from the functioning of economic institutions in the historical, pre-industrial economies studied by Polanyi et al. (1957), except that they give ideas with which it is possible to think about further and new forms of the economy that did not exist before. The future economy cannot be a return to older forms of economic systems and modes of production before industrialisation, although some elements of the new economy are similar to older forms, for example: the reduction of global value chains, exchange and the strengthening of local and regional economic systems, shorter transports and local recycling and reuse of materials. The new economy is to be built from single institutions and experiments; its systemic forms are unknown today. Terms such as post-capitalist, post-industrial, post-growth and degrowth economy and society have one common quality: they signal a lack of ideas and knowledge about the future society that is now discussed under another name as sustainable development.

In ecological economics and in the degrowth movement there is insufficient knowledge and experience available to design strategies for transforming the global economy; ideas are borrowed from heterogeneous sources and theories and realised in a variety of eclectic practices and experiments. Some of these ideas are based more on theories of economy and society, others purely pragmatic and experimental, as the suggestions for political action to achieve degrowth show: creating and introducing local currencies, fiat money, small and non-profit companies, local commons, participatory decision-making, reduction of working hours, voluntary work, basic income and income ceilings, limitation of exploitation of natural resources, maintenance of biodiversity, minimising waste, regulating and taxing environmental pollution, dismantling large-scale infrastructure systems, abolishing automobile-based transport and introducing new local and public transport systems, and so forth.

The degrowth debate is dispersed and incoherent, lacking theoretical underpinning and a critical system analysis of the modern capitalist system that is criticised because of its socially and environmentally destructive consequences. Some systematisation of the debate and theoretical reflection has occurred, especially with the version of "sustainable degrowth" (Martinez-Alier et al. 2010), as an attempt to integrate degrowth in strategies of sustainable development, allowing under the name of degrowth a variety of approaches and experiments that seek ways out of the growth path of the modern economy. The authors published their ideas at a time when the appeal of sustainable development was weakening (Martinez-Alier et al. 2010: 1745). To support a renewal and revitalisation of the sustainability discourse the authors formulated several conclusions (Box 6.4).

Box 6.4 Renewing the Sustainability Discourse with the Idea of Degrowth
Martinez-Alier et al. (2010: 1745) suggest introducing the idea of degrowth in the sustainability discourse to strengthen the perspective and the coherence of ecological economics for the fomulation of strategies for sustainability transformation:

1. In scientific, joint and practical knowledge production the conditions for achieving the aims of sustainable degrowth and sustainability should be in focus.
2. It is time to reflect on the profile of a future society with much less material consumption, under the different conditions of industrialised and newly industrialising countries, and without deteriorating quality of life in the Global North or the Global South (which requires, for example, reducing the legal working day, delinking revenues from wage employment, and, at the global level, redistributing and sharing resources between countries).
3. It is time to promote the degrowth idea and leave behind the outdated early ideas of sustainable development that oscillate in unclear ways between ideas of growth and limits to growth but have not brought progress to reducing the economic throughput of natural resources.
4. When with the degrowth paradigm the illusion of achieving sustainability through growth is finally shed, it is also necessary to clarify the concept of economic growth, which should not be reduced merely to GDP growth. The concept of growth is vague, polymorphic and ambiguous, and needs to be differentiated with regard to the various forms of growth processes that occur in coupled ecological-economic systems (Martinez-Alier et al. 2010: 1746).

The arguments of the authors can help to develop degrowth scientifically and theoretically, and by improving transformative agency and governance through networking and cooperation between degrowth and other movements, it also reacts to some of the critique of the degrowth concept in scientific and political debates.

Source: Martinez-Alier et al. 2010

The ideas discussed in the degrowth discourse connect with ideas popular in the environmental discourse and movements of the second half of the twentieth century; for example, the idea of small-scale local production ("small is beautiful": Schumacher) or the various ideas that emerged in the global ecological discourse that was initiated with the first global Conference on the Human Environment, hosted by the UN in Stockholm in 1972, and the publication of the "limits to growth" report by the Club of Rome and the "A blueprint for survival" report (Goldsmith and Allen 1972). Ideas of local, decentralised economic systems are supported by anthropological and ecological research. Normative ideas of fairness, justice and equality are influential in the new social and environmental movements. The pooling of ideas is a step towards constructing new differentiated strategies for sustainability transformation at the local, national and global levels. Pooling is necessary in both scientific thinking and governance processes because the clear message of global limits to growth from ecological research is difficult to translate in concrete forms of a non-growing economy. Little can be learned from the non-growing society before the industrial revolution, when societies did not have to struggle with global problems and limits. Ideas about green or sustainable growth are not generating the ecologically necessary reduction of material and energy use, nor are they creating more equality and solidarity. Rather, the green economy is more a part of "green capitalism" (Candejas in Brie 2014: 308ff) than it is of a transformation to a post-industrial and post-capitalist society. It seems unlikely that green growth can be achieved in the global economy (Stoknes and Rockström 2018; see Box 6.5).

Box 6.5 Green Growth and the Planetary Boundaries of Resource Use

Stoknes and Rockström (2018) discussed green growth with this question, reacting to the critique that green growth is not contributing to reducing natural resource use or observing ecological limits. The definitions of green growth by the UN, the Organisation for Economic Co-operation and Development (OECD) and the European Union (EU) are not clear nor operational: they have no science-based targets and are not related to global limits. With these concepts one can only compare how countries are faring in reducing emissions; one cannot assess whether and how far this is sufficient to achieve an economy within planetary boundaries. By specifying growth concepts Stoknes and Rockström differentiate between green growth in the inexact sense of these definitions, grey growth as inefficient or "greenwashing" and genuine green growth as staying within planetary boundaries. Genuine green growth needs to be measured in relation to science-based targets for environmental sustainability and it needs to specify whether zero emission goals can be achieved or whether similar effects like low or zero growth can be achieved. The authors argue that to measure the achievement of the targets,

(*continued*)

Box 6.5 (continued)

the measured economic growth needs to be connected with material and energy flow accounts that connect economic growth to planetary boundaries.

In their empirical study the authors need to limit the validity of their results more and more, so much so that in the end no conclusion on whether genuine green growth can be achieved globally is possible. The manifold difficulties begin with the inexact concepts and indicators presently available: planetary boundaries are not yet exactly defined and measuring all of them becomes methodologically difficult because of the complexity of the interacting economic and ecological processes. The authors do not discuss critically GDP as an indicator but accept it with pragmatic arguments: it is the established indicator to measure economic growth with which one must live. To make genuine green growth empirically measurable they select a limited number of countries—the Scandinavian countries (an exceptional case with regard to their position in the global economy)—and introduce specific definitions, concepts and measurements, limiting the ecological indicators for planetary boundaries to carbon emissions and the climate boundary (with the reasoning that it is already transgressed and thus this boundary change is already necessary). The results, that the Nordic countries have achieved genuine green growth at the country level, over longer periods, with different industrial structures, is devalued by the fact that the largest emitting countries and country groups—the United States, China, India, EU member states and the OECD countries—have not achieved the necessary levels of reduction. The authors see genuine green growth as highly unlikely for large and culturally diversified countries (Stoknes and Rockström 2018: 47).

Sources: mentioned in the text

The analyses of Stoknes and Rockström (2018) and Sandberg et al. (2019) are complementary. They bring the debate about the future path to sustainability closer to solutions by broadening and opening up the growth discourse to encompass ideas of a post-industrial and post-growth society where the economy has various institutions forms, including non-market forms (see Box 6.6).

Box 6.6 Comparing Green Growth and Degrowth

Sandberg et al. 2019 broaden the analysis of green growth and degrowth through the components of ethical reasoning and theoretical framing.

Green growth as a form of sustainable growth has become a dominant paradigm in the economic and political discourses of sustainable development, supported by such powerful global players as the UN, the OECD and the EU, but it receives increasing criticism for being powerless to prevent continued environmental degradation (Sandberg et al. 2019: 137, with further references). In this situation the comparison with degrowth can demonstrate alternatives. Degrowth is presently a marginal discourse, more of a critical movement, less influential in politics and in academic science—in spite of the transformation into an academic paradigm observed by Weiss and Cattaneo (2017). To compare ideas and strategies for growth and degrowth, the concepts need to be connected with further knowledge and arguments—which theoretical reasoning, which normative ideas, which empirical research can support them? Sandberg et al. argue, following DesJardins (2013), that scientific analysis and research alone cannot determine how environmental problems should be solved; the ethical and the normative assumptions that guide concepts and ideas for the solution need to be reflected upon and assessed. Furthermore, a theoretical framework is required that can connect concepts and broaden the knowledge basis for their application. As a framework for the analysis of environmental sustainability the authors describe four cognitive elements: explanation, diagnostics, anticipation and utopian vision (Sandberg et al. 2019: 134). The authors argue that the utopian or normative element needs to be connected with the explanatory, diagnostic and anticipatory elements of a critical theory to be anchored in the social reality and supported by empirical research.

In spite of the refined analysis and empirical data, the analysis of Stoknes and Rockström (2018), taking into account the complexity of the economic systems and relations between the countries in the Global North and South, does not provide arguments for the possibility of a successful transition to global sustainability on the basis of genuine green growth. Sustainable growth in the Scandinavian countries is only verified with regard to the climate boundary. The possibility of achieving genuine green growth in all countries, not yet taking into account all other planetary boundaries, is unrealistic, and, according to the estimation of the authors, not achievable. The results are weakened further insofar as the authors ignore, as is often the case in economic discourse, the importance of political agency in achieving

(continued)

Box 6.6 (continued)

environmental sustainability (a point highlighted by Sandberg et al. 2019). Their discussion is limited to scientifically based targets, available indicators and statistical data. The arguments for communication-friendly terms in science communication (based on psychological knowledge) do not help to understand the complexity of systems analysed and the transformation processes towards sustainbility. A positive terminology alone does not help to change the destructive processes of growth and global environmental change. Avoiding negatively perceived terms and dystopian connotations is of limited use when facing the practical difficulties of achieving sustainable development represented by the lack of political or economic institutional will or engaged action by citizen and consumers.

Sources: mentioned in the text

The findings of the analyses of green growth above are not contradicted analyses of biophysical growth. The work of Krausmann et al. (2018) identifies a key difference in its presentation of a more complete picture of the use of biophysical resources in the global economy for the period from 1900 to 2015, with a scenario showing potential growth until 2050. In this big picture of global resource flows the moderate reduction in recent years in some small countries that employ the strategy of genuine green growth shows limited effects; these small positive trends are overrun by the simultaneous acceleration of biophysical growth that is linked with the industrialisation of some countries in the Global South. In fifteen years this process has accelerated the growth of material and energy flows from ecosystems to the global economy and achieved the same volume of global flow as did the entirety of the twentieth century. It is possible that strategies of genuine growth can become part of the first phase of a transformation of the economic growth mechanism that requires in the long term a non-growing global economy that is ecologically sustainable. Despite the gains made in the exceptional case of the Scandinavian frontrunner countries, the historical dis-simultaneity of the transition to an industrial society in some countries and phasing out of industrial society in others, and the slow changes that come with green growth, means that, even if other countries follow the Nordic countries (which Stocknes and Rockström do not expect to happen), genuine growth will be overwhelmed by the scale of accelerated biophysical growth that is in full speed in the global economy, which will not be slowed down by the transition to a new bioeconomy and post-industrial or information society in the Global North. In the final analysis, it is necessary to show the widening gap between the ideas and effects of green growth and the accelerating exponential growth in the global economy since the beginning of the twenty-first century (Box 6.7).

Box 6.7 Economic Growth and Biophysical Growth of Resource Use
Examples of systematic analyses of the socioeconomic metabolism of the global economy (material and energy flows over longer periods in the modern and earlier human societies):

Schaffartzik et al. (2014) and Krausmann et al. (2018) offer another picture to the selective results of Stoknes and Rockström. For the period from 1900 to 2015 the data of Krausmann et al. (2018) show that the moderate progress towards a reduction in emissions through genuine green growth in a few small OECD countries happens amid an acceleration since the beginning of the twenty-first century of biophysical growth in terms of material and energy flows.

The global societal metabolism shows a multiplication of natural resource use, a massive extraction of material resources from the earth system and vast outflows of emissions and waste, all accelerating with the beginning of industrialisation in the Global South. This process, if it continues at the current rate, means that extraction and waste may double again before 2050, which is beyond that which can be assumed to be environmentally sustainable (though exact and agreed limits and criteria are not yet available) and which does not comply with the necessary phasing out of fossil fuels to mitigate climate change. The incremental changes and limited efficiency gains that sustainability policies have achieved up to now will not be sufficient to cope with the consequences of this new biophysical growth (Krausmann et al. 2018: 139).

The conclusions of the authors are compatible with that of Lenton et al. (2016: 353): achieving global environmental sustainability requires "scaling up new renewable and decarbonized energy technologies and the development of much more efficient material recycling systems"; it requires, furthermore, social changes in forms of consumption and ways of life that prevent deterioration of the quality of life in affluent countries while creating improvements in quality of life, without the present economic growth, in the Global South.

Sources: mentioned in the text

Beyond the scientific discussion and critique of growth and degrowth, there have been discussions of the practical aspects and problems facing the degrowth discourse and movement; for example, racial and ethnic discrimination (Gilmore 2013). A critical reconstruction of degrowth on the basis of a critical theoretical analysis of the modern economic world system has not yet been advanced. The degrowth idea develops from comparisons and confrontations with other economic ideas to reduce the throughput of natural resources in economic processes. Degrowth cannot remain solely a critique of growth and its consequences: it requires more systematic theoretical reflection and analysis and embedding in broader strategies of sustainability transformation. Such critique does not come mainly from orthodox economics but from authors of critical theory (Exner et al. 2008; Schwartzman 2012; Exner 2014). Exner (2014) summarises the main critical points: the degrowth idea lacks theoretical interpretation and elaboration; it remains a collection of diffuse, fragmented and heterogeneous, incomplete and contradicting, concepts; it criticises only symptoms, not the inherent destructive structures of the modern economy. Strengthening degrowth strategies requires analysis and critique of further elements of the economic system, which are currently taken up in the degrowth discourse only in partial and selective forms: in heterogeneous ideas and approaches to local economy, solidarity economy, demonetarisation and decommodification of the economy, critique of interest and credit. A critical question is: how should other components and institutions of the modern economy beyond market and growth be understood and addressed in degrowth strategies, especially capital, which, in the modern economy, is inseparable from money, or the state, or the institutions that were analysed in historical comparative studies of economics by Polanyi and in economic anthropology. Exchange, distribution, reciprocity and subsistence, the commons and other forms of property rights, all discussed in previous chapters, need to be taken up in connection with the degrowth discourse.

Schwartzman (2012) shares much of the critique of degrowth with other authors from critical theory with regard to the necessity of a supporting theoretical analysis and programme of political action at all levels, form the local to the global. He criticises the degrowth movement insofar as it focuses on the quantitative dimensions of growth and excessive resource use, but neglects qualitative dimensions, including specific social and ecological aspects of growth, quality of life and work (Schwartzman 2012: 119). He sees a lack of analysis and critique of corporate-driven globalisation and of action at global level in the degrowth movement, with its focus on localisation of the economy, especially of Latouche's ideas of exiting the economy instead of building a new global economy. The discussion about exiting is, however, another point of the degrowth debate that causes misunderstanding and polemical responses. Exiting can be interpreted in different forms that should not be lumped together:

1. Exit in the physical sense to which Schwartzman refers (creating local food cooperatives or local exchange and trading systems, emigrating into isolated ecovillages, other forms of isolated communities) is only one component of degrowth.

2. Exit in the form of changing economic minds and worldviews, which Latouche and Martinez-Alier refer to as "decolonizing minds from economism", is a condition of downsizing or reducing the throughput of energy and material in the economy and of changing ways of life (Martinez-Alier et al. 2010: 1744). How deeply rooted the orientation to growth in economics and the "growthmania" in the modern economy are can already be seen in the terminology prevailing in economics and policy discourses: in economic discussions of growth rates from the perspective of a growth-based economy, degrowth and reducing resource uses are not seen as processes towards another economy without growth, but rather as "negative growth".
3. Exit in the form of leaving the modern economy by way of transforming it to a post-industrial economy is the broader discussion in the sustainability discourse and process. Degrowth brings new ideas to this discussion, but it requires further clarification and theoretical framing.

The controversy about exiting could be clarified further with the distinction between "stepping out of economics" and "stepping out of economism" as a mindset (Martinez-Alier et al. 2010: 1744). The missing global action components that Schwartzman (2012) criticises in the "degrowth as the recreation of local economies" argument are discussed by other degrowth authors, especially Martinez-Alier, who unfolds a multi-scale action strategy for the Global North and South, supported by research on environmental, technological, demographic and social aspects of degrowth, including ecological distribution conflicts. Schwartzman highlights the necessity of an eco-socialist movement to develop degrowth strategies and sees climate change as a huge opportunity to change the global economy, especially through transforming the energy system and using solar energy.

Whereas Schwartzman's critique focuses on collective action and multi-scale agency to change the global economy, Exner (2014) directs the discussion to new economic institutions that should be developed further to strengthen the degrowth movement. The degrowth discourse has identified systemic risks that can cause the failure of modern capitalism, but it has not formulated how the system can be transformed and what elements could be retained from which to build a new economic system of the future. Solidarity economy (Poirier 2013) and the commons (discussed in social and human ecology: Ostrom 2005) are two of these potential elements as forms of economic production, distribution and consumption that can replace market and money to some degree, and can be developed further based on their principles of reciprocity and cooperation. Ultimately the transition to sustainability does not require a new utopia but the long project of building a new economic system in which the unleashed forces of the market in the globalised economy and the attendant capital-based excesses of growth, profit and speculation are controlled and "domesticated" once again, and with the market also the capital-based excesses of growth, profit, speculation (Exner 2014: 23f).

Beyond concepts—the trends of global growth: as Jackson (2019) summarises, growth is low in advanced economies and there is no prospect that it will rise again in the long run. Low growth is the new normality, and the outlook that the OECD provides for the coming decades. Assuming the exponential growth processes in the global economy come to an end because of the multiples crises in the global economy and growing resource scarcity, would this mean a sustainable situation? Rather it may become a long stagnation, the economic and ecological crises becomes permanent, without solutions in terms of social, economic and environmental sustainability, a long decline which could be managed only with increasing inequality, new poverty for large parts of the global population and increasing conflicts. What would still be missing is a transformation to a new economy that can deal with limits to growth by introducing new forms of development: sharing, redistribution of resources, reducing poverty and inequality. To realise a post-growth society requires empirical knowledge and practical working examples of transforming the modern economy, in which ideas from past and present debates about the non-market based exchange of goods and services, local economies and the commons can be developed.

The debate about whether degrowth is a good term for public and policy discourse tends to come to a similar dead end to the discussion about the many views and definitions of sustainable development. Rather, it is possible to use a plurality of terms that indicate the search for different, co-existing forms of development under the unifying idea of sustainability. The alternatives are not simply growth or no growth, but transforming growth so that its beneficial effects can be achieved in other forms of development. The transition to sustainability will be difficult but necessary to avoid the catastrophic collapses discussed in the "limits to growth" discourse.

6.3 Requirements of Social-Ecological Transformation Towards Global Sustainability

Social-ecological transformation to achieve sustainability is not yet accomplished as a concept and process. And it cannot be accomplished by following the development of ecological economics and its history of ideas. Costanza et al. (2010) discuss a large number of historical sources of ecological economics, but the sustainability discourse and process requires more non-economic knowledge and greater interdisciplinary knowledge integration. Improvements in policy, governance and transformation practices require two complementary steps of knowledge work: (a) the translation of economic knowledge in knowledge practices and regimes for sustainability transformation, applied in concrete projects, practices and experiments of economic transformation; (b) in these processes of formulating interdisciplinary and transdisciplinary knowledge practices, economic knowledge needs to be connected with other knowledge through interdisciplinary and transdisciplinary knowledge integration. The adoption of non-economic knowledge includes knowledge of the following components of sustainability transformation.

Knowledge about the non-sustainability of the global economy: the slow progress and challenges of transformation towards sustainability has so far been created through the development of economic, social and ecological indicators to measure improvements or deteriorations in welfare, well-being and the state of the environment. Integrating the available knowledge from such indicators in global assessment projects since the Millennium Ecosystem Assessment, the overall picture remains that of the UN sustainability reports until today: the necessary changes for global sustainability are not yet underway. The ways to change this situation cannot be found through new indicators alone, although indicators are necessary to measure the progress of the transformation.

The sustainability transformation includes political and power struggles, conflicts and complicated forms of building transformative agency and collective action. The concepts of environmental justice, reintegration of ecology and economics, adaptive management (Holling) and co-evolution of ecological and economic systems (Norgaard), such as they are discussed in ecological economics (Costanza et al. 2010), do not yet show the complicated nature of the transformative changes required. Regarding the creation of social agency, social subjects and transformative action groups for social-ecological transformation, it is necessary (a) to broaden the transformation debate, such as happens in transdisciplinary transformation research, and (b) to deal with the vested interests (Moe 2015) of actors who block the transformation process.

The critical reflections and assessments of economic growth, valuation and valorisation of nature translates in two variants of sustainable development: (a) relying on growth, technological change, innovative technologies and geoengineering as transformative strategies to combat climate change and simultaneously accepting forms of industrial growth in the newly industrialising countries of the Global South, with the use of fossil fuels and greenhouse gas emissions; (b) seeking alternative development paths towards a steady-state economy with low growth or degrowth in affluent societies and simultaneously developing economies of sharing and redistribution that allow countries of the Global South to escape the trap of polluting industrialisation, thus improving the welfare and well-being of the populations (Martinez-Alier and Muradian 2015: 10f). The first strategy is the riskier due to unforeseeable consequences for the environment and the natural resource base. To dissolve the dilemma of manoeuvring between two irreconcilable strategies, with insufficient knowledge about both of them, a more systematic reflection of the institutional components of growth-based and non-growth economies is necessary, for which Victor and Dolter (2017) provide some ideas.

The handbook on growth and sustainability edited by Victor and Dolter (2017) shows the present state of the growth discourse and its relevance to the sustainability discourse and strategies for sustainability transition. It describes the history of the growth debate, the reasoning for and against economic growth with regard to sustainability, the refining of the debate and how it can continue. Knowledge from economics and other social-scientific and natural-scientific disciplines is reviewed for this purpose, achieving some clarification of the dilemma of sustainable development. Whereas this debate between economists tends to come to a dead end, its opening up through interdisciplinary knowledge exchange and integration has brought ideas forth that dissolve the dilemma. The authors do not offer a simple conclusion for or against growth. However, the discussion between proponents and opponents of growth shows how the arguments for and against growth become clearer, and the critique of growth is becoming more convincing for creating strategies and agency for sustainability transformation. In strategies of transition to sustainability it is necessary to provide opportunities to experiment with both variants of sustainable development through growth and through transformation. The relevance of the progress in the growth debate to the sustainability discourse and process can be summarised as follows (Victor and Dolter 2017: 1ff):

1. The Brundtland Report of 1987 and its view of sustainable development is no longer describing the present situation; it was only the beginning of a debate that developed through the controversy about economic growth. The report itself remained unclear about the possibilities to sustain economic growth in the long run; its reasoning for growth is based on the policy not the scientific context. The report argues for growth in other forms and primarily for the countries in the Global South. However, it continued the controversy and scepticism about unlimited growth that had prevailed since the limits to growth report from 1972.
2. Alongside the introduction of sustainability policies in the 1990s two broad schools of thought developed. Proponents of pro-growth schools adapted ideas from orthodox economics, environmental economics and ecological modernisation theory and combined them with arguments about the environmental Kuznets curve, green growth and decoupling growth from resource inputs, pol-

lution and waste. These ideas strengthened the case for mainstreaming sustainable growth in political and economic policy. Proponents of the no-growth schools developed ideas from ecological economics and other variants of critical economics such as institutional and feminist economics, human, social and political ecology, ecological anthropology and social and environmental movements into a more interdisciplinary discourse.

3. The arguments of economists for growth are more social-scientific than ecological, referring to prosperity, technology, "the monetary growth imperative" and the practical impossibility of transitioning the modern economy to a non-growing economy. This is one of the biases of the growth debate in orthodox economics, where environmental effects are reduced to social ones and calculated in terms of costs and benefits. Also, the arguments for sustainable growth are mainly based on the core points of orthodox economics, technological progress and substitution of natural with human-made capital, but the reasoning became more refined. For the arguments with the environmental Kuznets curve—that growth increases pollution temporarily and brings it down in developed stages of the economy—it was shown that they are valid under specific conditions where environmental flows have local impacts, but not with regard to the reduction of CO_2-emissions that becomes more and more difficult in a growing economy. Green growth is working under specific conditions such as in the Scandinavian countries (discussed above), but not globally.
4. The arguments against sustainable growth developed from the debate about social and ecological limits to growth. Social limits refer to the negative social consequences and unequal distribution of costs and benefits through economic growth—including the conditions downplayed by growth protagonists: concentration of economic power and capital, oligopolistic and monopolistic markets, extreme forms of social and economic inequality and development of an uncontrolled and crisis-prone international financial economy. Ecological limits refer to population growth, growth of natural resource use (throughput of natural resources in economic growth) and exceeding the carrying capacity of ecosystems and planetary boundaries. One of these boundaries, not discussed by Rockström et al. (2009), is land (Haberl and Erb in: Victor and Dolter 2017: 277ff). Several authors, especially Jackson and Victor, showed that growth is not necessary for stability and well-being in developed economies. Also, the arguments that the costs and negative effects of growth outweigh the benefits (Daly, research on defensive expenditures) become more convincing through the statistical data available.
5. The arguments for and against growth no longer develop through exclusively economic or expert-based reasoning; they have become part of an interdisci-

plinary discourse in which the social and ecological conditions and contexts of growth are analysed more systematically. The different normative premises, assumptions, worldviews and paradigms that support the arguments have become clearer. For those advancing pro-growth arguments it can be shown that they are connected with a worldview that has been described under different names as a dominant Western worldview, the human exceptionalism paradigm or the "unlimited self" that describes the dominant "social character mask" of modernity.

6. The arguments for and against growth do not totally exclude each other; some are similar but formulated under different assumptions and conditions. The ecological modernisation reasoning, for example, is connected with the argument that the dominant economic rationality in production and consumption needs to be connected with a new ecological rationality that takes into account the scarcity of resources and negative environmental effects (similar to green growth with the formula "growth without environmental damages"). With such reasoning growth and non-growth arguments come closer together in the aim to observe ecological limits and to avoid environmental disruption. The conditions under which growth and sustainability go together (at which level of the economy, in which countries, under which social conditions) can be specified in further research and debate. Reasoning in the orthodox economic tradition, such as that by Samuelson and Nordhaus (2005: 563), that the limits to growth debate is that of a pessimistic "ominous-sounding group" which will be silenced by new environment-friendly capital and technologies replacing resource-intensive and polluting technologies, loses its persuasiveness. The contradictory conditions of growth and their change become gradually visible, and the formerly positive feedback loop between population growth, economic growth and welfare is weakening (Fischer-Kowalski and Steinberger in: Victor and Dolter 2017: 372 ff).
7. To move beyond economic growth without reducing welfare and prosperity is, in the last analysis, not an argument for sustaining the present growth-based economic system. Cultural, social and ecological transformations have become a challenge in modern society to move beyond "growth culture" and "growth addiction". Arguments in the broader contexts of growth can be double-edged: under certain premises they can be for growth, under others against growth. Arguments for possible cultural change towards non-growth, especially changing lifestyles, face counterarguments, for example in favour of human creativity and innovation as reasons for sustained growth. To achieve prosperity without growth Jackson (in: Victor and Dolter 2017: 522ff) argues that certain conditions for shifting the economy need to be specified: the nature of enterprise, the value of work, the structure of investment, the role of money. Changing these institutional practices requires collective learning along the road towards sus-

tainability. The final conclusion of Victor and Dolter (2017: 12) is that the question of whether the burdens which the global economy places on the biosphere can be reduced without reducing growth is still open, but it becomes more doubtful that growth can be maintained.

Starting from this review of the debate about growth and sustainability the further discussion of a transformation towards sustainability can be imagined as a long, complicated and conflict-full process in which economic growth, in the rich countries of the Global North and later in the poor countries of the Global South, will gradually come to an end. The OECD expects continued low growth throughout this century. This slow-down will be part of the painful and conflicting processes to deal with in the further sustainability discourse, when the prior attempts to stimulate new growth, for example by sustaining the fossil energy regimes, will become less and less effective and possible.

The hope of Mill (2004: 692), that growth is replaced by better distribution to reduce poverty and people "will be content to be stationary, long before necessity compels them to it", does not seem to have come true. The time has already come where it is necessary to deal with growing and irreversible environmental destruction. A realistic estimation and probability in terms of scenarios for a transformation to global sustainability in the twenty-first century is that this process will last longer and will not be achieved through limited processes of "greening" in certain countries nor through the present national and global climate and sustainability policies (with unclear targets for which the process of achieving them cannot be measured), in spite of technological change and some progress being made in rebuilding energy systems. More than the positive changes that have been achieved through policies will negative environmental changes, disasters and catastrophes conflicts due to unequal economic development define the course towards global sustainability. However, this course of conflict and catastrophe is not a dystopian view of the distant future; it describes economic and environmental disasters that are already a reality: climate change, biodiversity loss, desertification, global economic crises. Conflicts and environmental disasters are gradually becoming stronger and are being discussed as the new reality of sustainable development, as the rhetoric of the early policies of "win-win situations" fades away. According to Martinez-Alier et al. (2010: 1745) "intertwined social, economic and environmental crises such as the ones in which we are immersed, in contrast to the idea of 'soft landing' and convergence dominant in the mindset of sustainable development advocates" will become the reality of sustainable development. The discussion about ecological distribution conflicts has already been with us for some time (Martinez-Alier 1995; Scheidel et al. 2018). Such conflicts develop in the globalised economy, with the competition for scarce natural resources, in the transition to sustainability; they have a similar importance and function to the economic distribution conflicts in the form of class conflicts in earlier phases of modern industrial society. In a broader view Newig et al. (2019) discuss these themes too. *Failures of markets and policies and postponed social-ecological transformation will do their own to make the achievement of global sustainability a more conflicting and catastrophic process*. This is not meant to belittle the efforts to achieve sustainable development, which are necessary at all levels from the local to the global; nor is it meant to repeat the doomsday prophecies of the earlier environmental discourse or nourish environmental scepticism: it simply gives a more realistic picture of what to expect in the transformation process. The global transformation to sustainability requires dealing with different dilemmas that can be described as economic and political lock-in situations (Table 6.1).

Table 6.1 "Locked into growth": dilemmas of economic transformation

Name	Dilemma	Potential solutions	Sources of the debate
"The dilemma of growth"	The growth-based capitalist economy has no alternative to "grow or collapse". In this system: (a) economic growth is unsustainable, undermines ecological stability and creates disparities in social well-being; (b) degrowth is unstable, reducing consumer demands and causing unemployment and recession	To change economic structures and the concomitant social logic of consumerism requires either making growth sustainable or making degrowth stable through strategies such as reducing working hours and changing consumer cultures (creating alternatives to material consumption and status competition). The three conditions are: (a) ecological: establishing limits of resource use, (b) economic: changing the structure of the economic system, (c) social: changing the culture of consumerism (imperial lifestyles)	Jackson 2009; D'Alisa et al. 2015; Jackson 2019

(*continued*)

Table 6.1 (continued)

Name	Dilemma	Potential solutions	Sources of the debate
"Lock-in" situations	The dilemma of growth described by Jackson (2009) is a form of structural lock-in: in the present economy systems there are only limited possibilities of change and development	Systemic lock-ins are not of the kind of simple lock-in situations discussed in economics (customer dependence on sellers, products or services (customer lock-in, proprietor lock-in) or technology lock-in; the systemic lock-in of the growth-based market economy becomes manifest through the natural and social limits to growth, connected with path dependency	Erickson et al. 2015; McLellan et al. 2016
Path dependency and path innovation: complex systems and global change	The present and with that the future depend on the past: in path-dependent processes development happens as a product of the process history; small, contingent events can create self-reinforcement and processes that "lock-in" particular structures and pathways of development; innovations and ruptures of development paths are connected with crises and conflicts	The transformation to global sustainability is a path-breaking process where the development path of the global economy and world system is transformed; the complexity of the processes and the changes with sustainable development makes processes of path-breaking and innovation conflicting, unplannable and contingent; they become part of the combined processes of sustainability transformation and transformation of global change processes	Spangenberg 2005; NSF 2005; Ehlers and Krafft 2006

(*continued*)

Table 6.1 (continued)

Name	Dilemma	Potential solutions	Sources of the debate
Vested interests	Vested interests of persons, social groups or organisations: strong interests to maintain a given social, political or economic order because of the benefits gaining from it, although it is detrimental for the long-term development of social, economic or ecological systems	Vested interests are among the main difficulties to deal with in sustainable development, creating conflicts and requiring strategies of conflict mitigation	Moe 2010; Brette 2017
Discounting the future	Discounting the future (or temporal discounting) implies the preference for spending or consuming presently, not in future; it is based on the subjective assessment and comparing of present and future benefits achieved from spending	Discounting the future is based on the assumption that economic growth continues because of investments and innovations made today, which implies that future generations will be richer than the present. Discount rates do not show objective measurements, but subjectively chosen assumptions in economic or policy decisions	Groom et al. 2005; Atkinson et al. 2007

Sources: mentioned in the table

The progress in the discourse of ecological economics by way of discussing limits to growth and possible forms of degrowth can be summarised with regard to the interdisciplinary knowledge requirements in the sustainability process as progress from simple visions of sustainability, as they dominate still in the policy and governance practices and in parts of the scientific discourse, to more inclusive visions of a sustainable society and economy. These do not end with some vague normative goals but can be translated into strategies for sustainability transformation. In exemplary form such visions have been formulated by Jackson (2009) and Klitgaard (2013; see Appendix, learning exercises). In the development of global scenarios further improvements of sustainability strategies can be expected.

6.4 Discussion and Conclusions

In spite of their critique of orthodox economics ecological economists argue for methodologically pluralistic reasoning and research, also accepting neoclassical economics as a potentially meaningful approach from which only its misleading assumptions need be corrected (Costanza et al. 2010). Ecological economists criticise, in consensus with other critical economists, the consequences of the extreme social and economic inequality in the global economy and the functional degradation of ecosystems. The forms of inequality are more preventing than supporting the global redistribution and intragenerational and intergenerational solidarity in natural resource use. The changes in the sustainability discourse since the 1970s, when the global growth-critical debate began, showed some progress in terms of knowledge for improving sustainable development:

1. *The "limits to growth" discourse* since the 1970s has brought some concretisation of the concept of a steady-state economy and of ecological limits to growth. Ophuls (1977) discussed intensively the idea of the steady-state society and the concept of ecological scarcity to see how they can be translated in a political theory and policy concept for the transformation of the modern economy. But not much more in terms of concrete ideas of a transformation to a steady state is formulated by him than "learning to live with scarcity", except abstract principles, ethical ideas, values and norms for the future society. He speculates about potential forms of a steady state and its political organisation in terms of modesty, diversity and stewardship. Regarding the sources of knowledge for the transformation process Ophuls came back to the wisdom of political philosophy and ultimate values, seeking more in past thinking and philosophy for knowledge sources than creating new knowledge about economic and ecological transformation (Ophuls 1977: 222ff). After Ophuls the relevant discussion became about Daly's concept of the steady-state economy and the degrowth discourse.

2. *Connecting economics with the sustainability discourse*: the problems of inequality and unequal development that require further discussion in the sustainability discourse and process resulted in the broadening of the interdisciplinary knowledge bases of ecological, institutional and feminist economics, with knowledge from human, social and political ecology. Additional themes to develop in the interdisciplinary sustainability discourse include the reflection of the normative and ethical values and premises of economic development and their connection with ecological principles of fairness, solidarity, sharing and redistribution of resources for sustainable development. To understand the interaction of material and symbolic factors in the sustainability process, the physical economy (material and energy flows) needs to be analysed to be able to influence changes in social institutions, worldviews, ethics and value systems.
3. *How will the sustainability process change the discourse in economics?* The paradigm change of rethinking and transforming the modern economy from a growth-based market economy to a non-growing economy, necessary in dealing with the limits to growth discussed in ecology, is one of the most difficult in the history of the modern economy and economic science, where economic knowledge comes to its limits. *The progress achieved with the discourse of ecological economics is not solely the result of research in ecological economics but of the joint efforts of the critical interdisciplinary coalition of researchers and other actors developing with the sustainability process. The epistemic community of interdisciplinary sustainability researchers made possible a breakthrough in the deadlocked economic growth debate that was rotating around the narrow question of whether growth is necessary or not in the modern economy: opening up the debate about the future of modern society that is no longer a growth-based industrial and capitalist society.*

The knowledge about social and ecological limits to growth is growing. The growth paradigm, still the mainstream in the academy and in policy, can come to an end with the next phase of the sustainability process, where it needs to be shown that the future global economy is not possible with primitive industrial growth and new ideas of green growth. The concept of social-ecological transformation opens up possibilities to direct the discussion away from constrictions of the growth concept to a broader view of development and change where growth can gradually dissolve in more encompassing processes. When global limits of growth need to be observed in sustainable development, growth is on the way to ending and alternatives of economic development in a new economy can be sought. The present knowledge problem in the sustainability discourse is about this new economy. One knows that it is not an industrial or capitalist economy but one does not know scientifically and practically how to transform the global economy into one of sustainable growth, or one of sustainable degrowth and a post-growth society. The knowledge required develops only in the future, through new research and collective learning as the transition unfolds.

The concept of a non-growing economy, discussed since nineteenth-century classical political economy, is that of a steady state. This concept serves as a starting point for further thinking and research about a future sustainable economy. What can be learned from the discussion of a stationary or steady-state economy was already learned once in classical political economy: the future of the modern economy is not the continuation of the industrial society forever, as some evolutionary theorists in early sociology, especially Spencer, assumed; economic growth cannot continue forever with the limited resource base on the earth; after the industrial epoch of growth will come one where economic growth has ended. It is an open question whether the end of growth happens in the form of social and ecological catastrophe or in the forms of joint learning, controlled development through global governance and sustainability transformation, and with new ideas of a non-growing economy that is both prosperous and advances human well-being.

The controversy about economic growth will not be solved in the near future, when the policy option for green growth in the name of sustainability is realised with the support of influential global players in economy and politics. The epistemic community of ecological economists, ecologists and environmental movements arguing for a transformation towards a non-growing economy that observes the natural and social limits to growth is still in a weaker position in terms of influence and power in science and politics, even as doubts about continued growth grow and spread in economics and other discourses. Yet the growth-critical epistemic community develops ideas and arguments for an environmentally and socially sustainable economy. In the near future the sustainability process requires experimenting for some time with different, even contradictory, ideas and strategies, that of growth and degrowth, and with different forms of research, in interdisciplinary and transdisciplinary knowledge production. The social-ecological transformation towards sustainability begins without knowledge of how to build the new society; this knowledge needs to be found and learned successively on the long road towards sustainability.

Summary of Chap. 6: Ecological Economics: Critical Perspectives
The ideas, concepts and models from ecological economics relevant for the interdisciplinary sustainability discourse and process include: ecological indicators and knowledge about ecosystems used together with knowledge and principles for economics; the critique of economic growth (which implies also a critique of the idea of substitution of natural capital from neoclassical economics); and ideas to build a future economy that qualifies as post-industrial, post-capitalist and post-growth, with a diversified set of partially competitive, partially complementary, institutions (including elements of market, sharing, solidarity and redistributive economy).

The transformation towards the future sustainable economy requires manifold activities and cannot develop only from disciplinary research. Interdisciplinary and transdisciplinary research is required and in-depth research on the systemic functions and components of social and ecological systems, their interaction and coupling. Furthermore, joint learning of the actors and decision-makers, integrating of economic and ecological knowledge in the processes of sustainability governance, building of transformation action groups and development of long-term strategies for the sustainability transformation are required in the future sustainability process.

Appendices: Further Information and Material

1. Questions and Individual Exercises

Learning Exercise 1: Development of an Interdisciplinary Economic Theory for Sustainability Transformation

Discuss in a group the following draft of an idea for an interdisciplinary economic theory of sustainability (Klitgaard 2013: 278ff) and answer the following questions: can the sustainability discourse solve the many problems formulated in the sustainable development goals of the UN with the help of a complex theory, as Klitgaard suggests? What else is required to achieve a sustainable economy and society?

Klitgaard summarises the requirements of sustainability transformation as follows:

1. To deal with "the real world in which we live" the economy cannot be treated solely as an abstract and isolated system of exchange, but as a system of production and exchange embedded in both a finite and non-growing biophysical system and in a social and institutional context. The theory to develop must recognise the prevalence and persistence of monopoly or concentrated industry in the economy. It cannot rely on "perfect competitive" markets that result in efficient allocation and equity as outcomes. A sustainability model cannot treat the planet's ability to support life as an externality that can be internalised by somehow finding a price for nature's assets and services.
2. The "evolution of societies must be placed in a historical context". Human interaction with nature has changed significantly in the transition from hunters and gatherers to agriculture to industrial producers. To understand present changes, the theory to develop needs to understand prior transitions. In the long course of human history before industrialisation society had three main characteristics: (a) dependence on the solar flow as an energy source; (b) most production was for direct use, with only a small economic surplus; (c) basic needs were satisfied by production for use and reciprocity, market activity was sporadic and the global population small.
3. With the industrial era the situation changed quickly through the use of fossil fuels. To achieve sustainability, energy, population and the nature of exchange must be addressed. Sustainability cannot be achieved by "fixing" one of these elements; innovation, technological change, increases in efficiency and relative decoupling of GDP from material production, or smart growth, cannot be expected to bring without further changes the requisite degrowth to live well within nature's limits and reduce the global overshoot. The institutional context of resource use becomes part of larger changes in production and consumption. Mainstream economics does not differentiate between exchange value and use value as in the era of classical political economy. The distinction needs to be introduced again to understand the historical context, the social and institutional frameworks of the economy and of technological change in the transition to sustainability. The internal forces that create depression and stagnation need to be understood, how biophysical limits will affect the traditional internal limits to growth and how the global economy has changed since the last great depression through globalisation and financialisation, and the pressures through the global economic and ecological changes. Sustainability economics needs to understand the consequences of the rise of global finance for growth and how the limits to growth, such as peak oil and climate change, will affect the financial economy and the real economy.

4. A sustainability theory needs to address many themes: including the quality of work in terms of meaningful work, not only its presence or absence. Meaningful work includes different aspects of qualification and responsibility and should be meaningful for the worker, the community and the society. In a steady-state economy, ever-expanding consumption simply cannot be the means by which people seek fulfilment. Like economic theories before, a theory of sustainability needs to deal with labour productivity to change the dysfunctional connection between increasing productivity and degradation of work in modern capitalism towards a situation where rising productivity through technological change is spreading throughout the economy to achieve a society rich in culture and community, with fewer working hours and more leisure time.
5. A realistic vision of the role of energy in economic activity and growth of productivity (see Hall and Klitgaard 2012): the interactions of energy quality, technological change, the labour process, the global financial economy and the evolution of human society need to be understood better—impending absolute scarcity, the Energy Ret and the consequences of its use, especially for climate change. As with the other parts of the theory, energy analysis requires transdisciplinary efforts to understand biophysical limits, the functions of the economy and the political processes, and the domestic and international political contexts.

To discuss the vision of a sustainable society more systematically see the following learning exercise.

Learning Exercise 2: An Economic Vision of a Sustainable Society and Economy

Discuss in a group the following vision by Klitgaard (2013: 294) *and the arguments for it critically. If you do not accept certain arguments, try to find other or further ones, but do not reduce the vison to a selective picture—try to think about all necessary changes to achieve global sustainability.*

Keep your ideas about the sustainable economy and society in the form favoured by Klitgaard: as a realistic description of the changes required to solve the social, economic and environmental problems existing today. Do not just describe a "good society" in an ideal or utopian form without thinking how it can be achieved.

Main requirements in terms of social sustainability:

1. It must meet the needs of people for income, healthcare, employment and chances to increase their capabilities and provide opportunities for the next generation.
2. Basic social institutions, technologies and human behaviour (lifestyles) need to change to achieve a sustainable balance with nature.

Main requirements in terms of ecological sustainability:

3. To reduce the overshoot of nature's limits it must be a non-growing society with a more equal income distribution.
4. Energy problems will necessitate more local production and distribution—in the end we will be dependent once more upon the solar energy, harvesting it more effectively with modern technologies.

Main requirements of change of the economy—impossibility of sustainable capitalism:

5. Entrepreneurial innovation will not lead to sustainability—it is dependent upon growth as any other form of capitalism. Financial innovation brought not much more than the complex derivative securities that caused the recent financial debacle and a rapid increase in income inequality.
6. If we desire a small-scale local economy, then this economy needs to be one based upon a community of associated producers. While a few small businesses may function perfectly well in a non-growing environment, the whole capitalist system, with large-scale enterprises and global markets, does not work without growth.

Main requirements of theoretical knowledge:

7. An economic theory of the sustainable economy must integrate production, consumption, work, finance, energy and policy, to name but a few characteristics.
8. The theories of heterodox political economy are a starting point for a theory of sustainability—they include economic activity and policy, as well as technology, in institutional and historical contexts. Without historical and institutional analyses, we may become victims to greenwashing and technological quick fixes.

Learning Exercise 3: Transition to a Sustainable Economy

Discuss the following ideas of Jackson (2009: 171ff) *critically—are they coherent, clear, convincing? Correct or complement them with your own ideas.*

Not only a vision and contours of a theory, but a transition scenario to a non-growing society is necessary for the transition to a sustainable economy:

> Consumer society seems hell-bent on disaster; but dismantling the system doesn't look easy either. Overthrowing it completely could drive us even faster along the road to ruin. But incremental changes are unlikely to be enough. Faced with this kind of intractability it's tempting to retrench. To cling more tightly to existing tenets. Or to resort to a kind of fatalism. A place where we accept the inevitability of a changing climate, an unequal world, perhaps even the collapse of society. And concentrate all our efforts on personal security.

Impossibility theorems are omnipresent: economies can only survive if they grow, people do not give up materialism, the state is powerless. But axiomatic truths like these dissolve under more careful scrutiny. A different kind of macroeconomics is possible. People can live well without more stuff. A new vision of governance makes sense.

The economic crisis provides an opportunity to change: sweeping away short-term thinking, replacing it with policymaking capable of addressing the challenges of climate change, delivering a lasting prosperity. There are only two possibilities for change of this order: revolution or to engage in the work of social transformation.

The ideas of revolution imply ending capitalism, rejecting globalisation, undermining corporate power, overthrowing corrupt governments, dismantling the old institutions. There are risks here too: a new barbarism lurks in the wings, a world constrained by resource scarcity, threatened with climate change, struggling for economic stability: how long could we maintain civil society in such a world?

To reject revolution is not to accept the status quo. The scale of the required transformation is massive, but concrete steps through which to build change are needed, and for this governments and those able to make or influence policy are needed. These steps open up a public and policy dialogue on the issues. It is possible already to establish some clear directions of travel: establishing the limits, fixing the economic model, changing the social logic. Clarifying these points will help to find new ideas in the transformation process.

2. Further Reading Suggested: Deepening, Thematically Specialised

Textbooks of ecological economics:

Daly, H., & Farley, J. (2004). *Ecological Economics: Principles and Applications (2nd ed. 2011). Washington, Covelo, London: Island Press.* This classic textbook of ecological economics is oriented to the description of ecological economics in a critical discussion of the main themes of economics as a disciplinary science and its limits and failures. Its chapters include the thematic scope and guiding ideas of ecological economics, ecosystems and resources, microeconomics, macroeconomics, international trade and policy. The authors underline that ecological economics is not an academic discipline, arguing that real problems do not respect academic boundaries (for lack of a better term they call ecological economics a "transdiscipline").

Shmelev, S.E. (2012). *Ecological Economics. Sustainability in Practice. New York et al.: Springer.* This interdisciplinary textbook is for students with some knowledge of economics and environmental science, focusing on sustainability. Its chapters cover a variety of themes relevant to economics (economic valuation and decision-making, market failures and externalities, economic models and input-output analysis) and sustainable development (measuring progress of sustainable development, climate change and renewable energy, biodiversity loss, sustainable cities, regional waste management, business and sustainable development).

Costanza, R., Cumberland, J.H., Goodland, R., Norgaard, R., Kubiszewski, I., & Franco, C. (2015). *An Introduction to Ecological Economics (2nd ed.). Boca Raton, FL: CRC Press.* This textbook from mainly US-American authors of the founding generation of ecological economics is a revised version of the 1997 first edition. Differing from other textbooks mentioned here it keeps a thematic focus on economics and economic knowledge from a historical perspective, with chapters on humanity's current dilemma, the historical development of economics and ecology, the principles and objectives of ecological economics, environmental institutions, instruments and policies. This book shows how knowledge of environmental problems developed from partial and inadequate knowledge to improved knowledge of global problems that need to be dealt with in sustainable development.

Common, M., & Stagl, S. (2016). *Ecological Economics. Cambridge et al.: Cambridge University Press.* This introductory textbook provides a systematic overview of ecological economics from an interdisciplinary perspective, including economic and ecological sciences. It describes the interdependence of the modern economy and its environment, the environmental and sustainability discourse in economics, the national policy processes and instruments, the international institutions and the interaction between economy and sustainability with regard to climate change and biodiversity.

Handbooks of ecological economics:

*Martinez-Alier, J., & Muradian, R. (Eds.). (*2015*). The Handbook of Ecological Economics. Cheltenham, UK and Northampton, MA: Edward Elgar.* This handbook (from mainly European authors) offers a more advanced and differentiated analysis of the state of the discourse in ecological economics than the textbooks described above. It includes chapters on the economic core themes of ecological economics (philosophical sources, value deliberation, macroeconomic policies, degrowth, consumption, economic instruments), chapters on ecological core themes (social metabolism, ecosystem services, traditional ecological knowledge, common pool resources, water) and chapters on policy (global environmental governance, policy instruments). The chapters are exemplary for the development of the transdisciplinary field, but the history and current scholarship in ecological economics is not systematically reviewed.

Spash, C.L. (2018). *Routledge Handbook of Ecological Economics: Nature and Society. Abingdon and New York: Routledge.* This handbook describes the advanced state of the development of ecological economics since its beginnings in the 1980s. The early critique of mainstream economics and the discussion of environmental degradation and limits to growth developed into a rich interdisciplinary subject with a polyphony of critical voices, approaches and theories in a variety of research themes. Two salient perspectives include the broadening of the thematic scope of ecological economics to "social ecological economics", and the transformation to another future economy and society as a guiding perspective for which the development of ideas about a post-growth society is the signpost. The subject develops through interdisciplinary knowledge gathering and integration, with knowledge from social- and natural-scientific disciplines and interdisciplinary approaches, especially heterodox economics and political ecology. This provides a continuously improved understanding of biophysical and social reality and their interrelations.

Victor, P.A., & Dolter, B. (Eds.). (2017). *Handbook on Growth and Sustainability*. Cheltenham, UK and Northampton, MA: Edward Elgar.

Scientific journals (international, peer-reviewed) with research relevant for the themes of ecological economics: "Ecological Economics", "Ecology and Society", "Environmental Sustainability", "Environment, Development, Sustainability", "Feminist Economics", "Global Environmental Change", "International Journal of Ecological Economics and Statistics", "International Journal of Sustainable Development and World Ecology", "International Journal of Environment and

Sustainability", "International Journal of Social Ecology and Sustainable Development", "Journal of Development Economics", "Journal of Environmental Sustainability", "Journal of Cleaner Production", "Journal of Political Ecology", "Resource and Energy Economics", "Society and Natural Resources", "The Journal of Economic Issues" (institutional economics), "The Journal of Institutional and Theoretical Economics", "GAIA", "Natures, Sciences, Sociétés", "Sustainability Science", "Sustainability".

References

Altvater, E. (1992). *Der Preis des Wohlstands oder Umweltplünderung und neue Welt(un) ordnung*. Münster: Westfälisches Dampfboot.

Atkinson, G., Dietz, S., & Neumayer, E. (Eds.). (2007). *Handbook of Sustainable Development*. Cheltenham, UK and Northampton, MA: Edward Elgar.

Banerjee, A. V., & Duflo, E. (2011). *Poor Economics: A Radical Rethinking of the Way to Fight Global Poverty*. New York: PublicAffairs.

Barnett, H. J., & Morse, C. (1963). *Scarcity and Growth. The Economy of Natural Resource Availability*. Baltimore: Johns Hopkins University Press.

Boulding, K. E. (1966). The Economics of the Coming Spaceship Earth. Environmental Quality in a Growing Economy. In H. Jarrett (Ed.), *Essays from the Sixth RFF Forum* (pp. 3–14). Baltimore: Johns Hopkins University Press.

Brette, O. (2017). The Vested Interests and the Evolving Moral Economy of the Common People. *Journal of Economic Issues, 51*(2), 503–510.

Brie, M. (Ed.). (2014). *Futuring: Perspektiven der Transformation im Kapitalismus über ihn hinaus*. Münster: Westfälisches Dampfboot.

Bromley, D. (1991). *Environment and Economy: Property Rights and Public Policy*. Oxford: Blackwell.

Bromley, D. (Ed.). (1992). *Making the Commons Work: Theory, Practice, and Policy*. San Francisco: ICS Press.

Bromley, D. (Ed.). (1995). *Handbook of Environmental Economics*. Oxford: Blackwell.

Bromley, D. (1999). *Sustaining Development: Environmental Resources in Developing Countries*. Cheltenham, UK: Edward Elgar.

Bromley, D. (2006). *Sufficient Reason: Volitional Pragmatism and the Meaning of Economic Institutions*. Princeton, NJ: Princeton University Press.

Bromley, D. (Ed.). (2014). *Institutions and the Environment*. Cheltenham: Edward Elgar.

Bromley, D., & Anderson, G. (2012). *Vulnerable People, Vulnerable States: Redefining the Development Challenge*. London: Routledge.

Bryant, R.L. (Ed.). (2015). *The International Handbook of Political Ecology*. Cheltenham, UK and Northampton, MA: Edward Elgar.

Burkett, P. (2006). *Marxism and Ecological Economics. Toward a Red and Green Political Economy*. Leiden: Brill.

Bruckmeier, K. (2018). The Significance of Economic Knowledge in the Environmental Sustainability Discourse. In N. Soukupová (Ed.), *Proceedings of the 12th International Scientific Conference INPROFORUM: Innovation, Enterprises, Regions and Management* (pp. 11–24). Ceské Budejovice: Faculty of Economics, University of South Bohemia.

Cardoso, F. H., & Faletto, E. (1979). *Dependency and Development in Latin América.* Berkeley and Los Angeles: University of California Press.

Common, M., & Stagl, S. (2016). *Ecological Economics*. Cambridge et al.: Cambridge University Press.

Costanza, R., Cumberland, J., Daly, H., Goodland, R., & Norgaard, R. (2010 [1997]). *An Introduction to Ecological Economics*. Enyclopedia of Earth.

Costanza, R., Cumberland, J. H., Daly, H., Goodland, R., Norgaard, R., Kubiszewski, I., et al. (2015 [1997]). *An Introduction to Ecological Economics* (2nd ed.). Boca Raton, FL: CRC Press.

D'Alisa, G., Demaria, F., & Kallis, G. (Eds.). (2015). *Degrowth: A Vocabulary for a New Era*. Abingdon, UK and New York: Routledge.

Daly, H. (1980). *Economics, Ecology, Ethics: Essays Toward a Steady-State Economy*. San Francisco: Freeman and Company.

Daly, H. (1994). Steady State Economics. In C. Merchant (Ed.), *Key Concepts in Critical Theory: Ecology*. Amherst and New York: Humanity Books.

Daly, H. (1997 [1996]). *Beyond Growth. The Economics of Sustainable Development.* Boston: Beacon Press.

Daly, H. (2008). On a Road to Disaster. *New Scientist*, Special Issue "The Folly of Growth", 18 October: 54–46–47.

Daly, H., & Farley, J. (2004). *Ecological Economics: Principles and Applications.* Washington, Covelo, London: Island Press.

DesJardins, J. R. (2013). *Environmental Ethics: An Introduction to Environmental Philosophy*. (5. Edition). Boston et al.: Wadsworth: Cengage Learning.

Drews, S., & van den Bergh, J. C. J. M. (2017). Scientists' Views on Economic Growth Versus the Environment: A Questionnaire Survey Among Economists and Non-Economists. *Global Environmental Change, 46*, 88–103.

Easterly, W. (2002). *The Elusive Quest for Growth. Economists' Adventures and Misadventures in the Tropics*. Cambridge MA and London, England: MIT Press.

Ehlers, E., & Krafft, T. (Eds.). (2006). *Earth System Science in the Anthropocene*. Berlin et al.: Springer.

Erickson, P., Kartha, S., Lazarus, M., & Tempest, K. (2015). Assessing Carbon Lock-In. *Environmental Research Letters, 10*. https://doi.org/10.1088/1748-9326/10/8/084023.

Exner, A. (2014). Degrowth and Demonetization: On the Limits of a Non-Capitalist Market Economy. *Capitalism Nature Socialism, 3*, 9–27.

Exner, A., Lauk, C., & Kulterer, K. (2008). *Die Grenzen des Kapitalismus. Wie wir am Wachstum scheitern*. Vienna: Ueberreuter.

Faber, M. (2008). How to Be an Ecological Economist. *Ecological Economics, 66*(1), 1–7.

Fischer-Kowalski, M., & Haberl, H. (Eds.). (2007). *Socioecological Transitions and Global Change: Trajectories of Social Metabolism and Land Use*. Edward Elgar.

Foster, J. B. (2000). *Marx' Ecology: Materialism and Nature*. New York: Monthly Review Press.

Foster, J. B. (2009). *The Ecological Revolution: Making Peace with the Planet*. New York: Monthly Review Press.

Foster, J. B., & Burkett, P. (2004). Ecological Economics and Classical Marxism: The "Podolinsky Business" Reconsidered. *Organization & Environment, 17*(1), 32–60.

Foster, J. B., & Burkett, P. (2016). *Marx and the Earth: An Anti-Critique*. Boston: Brill.

Foster, J. B., & McChesney, R. W. (2012). *The Endless Crisis*. New York: Monthly Review Press.

Franz-Bahlsen, A. (2014). Gender and (UN)Sustainability – Can Communication Solve a Conflict of Norms? *Sustainability, 6*, 1973–1991.

Gardiner, J. (1997). *Gender, Care and Economics*. London: Palgrave Macmillan.

Georgescu-Roegen, N. (1975). Energy and Economic Myths. *Southern Economic Journal, 41*(3), 347–381.

Giampietro, M., Mayumi, K., & Sorman, A. (2011). *Metabolic Patterns of Societies: Where Economists Fall Short*. London: Routledge.

Giampietro, M., Mayumi, K., & Sorman, A. (2012). *Energy Analysis for a Sustainable Future*. London: Routledge.

Gilmore, B. (2013). The World Is Yours: "Degrowth", Racial Inequality and Sustainability. *Sustainability, 5*, 1282–1303.

Goldsmith, E., & Allen, R. (1972). *A Blueprint for Survival*. Penguin Books.

GreenCom and Academy for Educational Development. (Eds.). (2000). *Environmental Education and Communication for a Sustainable World: Handbook for International Practitioners*. Washington, DC: Academy for Educational Development.

Groom, B., Hepburn, C., Koundouri, P., & Pearce, D. (2005). Declining Discount Rates: The Long and the Short of It. *Environmental & Resource Economics, 32*, 445–493.

Hall, C. A. S., & Klitgaard, K. (2012). *Energy and the Wealth of Nations: An Introduction to Biophysical Economics*. Cham, Switzerland: Springer and Springer Nature.

Held, D., & Koenig-Archibugi, M. (Eds.). (2003). *Taming Globalization: Frontiers of Governance*. Cambridge et al.: Polity Press.

Jackson, T. (2009). *Prosperity Without Growth: Economics for a Finite Planet*. London and Sterling, VA: Earthscan.

Jackson, T. (2019). The Post-Growth Challenge: Secular Stagnation, Inequality and the Limits to Growth. *Ecological Economics, 156*, 136–146.

Jo, T.-H., Chester, L., & D'Ippoliti, C. (Eds.). (2019 [2018]). *The Routledge Handbook of Heterodox Economics*. Abingdon, UK and New York, NY: Routledge.

Kallis, G. (2011). In Defence of Degrowth. *Ecological Economics, 70*, 873–880.

Kallis, G. (2017). Radical Dematerialization and Degrowth. *Philosophical Transactions of the Royal Society, A, 375*, 20160383. https://doi.org/10.1098/rsta.2016.0383.

Kallis, G. (2018). *Degrowth. The Economy. Key Ideas*. Newcastle upon Tyne: Agenda Publishing.

Kapp, W. (1950). *The Social Costs of Private Enterprise*. Cambridge MA: Harvard University. Press.

Kapp, W. (1968). In Defense of Institutional Economics. *The Swedish Journal of Economics, 70*(1), 1–18.

Kapp, W. (2011). *The Foundations of Institutional Economics*. New York and London: Routledge.

Klitgaard, K. (2013). Heterodox Political Economy and the Degrowth Perspective. *Sustainability, 5*, 276–297.

Krausmann, F., Lauk, C., Haas, W., & Wiedenhofer, D. (2018). From Resource Extraction to Outflows of Wastes and Emissions: The Socioeconomic Metabolism of the Global Economy, 1900–2015. *Global Environmental Change, 52*, 131–140.

Langley, P., & Mellor, M. (2002). 'Economy', Sustainability and Sites of Transformative Space. *New Political Economy, 7*(1), 49–65.

Latouche, S. (2009 [2007]). *Farewell to Growth*. Cambridge: Polity Press.

Lenton, T. M., Pichel, P.-P., & Weisz, H. (2016). Revolutions in Energy Input and Material Cycling in Earth History and Human History. *Earth System Dynamics, 7*, 353–370.

Martinez-Alier, J. (1987). *Ecological Economics: Energy, Environment and Society*. Oxford: Basil Blackwell.

Martinez-Alier, J. (1990). *Ecological Economics, Energy, Environment and Society*. Oxford: Basil Blackwell.

Martinez-Alier, J. (1995). Political Ecology, Distributional Conflicts, and Economic Incommensurability. *New Left Review, 211*.

Martinez-Alier, J., & Muradian, R. (Eds.). (2015). *The Handbook of Ecological Economics*. Cheltenham, UK and Northampton, MA: Edward Elgar.

Martinez-Alier, J., Pascual, U., Vivien, F.-D., & Zaccai, E. (2010). Sustainable De-growth: Mapping the Context, Criticisms and Future Prospects of an Emergent Paradigm. *Ecological Economics, 69*(9), 1741–1747.

Mayumi, K. (2001). *The Origins of Ecological Economics: The Bioeconomics of Georgescu-Roegen*. New York and London: Routledge.

McCay, B. J., & Acheson, J. M. (Eds.). (1987). *The Question of the Commons: The Culture and Ecology of Communal Resources*. Tucson: University of Arizona Press.

McLellan, B. C., Chapman, A. J., & Aoki, K. (2016). Geography, Urbanization and Lock-In – Considerations for Sustainable Transitions to Decentralized Energy Systems. *Journal of Cleaner Production, 128*, 77–96.

Meier, G. M. (2005). *Biography of a Subject: An Evolution of Development Economics*. Oxford et al.: Oxford University Press.

Mellor, M. (1997). *Feminism and Ecology*. Cambridge: Polity Press.

Ménard, C., & Shirley, M. (Eds.). (2005). *Handbook of New Institutional Economics*. Dordrecht: Springer.

Mill, J. S. (2004 [1848]). *Principles of Political Economy*. Amherst, NY: Prometheus Books.

Moe, E. (2010). Energy, Industry and Politics: Energy, Vested Interests, and Long-Term Economic Growth and Development. *Energy, 35*(4), 1730–1740.

Moe, T. (2015). Vested Interests and Political Institutions. *Political Science Quarterly, 130*(2), 277–318.

Moore, J. W. (2000). Environmental Crisis and the Metabolic Rift in World Historical Perspective. *Organization and Environment, 12*, 123–156.

Moore, J. W. (2015). *Capitalism in the Web of Life. Ecology and the Accumulation of Capital*. London: Verso Books.

Newig, J., Derwort, P., & Jager, N. W. (2019). Sustainability Through Institutional Failure and Decline? Archetypes of Productive Pathways. *Ecology and Society, 24*(1), 18. https://doi.org/10.5751/ES-10700-240118.

Norgaard, R. (1994). *Development Betrayed: The End of Progress and a Coevolutionary Revisioning of the Future*. London: Routledge.

NSF. (2005). Complex Environmental Systems: Pathways to the Future. Retrieved from www.nsf.gov/geo/ere/ereweb.

O'Connor, J. (1988). Capitalism, Nature, Socialism a Theoretical Introduction. *Capitalism, Nature, Socialism, 1*, 11–38.

O'Connor, J. (1998). *Natural Causes: Essays in Ecological Marxism*. New York and London: Guilford Press.

O'Hara, S. (1999). Economics, Ecology and Quality of Life: Who Evaluates? *Feminist Economics, 5*(2), 83–89.

O'Hara, S. (2010). Feminist Ecological Economics in Theory and Practice. In A. Salleh (Ed.), *Eco-Sufficiency & Global Justice – Women Write Political Ecology*. London, New York, Melbourne: Pluto Press.

O'Hara, S. (2016). Production in Context: The Concept of Sustaining Production. In J. Farley & D. Malghan (Eds.), *Beyond Uneconomic Growth* (Festschrift in Honor of Herman Daly) (Vol. 2). Burlington, VT: UVM.

Ophuls, W. (1977). *Ecology and the Politics of Scarcity*. San Francisco: W. H. Freeman and Company.

Ostrom, E. (2005). Doing Institutional Analysis: Digging Deeper than Markets and Hierarchies. In C. Ménard & M. Shirley (Eds.), *Handbook of New Institutional Economics* (pp. 819–848). Dordrecht: Springer.

Pearce, D. (2002). An Intellectual History of Environmental Economics. *Annual Review of Energy and the Environment, 25*, 57–81.

Podolinsky, S. (1880). Le socialisme et l'unité des forces physiques. *Revue Socialiste, 8*, 353–365.

Poirier, Y. (2013). *Social Solidarity Economy and Related Concepts – Origins and Definitions: An International Perspective*. Asia Solidarity Economy Council: Developments in Solidarity Economy in Asia, 71–86. Retrieved from http://www.socioeco/org.bdf_fiche-document-1873_en.html.

Polanyi, K. (1944). *The Great Transformation*. New York: Rinehart.

Polanyi, K., Arensberg, K., & Pearson, H. (1957). *Trade and Market in the Early Empires*. New York: Free Press.

Rees, W. (2014). *Avoiding Collapse: An Agenda for Sustainable Degrowth and Relocalizing the Economy*. Vancouver: Canadian Centre for Policy Alternatives.

Rice, J. (2007). Ecological Unequal Exchange: Consumption, Equity and Unsustainable Structural Relationships within the Global Economy. *International Journal of Comparative Sociology, 48*(1), 43–72.

Rockström, J., Steffen, K., Noone, K., Persson, A., Chapin, III, F. S., Lambin, E., Lenton, et al. (2009). Planetary Boundaries: Exploring the Safe Operating Space for Humanity. *Ecology and Society 14*(2), 32. http://www. ecologyandsociety.org/vol14/iss2/art32/.

Ropke, I. (2004). The Early History of Modern Ecological Economics. *Ecological Economics, 50*, 293–314.

Ropke, I. (2005). Trends in the Development of Ecological Economics from the Late 1980s to the Early 2000s. *Ecological Economics, 55*, 262–290.

Sachs, J. D. (2005). *The End of Poverty: Economic Possibilities for Our Time*. New York: Penguin Books.

Samuelson, P. A., & Nordhaus, W. D. (2005). *Economics* (18th ed.). New York: McGraw-Hill.

Sandberg, M., Klockars, K., & Wilén, K. (2019). Green Growth or Degrowth? Assessing the Normative Justifications for Environmental Sustainability and Economic Growth Through Critical Social Theory. *Journal of Cleaner Production, 206*, 133–141.

Schaffartzik, A., Mayer, A., Gingrich, S., Eisenmenger, N., Loy, C., & Krausmann, F. (2014). The Global Metabolic Transition: Regional Patterns and Trends of Global Material Flows, 1950–2010. *Global Environmental Change, 26*, 87–97.

Scheidel, A., Temper, L., Demaria, F., & Martinez-Alier, J. (2018). Ecological Distribution Conflicts as Forces for Sustainability: An Overview and Conceptual Framework. *Sustainability Science, 13*, 585–598.

Schnaiberg, A. (1980). *The Environment: From Surplus to Scarcity*. New York: Oxford University Press.

Schwartzman, D. (2012). A Critique of Degrowth and Its Politics. *Capital Nature Socialism, 23*(1), 119–125.

Sen, A. (1992). *Inequality Reexamined*. New York and Oxford: Russell Sage Foundation and Clarendon Press.

Sen, A. (1995). Gender Inequality and Theories of Justice. In M. Nussbaum & J. Glover (Eds.), *Women, Culture and Development: A Study of Human Capabilities* (pp. 259–273). New Delhi et al.: Oxford University Press.

Shmelov, S. E. (2012). *Ecological Economics: Sustainability in Practice*. New York et al.: Springer.

Söderbaum, P. (2008). *Understanding Sustainability Economics*. London: Earthscan.

Spangenberg, J. (2005). Economic Sustainability of the Economy: Concepts and Indicators. *International Journal of Sustainable Development, 8*(1–2), 47–64.

Spash, C. L. (2013). *The Ecological Economics of Boulding's Spaceship Earth*. Discussion Paper SRE 2013/02. Vienna University of Economics and Business, Institute for the Environment and Regional Development.

Spash, C. L. (2018 [2017]). *Routledge Handbook of Ecological Economics: Nature and Society*. Abingdon and New York: Routledge.

Spash, C. L., & Ryan, A. M. (2010). *Ecological, Heterodox and Neoclassical Economics: Investigating the Differences*. München: Munich Personal RePEc Archive (MPRA), paper 26292.

Stoknes, P. E., & Rockström, J. (2018). Redefining Green Growth Within Planetary Boundaries. *Energy Research & Social Science, 44*, 41–49.

Streeck, W. (2016). *How Will Capitalism End? Essay on a Failing System*. London and New York: Verso.

Vatn, A. (2005). *Institutions and the Environment*. Cheltenham: Edward Elgar.

Vatn, A. (2016). *Environmental Governance: Institutions. Policies and Actions*. Cheltenham: Edward Elgar.

Victor, P. A., & Dolter, B. (Eds.). (2017). *Handbook on Growth and Sustainability*. Edward Elgar.

Wallerstein, I. (1974). *The Modern World System: Capitalist Agriculture and the Origins of the European World-Economy in the Sixteenth Century*. New York et al.: Academic Press.

Wallerstein, I. (1980). The Modern World System II. In *Mercantilism and the Consolidation of the European World-Economy, 1600–1750*. New York et al.: Academic Press.

Weiss, M., & Cattaneo, C. (2017). Degrowth – Taking Stock and Reviewing an Academic Paradigm. *Ecological Economics, 137*, 120–130.

Wolf, E. (1982). *Europe and the People Without History*. University of California Press.

7 Conflict, Relapse and Failure in the Sustainability Process: Neglected Problems

This chapter discusses neglected problems and obstacles in the sustainability discourse and process: conflicts, setbacks, failures, crises, unexpected disturbances and catastrophes. These difficulties met in sustainability governance and collective action result from manifold social and environmental problems that need to be solved: social inequalities, social exclusion, poverty and hunger, overuse of the natural resource base, climate change, biodiversity loss, land use change, human modification of ecosystems, and so forth. Integrating global governance processes across cultural and political boundaries implies mitigating many environmental and resource use conflicts (violent and non-violent), failures and delays in the policy and governance processes, the slow change towards environment-friendly ways of life and the risks of a global economic and societal collapse, as discussed in the "limits to growth" discourse. Most of these problems are not caused by failures in the sustainability process but through other national and international economic and political processes (unequal development, economic globalisation, competition, global exchange and trade, national and international policies), where they could not be solved and now wait for sustainable development as the last way out. The unfavourable conditions under which the sustainability transformation began will continue to appear in these postponed and often unsolvable problems and conflicts.

Figure 7.1 shows in simplified form how these problems and obstacles interact and reinforce each other in sustainable development as a process overloaded with conflicts.

K. Bruckmeier, *Economics and Sustainability*,
https://doi.org/10.1007/978-3-030-56627-2_7

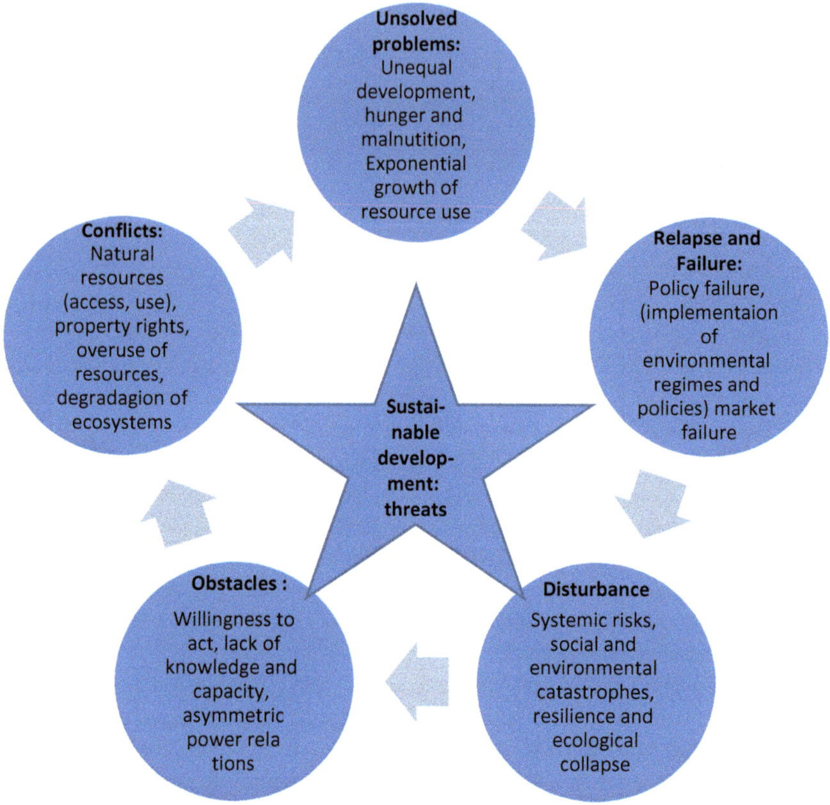

Fig. 7.1 Problems, obstacles and conflicts in sustainable development

From the analyses in the preceding chapters the unfavourable conditions of the sustainability transformation can be summarised as follows:

1. *incompatibility of continued (exponential) economic growth and sustainable use of natural resources* that appears in the contrasting processes of economic globalisation and sustainable development;
2. *continuing inequality between the Global North and South* through the economic and social divides in the modern world system, reinforced by increasing inequality within all countries;
3. *contrasts between governmental sustainability policies and social and economic realities of intensification of resource use and environmental degradation*

that show the lack of progress towards social, environmental and economic sustainability in most countries;
4. *increasing violent conflicts* in the globalising economy and society, with economic globalisation as the core process of "managed decline" in rich and poor countries;
5. *contrasts between harmonistic global ethics of "our common future" and divided realities of sustainability policies* and practices (e.g. environmentalism of the rich, environmentalism of the poor, continued growth on the basis of unequal development and exchange, ecological distribution conflicts);
6. *the complexity of the interconnected problems and changes in social and ecological systems* that create problems for managing and maintaining the sustainability process in the long run;
7. *the lack of agency and deficits of organised social subjects of transformation* (social or political groups, institutions, transformative action groups) and the difficulties of building global environmental governance or earth system governance (global policies and governance are lagging behind the rapid social and environmental changes).

In abstract, system-theoretical language, the transition to sustainability can be seen as creating emergent properties with non-intended consequences (Hannigan 2006: 136ff): emergent properties or capacities are generated through the aggregated capacities and collective functioning of a whole system, not through the parts of the system or the individuals that belong to a system, group or community. Emergence is a result of cooperation, collective action, integration and coordination of activities and agents. In the sustainability transformation the global economy and the politically and culturally divided world of states, national economies and cultures are involved in creating transformation capacity and agency and multi-scale forms of social and environmental change, especially global change processes (climate change, biodiversity loss, land use change and urbanisation). The dilemma of the sustainability process appears then in the form:

1. The sustainability process cannot be reduced to simple systems, local processes, or specialised systems (e.g. the economy, politics); it encompasses coupled social-ecological systems at all scale levels from the local to the global.
2. In practice the sustainability process needs to be driven and managed through small, fragmented and diverse systems at different levels from the local to the global; it requires the continued networking and cooperation of more actors and institutions than other, more sector-specific global processes in the economy, politics and culture.

When searching economic and ecological research knowledge for conflict resolution or mitigation not much is found. The subject areas discussed in this book—economics, ecology and interdisciplinary sustainability research—are weak on knowledge, methods and tools for conflict resolution. Efforts to develop conflict research, typologies and solution methods in these fields are exceptional: the typology of ecological distribution conflicts with ecological economics (Martinez-Alier). The underdeveloped subdiscipline of conflict economics (Croson and Bolton 2012; Wärneryd 2014; Anderton and Carter 2019) is oriented towards monetary assessment of conflicts and selective, economic methods and instruments for conflict mitigation. The larger part of conflict research and conflict resolution is done in policy research, in the interdisciplinary field of "peace and conflict research", and in specialised practices of conflict mitigation as mediation, often working with psychological knowledge, concepts and methods. The pattern of externalisation and shifting of conflicts is a symptom of the deficits of sustainability research and governance: conflicts are ignored as long as possible, and when addressing them becomes unavoidable they are left to the responsibility of the specialised experts of ten formal and legal institutions.

7.1 The Societal Context of Conflicts and Crises in the Sustainability Process

Sustainability appears in the scientific and political discourses as a diffuse idea that needs to be specified in the sustainability discourse and process, in science and politics. The idea of intragenerational and intergenerational solidarity remains abstract; it does not provide a conceptual basis for the mitigation of environmental conflicts that need to be solved in the sustainability process (Bruckmeier 2016: 125ff). This process is itself in crisis owing to the limited success of policies and weakening public support for the "great transformation" to sustainability in the decades since the enthusiastic start inaugurated by the global conference in Rio de Janeiro in 1992. With weaker public support came the voices of environmental scepticism, fatalism, the spreading of dystopian ideas, of doomsday prophecies and predictions of global collapse. Under these conditions the future sustainability process requires new efforts to build capacity and address complicated situations, conflicts, disturbances and environmental catastrophes, while maintaining the process and continually developing the cooperation between actors and the building of transformative action groups.

The obstacles on the path to sustainability are only partly known. Some are discussed and reflected upon in the process: mainly those identified in policy evaluations and review processes or discussed in science and research. Other obstacles are widely ignored and do not receive sufficient public attention. One of the most important reasons for this is that the problems are not seen as being part of the sustainability policy or process, or that they should not be solved there but in other policy processes or outside politics. The overlapping of policy fields in the sustainability process—environmental policies and regulations, global governance, development policies, economic policies, and others—can impede solutions being found for emerging conflicts. Who should do it, who can do it, in which form? Of these ignored conflicts an increasing number become manifest in the future sustainability process, such as the taboo themes of violence, conflict, catastrophe, relapse and market and government failure. Sustainable development, interpreted as a "win-win situation", does not take into account the big differences, inequalities and complicated relations between the many countries, institutions, actors and citizens in the transformation process. These actors are in different situations, with different and competing worldviews, interests, experiences and expectations; these differences thus appear as causes of conflicts, resulting in conflict-loaded processes of problem-solving.

The abstract and inexact terms of "conflict" and "problem" evoke further difficulties in identifying necessary and adequate mechanisms of conflict resolution and problem-solving (see Box 7.1). The conflicts influencing the sustainability process include:

1. a larger group of local and non-violent conflicts in the routine processes of natural resource use, between smaller groups of resource users, and, because of the changes connected with sustainable development, increasingly also between the actors involved;
2. a limited group of violent conflicts, civil and other wars, between larger social, ethnic and resource user groups, communities, institutions and countries, about access to and control and use of natural resources and solving environmental problems. Violent conflicts over natural resources receive more public attention and efforts towards conflict resolution (Bannon and Collier 2003), for the obvious reason that civil wars and wars about access to natural resources disturb and destruct much more than the directly involved groups and their lives, the local areas or communities where the conflicts happen: local conflicts also have repercussions for national economies, international trade, political systems and policies.

For both types of conflict there are mechanisms of conflict resolution. When they are applied in sustainable development, they bring new burdens, costs and difficulties in the process. Conflicts can escalate to armed and violent conflicts, discussed in the scientific literature as potential climate and resource wars of the twenty-first century (Welzer 2008; Dyer 2011): people will fight for access to land and resources in wars that gain public attention and cause military interventions or attempts at reconciliation through national and international political institutions. Some of this is already reality: the wars in the Near East, between Israel and its Arab neighbouring countries, have several reasons; an important one is access to drinking water in dry areas (specifically, access to the aquifers at the Golan Heights). Civil conflicts and wars in East Africa, for example in Sudan, are seen as early examples of climate wars. Conflicts over access to oil have been one cause of the wars in the Near East and in Iraq since the 1990s. In other conflicts that have not escalated into violence disputes over environmental and resource use have played a significant role. These conflicts will have to be addressed in the routines of sustainability governance as components of sustainability transformation (Box 7.1).

Box 7.1 The Conflict Concept

The abstract notion of conflict encompasses manifold forms of conflicts in the sustainability process, including:

(a) conflicts over material or energy resources
(b) symbolic conflicts between different values, interests and knowledge frameworks of persons and groups
(c) political, economic or cultural conflicts
(d) violent or non-violent conflicts, conflicts between a small number of persons or small groups or large conflicts between social and ethnic groups or between institutions and states

Definitions of environmental or natural resource use conflicts such as "*incompatible interaction between at least two actors over the use of natural resources or an environmental system, where one of the actors is damaged by the interaction, and the other actor intends or ignores this damage*" (Mason et al. 2007: 327) are too broad and vague to be useful in conflict analysis or conflict resolution. Such abstract terms, widespread in conflict research, cause controversies about their interpretation of the phenomena that should be seen as conflicts.

General definitions need to be specified through more concrete aspects of conflicts, such as the actors involved and their rights, interests, positions, values and worldviews, procedures of conflict mitigation, and so forth

(*continued*)

Box 7.1 (continued)

(Stepanova and Bruckmeier 2013). Environmental and resource use conflicts become ever more complex with the processes of globalisation and global environmental change, and conflicts in the sustainability process are examples of such complex conflicts between organisations, between several social systems, and crossing political and other boundaries. Some of the knowledge and practical requirements for the analysis and resolution of complex conflicts can be found in recent conflict research, dispersed in several areas of specialisation; for example, research on environmental and resource use conflicts (Bruckmeier 2005), political conflict research (Dudouet 2006; Varisco 2010), social-psychological research on conflicts and mediation (Bar-Tal 2009) and research in some fields of economic and management where the resolution of conflicts is described as "transformation" (Mason and Muller 2007) in the sense of a guided transformation of conflict towards cooperation.

Non-violent and local conflicts with few persons or small groups involved may not be perceived as conflicts at all, especially when they lack legal and political disputes and decisions, or public attention, are dealt with informally, without political decisions, or do not require formalised processes of conflict mitigation. Environmental and resource use conflicts often have several reasons and causes, are often connected with other problems and do not always appear in political processes but rather in lifeworld contexts or daily routines or professional work. Such conflicts need to be analysed in their development and in their social and natural contexts. Cultural perceptions of conflicts vary between views of conflicts as taboos and conflicts as necessary components of social life. Also in culturally similar areas such as in Europe exist significant differences in understanding, assessing and resolving conflicts, for example whether they should be solved through external intervention or by the conflicting parties themselves (Bruckmeier 2015). For a study of conflict resolution in marine governance see Vasconcelos and Flavia Silva (2015); for a recent publication with studies on conflicts related to nature resource use and sustainable development see Almered Olsson and Gooch (2019).

Sources: mentioned in the text

The sustainability process, a long process of social-ecological transformation, is still in an early phase after three decades of global sustainability governance. Routines for conflict mitigation have not yet been established. The process has so far shown more deficits and failures than successful policies and changes towards sustainability: more postponing of necessary debates and reforms, more sustainability diplomacy, political rhetoric and weak policies; delays in the social-ecological transformation and in earth system governance on the way to a sustainable economy and society. Changes in ways of life and consumption, of the global economic system, reduction of exponential growth of the economy and of resource use and regulation of global environmental and economic change are not yet advanced in sustainability governance. The United Nations (UN) institutions guiding global sustainability governance repeat this message in nearly all of their reports and assessments.

The limited success in national and international sustainability policies does not include these core problems in the sustainability process, only some preparatory steps, such as the creation of a global agenda for sustainability policy since 1992; the opening up of the policy to civil society actors; the formulation of national programmes for sustainable development in the Global North and South. Most of these programmes are not efficiently implemented. Advances in the economy, the main system to change through sustainability transformation, are limited to the creation of a global dialogue among economic enterprises and actors through the World Business Council for Sustainable Development (WBCSD) and other fora of sustainability communication and networks, and many voluntary activities to change economic production processes in enterprises towards sustainability. The voluntary activities by enterprises and other organisations are often "greenwashing" and symbolic company policies to support business success. Single enterprises voluntarily experiment with degrowth and can succeed in maintaining a continued business strategy of degrowth. But this has no transformative effects on an economic system that is programmed for growth. The relevant advances in civil society are more difficult to overlook; they happen mainly at local levels and would require reviewing local, urban and community programmes for sustainability transformation, of which many exist. The development of a global civil society is slow, presently limited to some global movements; it will develop only with further progress in global environmental and sustainability governance.

The most relevant breakthrough in the 2010s in global sustainability governance is that the idea of "another great transformation" is now on the agenda, outstripping the weak and vague forms of sustainable development through policy reforms and international diplomacy. It is the voluntary activities of enterprises and the piecemeal changes of everyday life and consumption processes, not the political routines, that are the core processes of sustainable development. These small behavioural changes are necessary and need to become parts of larger, more integrated and powerful strategies, but they cannot replace larger and longer term changes, ultimately, the transformation of the modern economy and society to a new system, for which the preliminary names “post-industrial society”, “post-capitalism”, “knowledge society”, or the more fuzzy “new world order” have been given.

7.1.1 Prevention and Creation of Conflicts through Sustainable Development

To deal with conflicts and obstacles in the sustainability process, the concept of sustainable development needs to be discussed again, beginning with the strange way in which conflicts in sustainable development are reformulated in the Brundtland Report. Sustainable development as intragenerational and intergenerational solidarity in resource use is not a peaceful process of redistributing resources; it requires a conflict-loaded building of a new world order, which is not discussed in the report, only indirectly referred to in the abstract terms of problem and solidarity. The relationship between sustainable development and conflict can be interpreted in two contrasting views, conventional and critical:

1. *Conventional view of sustainable development as preventing new conflicts by solving old problems*: the UN and the World Bank initiated the Humanitarian-Development-Peace Initiative, where sustainable development is seen as a process of preventing conflicts (Corral et al. 2020). Similarly the International Institute for Sustainable Development (IISD) is oriented in its work with the sustainable development goals of the UN to build peace and stability as prerequisites for sustainable development in countries that are socially and politically fragile, vulnerable to climate change and affected by violence (many countries in Africa). These conditions threaten or prevent improvements of education, health and other determinants of well-being. Under such conditions, in many countries in the Global South, sustainable development is confronted with a vicious circle of violence, poverty and blocked development, preventing the sustainable development goals from being achieved. This is an extreme situation, but even in affluent countries with more stable development the transformation process is not one of harmony. The UN, the IISD and the World Bank do not have strategies to break the vicious circle; their views and roles as representatives of the existing, unsustainable political and economic world order is to maintain this order according to the interests of the industrialised countries of

the Global North. From the perspective of these powerful global players sustainable development moves between peace and security policies, investment programmes and global environmental governance programmes (Allan 2020), without a strategy for transformation.

Sustainability policies and governance are not organised for the solving of resource use conflicts. The many violent and non-violent conflicts between resource users described in the scientific literature as economic and ecological distribution conflicts (Martinez-Alier 1995) are not seen as part of the policy programmes and governance strategies. The conflicts appear as a reality of their own kind, shifted to other fields of policy and governance, not in sustainability research and policy but in specialised forms of conflict research and conflict solution to the institutions of legal and public administrative systems: arbiters, courts, police and the military.

2. *Critical view of sustainable development—new conflicts emerge in the sustainability process when the old problems created by industrial society should be solved*: this does not mean that sustainable development is only and everywhere a process of conflict mitigation, but conflicts need to be seen realistically and solved in the transformation process. To do that requires policy programmes and strategies other than those presently in place, where mainly goals are formulated and the processes to achieve them remain unclear. Conflict mitigation, to become part of a transformation process, needs to be connected with other activities, such as mobilising citizens, consumers and resource users to participate in the process; creating fora to discuss local strategies for transformation; opening up the governance process to broader participation and developing more active forms of knowledge integration, application and joint learning; critically reviewing the failure of earlier, symbolic, sustainability policies; and setting neglected themes like the redistribution of resources between the Global South and North on the policy agenda. A discussion of the conflicts that need solving in the transition to sustainability, and the means to do that, should be part of sustainability agendas at all levels of the governance process.

7.1.2 Origin of the Conflicts Affecting Sustainable Develop ment

Most of the conflicts and the problems that will need solving in the transformation processes originate in the present unsustainable resource use practices that are shown in economic and ecological indicators of development.

Conflicts generated by the development of modern society and its resource-intensive ways of life and consumption: resource-intensive lifestyles are less the choices of individuals than they are a necessary part of an economic system that cannot develop and grow without high levels of consumption. The direct manifestation of the high consumption levels is not wealth in monetary terms and income levels but the physical indicators of material and energy flows in the modern economy, organised in more complex patterns of societal metabolism. The solution for the overconsumption of resources includes lifestyle changes towards more modest resource use. These changes may cause individual conflicts and resistance to change; however, these are not the main conflicts to deal with, and for these no political mechanisms of conflict mitigation are necessary. The main conflicts to deal with occur in the economic system, between different organisation, enterprises and resource user groups. To understand the nature of these conflicts and their origin in competing resource use and the changing availability of resources, and to find solutions, it is necessary to analyse the total resource flows in the economy, their interconnections and their changes through economic development. For the analysis and resolution of economy- and resource-related conflicts in the sustainability process, it seems useful to differentiate between various types of change that interact in economic resource use processes:

1. *Development* is a process of incremental social, economic technical change based on economic growth, as continuous and short-term changes.
2. *Structural change* is part of the dynamics of economic development in the modern economy (the market-based processes of capital accumulation, economic concentration, changing power relations between politics and the economy, state and market-based enterprises resulting in slow, growth-dependent changes in the concentration of capital and in technical modernisation).
3. *Modernisation* is the overarching process, including combined economic, technical and social changes, in which conflicts and crises appear periodically, connected with business cycles and "long waves" (Kondratiev waves). Long waves are innovation-based cycles within the industrial economy that began with the European industrialisation two hundred years ago. In the industrial epoch we are presently in the fifth cycle based on innovations in information technology and the digital economy. The sixth cycle is being initiated through presently discussed strategies of green and blue growth. A decisive part of sustainable development is innovation in the transition to a sustainable and knowledge-based economy based on green growth and genetic modification of living resources. However, this would be sustainability within the logic and limits of the growth-based economy—less a transformation, more an economic innovation

cycle, exposed to the systemic risks and crisis-prone development trends of the capitalist system. For this sixth cycle the big controversy in public policy and governance will be whether it remains a cycle of the existing system: if so, it would be a weak form of sustainability that does not solve the problems that sustainable development was conceived of as the solution. To seek for alternatives and new development paths, not only global scenarios should be discussed, but forms and possibilities of a social-ecological system transformation towards a post-industrial and post-capitalist economy and society.

4. *Transformation* is a complex change of social, economic and ecological systems into another qualitative state: a non-linear qualitative change of complex systems. Not all transformations are necessarily large-scale social-ecological transformations with the building of a new society, economy or societal metabolism. But this type of transformation was the decisive form of societal change in human history that brought, as the non-intended consequence of collective action, the long trend of intensifying natural resource use that comes to an end with the exponential growth triggered by the industrial revolution. A new social-ecological transformation is necessary to break the ruinous development path. Sustainable development is the only global process that can qualify as transformation towards a post-industrial, post-capitalist and post-growth economy.

Various forms of change are connected in the processes of economic, social and environmental change in modern society that can become part of sustainability transformation. To understand their interaction and dynamics, it is necessary to analyse the economic system theoretically, and to understand the conflicts in the change processes. It is not necessarily an exceptional situation or a dysfunction of institutions that causes the conflicts that environmental policy or sustainable development face: rather, the routine processes of economic development and resource use underlie these conflicts. A historical review of the development of the modern economic world system and its social, political and cultural contexts demonstrates how systemic mechanisms, deep structures, incoherencies and contradictions of systemic processes and the conflicts accompanying the economic modernisation processes have been ignored.

Conflict-generating systemic mechanisms and dynamics in the modern society and economy: from research in social and political ecology, world system analysis, ecological economics and critical political economy, the inter-systemic dynamics of societal and ecological systems can be described. The dynamics are conflict-prone, continuously evoking conflicts in the development of modern society and the global economic system (e.g. caused by unequal development, exchange of resources, socio-metabolic regimes and growth-related issues):

1. *Internal systemic incoherence*: contradictions and conflicts in the modern economic world system that are connected with the economically structured processes of natural resource use. Tensions in the social-ecological systems appear in conflicts of appropriation, access, distribution, ownership rights, accumulation regimes and through the valorisation and commercialisation of natural resources and their transformation into economic goods.
2. "*The domestication of space*": territorial expansion, globalisation and space-bound conflict dynamics in economic development and accumulation in the expansion and globalisation of the modern economy. Conquests such as colonisation, centre–periphery capitalist dynamics, postcolonial development and globalisation in the twentieth century are components of the expansive development of the global economy that has not ended. It continues with the blue growth strategy and the exploitation of the oceans and the deep sea entering a new phase. The domestication of cosmic space opens up speculation about humankind leaving the planet earth and embarking on interplanetary colonisation.
3. "*The domestication of time*": temporal structures of economic development, intensification of resource use and the dynamics of economic change. The short-term and long-term cycles and the system transformation towards a post-industrial economy create a build-up of conflicting processes that are expressed in cyclical processes of economic reproduction, accumulation and modernisation in the industrial society. The market-bound entrepreneurial processes of creation, destruction and innovation (Schumpeterian dynamics) and the Kondratiev waves as "long waves" of innovation-bound economic cycles and changes of accumulation regimes show the temporal dynamics of system-related conflicts.
4. *The inter-systemic social-ecological dynamics of contrasting and conflicting development processes in social systems and ecosystems in modern society*: unequal economic development and overuse of the resource base of the earth system as a core process of modernisation appear in the clashes between economic globalisation ("dis-embedding of markets") and sustainable development ("re-embedding the economy" through social-ecological transformation).

In the processes of economic globalisation and sustainable development the developing economies of the Global South are neglected, although the consequences of uneven development are the main reason for the contrasting transformations. The global sustainability policy up to now and in the foreseeable future is dominated by the interests and political and economic power of the affluent societies in the Global North, in a striking contrast to the meaning of sustainable development implied in the Brundtland Report: to overcome the Global North-South divide and create more equal and better conditions of life everywhere through strategies of redistributing and sharing natural and other resource globally. Such redistribution has not happened and is not yet foreseeable. The available global funds such as the Global Environmental Facility and other forms of international development aid renamed as part of global governance and sustainable development are not effective forms of redistribution. Rather, they maintain the dependence of the countries of the Global South on the economic interests of the Organisation for Economic Co-operation and Development countries. Sustainable development inherits the systemic conflicts of the industrial society and becomes part of their solution through social, economic and ecological transformation.

7.2 Knowledge-Related Conflicts in the Sustainability Discourse and Process

The scientific conflicts about sustainable development: when conflicts are understood as clashes of values and interests between persons and social groups, scientific conflicts also appear as knowledge-, value- or interest-based conflicts between experts—though they are usually called "controversies" in scientific discourse. In the sustainability discourse and process scientific controversies over economic growth, climate change and ways to achieve sustainability become part of the programmes and practices of sustainability governance. The discussion of economic knowledge and its application in sustainability governance shows that controversies over growth and sustainability have become a continuous "discursive battlefield of knowledge" (Lie 2012); they are part of the broader processes of development and modernisation described by Long (Long and Long 1992) as "battlefields of knowledge". This means that the important components of research become knowledge and power, rather than the production of new knowledge or the achievement of goals in scientific, political and economic organisations. Scientific controversies are not always guided by the rules of the discourse ethics described by Habermas (1991), where a convincing argument is enough to achieve consensus. At least for the proponents of growth—the influential global players in the sustainability discourse and the majority of economists in academic science—the practice of sustainable development is an example of a hegemonic knowledge battleground: the definitional power of an epistemic community is guiding and defining the global sustainability agenda with strategies of green or sustainable growth.

Ecological economics and critical political economy: controversies about the forms of transformation of the economy between critical and orthodox economists and between economists and ecologists are a driving force of the sustainability discourse. Although the critical system analyses of the global economy, modern world system and interaction between modern society and nature are theoretically and empirically more elaborate, detailed and convincing, the critical economists are less influential and powerful in sustainability governance. This can be said for the system analyses of modern capitalism in ecological perspectives: world system theory (Wallerstein et al.); ecological Marxism (O'Connor, Foster, Altvater et al.: Burkett 2006) and critical political economy (Schnaiberg 1980: "the treadmill of production"); critical environmental sociology (Rice 2007); capitalist world ecology (Moore 2014); ecological economics and political ecology (Martinez-Alier); institutional and ecological economics (Klitgaard 2013); critical development economics (Sachs 2005); feminist economics (Langley and Mellor 2002; O'Hara 2016); social ecology (Fischer-Kowalski and Haberl 2007); and sustainability science (Kasemir et al. 2003). In comparison, research in orthodox economics and conventional policy analysis is theoretically less developed, fragmented by specialisation, comprising mainly empirical research for specific cases and without significant interdisciplinary knowledge integration and generalisation. Such asymmetrical power relations in science are not exceptional; it was similar during the development of modern environmentalism. Although it is necessary to gain acceptance for a system transformation among more economists, the processes and practices of transformation remain controversial (Box 7.2).

Box 7.2 Knowledge Conflicts Between Economy and Ecology

Examples for critical discussions about the contrasting development processes in economic and ecological systems:

1. *An anthropologist's analysis of the ecological transition*—Bennett (1976: 298ff): adaptive and maladaptive changes have happened in human history, fluctuations between overuse and underuse of natural resources, and both can permit sustained yield. "Characteristically, as human society has become more complex, and economic growth has become a dominant feature, the number of conflicts between desired states of being has increased. Out of this situation competition for resources emerges, as well as human group conflict" (Bennett 1976: 299). The cumulative ecological damage to the environment and to human life may in future require giving up present levels of welfare and intensive resource use in the developed world.

(*continued*)

Box 7.2 (continued)

2. *An ecologist—two fundamentally different forms of ecosystems as a consequence of the ecological transition*—Southwick (1996): the only kind of sustainability that makes sense in the long run is ecological sustainability. Sustainable growth is not possible because all growth must come to an end at some point. Sustainable development implies long processes of change, but it has limits. The message of ecology is essentially one of limits—"planet earth is finite" (Southwick 1996: 332). This implies discussing sustainability in two forms: for natural ecosystems, and for human-dominated ecosystems in modern society. The resource use in the modern economy has a strong impact on the functions and services of ecosystems; sustainable development requires limiting growth and reducing environmental destruction (Southwick 1996: 332ff)
3. *Technology for a sustainable future*—Elliott (1997: 205ff): Technical fixes are suggestions to avoid basic changes to modern society and its use of energy resources. The envelope of sustainable development allows for different forms of social choice, for social negotiation of means and ends; there is "a variety of approaches to sustainability tailored to specific social, cultural and environmental contexts, but there are also overall limits to the range of social and technological choices—primarily environmental limits"; a "key task for the future is to try to reach agreements on the right mixes in each situation, locally, nationally and globally, while recognising that each of these levels of concern interacts to some degree" (Elliott 1997: 207).
4. *The controversy between economists and ecologists about the "limits to growth" report* (Meadows et al. 1972)—Hall and Klitgaard (2012: 213ff): "Economists particularly disliked the idea of absolute scarcity of resources, and they wrote a series of scathing reports against the scientists … most closely associated with the limits to growth", trying to re-establish the idea of the market as ultimate arbiter and optimal distributor, assuming that it reflects the choices of the consumers. Their arguments did, however, not hold over time, were "illusory at best, and generally based on incomplete information"; as later developments and improved analyses of economic resource use, especially energy use, have shown: the limits to growth report and its updates are not invalidated, but economists need to learn economics from a biophysical and a social perspective to understand how global resource use problems can be solved.

Sources: mentioned in the text

The controversies over a conflict between economy and ecology or between the knowledge provided in economics and ecology cannot be dealt with in abstract form; they need to be concretised and translated in debates about transformative action in policy and governance processes. For this purpose knowledge from research about environmental and resource use conflicts should be reviewed and adopted in the sustainability process.

7.3 The Development of Sustainability-Related Conflict Research

Sustainable development as a conflicting process: the abstract form of a conflict implied in the definition of sustainable development in the Brundtland Report is that of a global conflict of access to resources that requires for its solution a fair distribution of resources within the presently living generations and between present and future generations. This is too abstract and vague a conflict to be dealt with in the concrete form of conflict resolution. It cannot be mitigated in its global dimensions, only diagnosed as the general conflict-generating mechanism of unequal access to resources and asymmetrical power relations in the global economy and society that dissolves into manifold forms of concrete conflicts between groups of actors, resource users and resource owners, dispersed across place and time.

The generalised conflict appears finally in specific local conflicts between concrete groups of actors and resource users (Bruckmeier 2019). In this concretisation the conflicts lose, however, their appearance as a unique type of resource use conflict; they appear as conflicts with manifold causes, courses, consequences and views held by the actors involved and many forms of successful or failing conflict resolution. Solidarity is then difficult to interpret and practice, because the conflicts happen between limited social groups or individuals, whereas the solidarity meant in the context of sustainable development is one between presently living generations and between them and future generations. For local conflicts this implies that the conflicting parties should not only act in terms of solving their concrete conflict; they should take into account in their individual or group behaviour and interests the effects their resource use has for the global population and for future generations. Practically speaking, this is impossible: in most resource use conflicts and resolutions the global sustainability dimensions are ignored. To deal with such dimensions requires coordinated forms of global governance and social-ecological transformation.

7.3.1 Resource Use Conflicts and Ecological Distribution Conflicts

Local and non-violent conflicts over environmental issues, climate change and resource use affect policy and governance processes and the lifeworld, and these need to be solved in the sustainability process and other environmental policy and governance processes. Such conflicts often originate in the non-political process of economically organised resource use, production and management and in the lifeworld, but their solution occurs in politically structured processes of conflict mitigation. The sustainability process does not become only or mainly one of conflict mitigation, but conflicts happen more often and cannot be ignored or suppressed in the further phases of transformation that include changes of consumption behaviour and lifestyles.

In the "limits to growth" debate (Meadows et al. 1972), the situation with regard to global overuse and scarcity of key resources appears as a dramatic and disaster- and conflict-related process. Global environmental change and processes connected with it became part of environmental policy and governance, for example regulating and preventing the total overfishing of the oceans, the deforestation of tropical rainforests, the desertification and erosion of agricultural land and globally spreading pollution and environmental destruction. These processes to reduce exponential growth of resource use have had to date limited success in local projects, nor have they significantly reduced the global growth of resource use, as ecological indicators show (see Chap. 6). Neglecting and shifting problem-solving only creates further conflicts in the future sustainability process.

The research relevant for conflicts in the sustainability process is dispersed across various fields of thematically specialised conflict research; for example, environmental conflict research, research on conflicts in natural resource use and policy-oriented peace and conflict research in local, national and international conflicts. Other forms of conflict research include discipline-specific conflict research in economics (economics of conflict resolution), political science, sociology, psychology and cultural anthropology. In spite of this dispersion of conflict research, the conflicts related to sustainable development, manifesting as political, economic or civil society conflicts, can be identified without difficulty as environmental conflicts about pollution and environmental damages or resource use conflicts about access to or ownership and use of natural resources. A relevant theoretical concept related to these conflicts is that of "ecological distribution conflicts" (Martinez-Alier 1995; see Box 7.3).

Box 7.3 Ecological Distribution Conflicts: Martinez-Alier

Ecological distribution conflicts (Martinez-Alier 1995, 2003; Martinez-Alier and O'Connor 1996; Martinez-Alier and Naron 2004; Scheidel et al. 2018) are *social conflicts resulting from unequal access to natural resources and unjust distribution of benefits and burdens of pollution between economic actors (natural resource users, enterprises, etc.).* Such conflicts can be classified according to the phases of extraction, production and processing, transport, waste and pollution, as follows:

Conflicts about the extraction of materials and energy:

1. mining conflicts (siting of mines and smelters, their pollution)
2. biopiracy (illegitimate appropriation of wild or agricultural genetic resources by private firms without recognition of peasant and indigenous resource ownership)
3. land degradation (pressure of population or production)
4. wood plantations (for commercial or other use)
5. mangrove conservation or shrimp production
6. water conflicts (defence of rivers against dams or use of water for irrigation)
7. national/local fishing rights

Conflicts about transport:

8. for example, spills from tankers, pipelines, building of new motorways, harbours, airports

Conflicts about waste and pollution:

9. "toxic struggles" (about risks from heavy metals, dioxin, etc.)
10. waste dumping ("toxic imperialism")
11. transboundary pollution
12. equal rights to carbon sinks
13. consumer and citizen safety (burden of risks from nuclear technology, genetically modified organisms, included in the concept of "risk society" by Beck)

Sources: mentioned in the text

Ecological distribution conflicts have their origins in the economic processes of extraction, exchange and transport, production and processing, but they appear as environment-related conflicts that require solutions in public policy and governance processes. A specific feature of ecological distribution conflicts is that they include conflicts over long distances, between social and resource user groups that are not in local and direct contact with each other; for example, in conflicts between the local population and resource users in the Global South, where extraction of resources occurs, and the consumers in the Global North, where most of these resources are used. Because the conflicts have manifold reasons and causes, it is not possible to classify them exactly as conflicts that emerge outside the process, as conflicts that cause sustainable development, as conflicts that emerge in the sustainability process itself (through economic and political failure in the transformation process) or as conflicts that can and need to be solved through sustainability transformation. It can only be seen that the sustainability process is, more than other policy and governance process, and in contrast to the harmonistic view by many participants in the process, burdened with the solution of environmental and resource use conflicts.

Specialised research on conflicts in the sustainability process is limited, but in sustainability research more and more conflict studies are carried out. Recent examples include a special issue on conflicts of the journal "Sustainability" from 2019, and the conflicts studies in Almered Olsson and Gooch (2019). In future sustainability research mapping and integration of the different thematic facets of multidisciplinary and interdisciplinary sustainability research is required to show the connections between the research fields, both the overlaps and the gaps. Such mapping is done in forms established in the scientific process, through databases, metadata, handbooks and reviews that describe the continuing research and the state of research in specialised themes. Such mapping is difficult for the broad subject of sustainable development and for sustainability research, and a task of the sustainability process will be to address the many challenges in gathering, transferring and applying knowledge practically. Several sustainability-related handbooks referred to in the chapters of this book show that sustainability research—in contrast to climate research—advances in uncoordinated research.

The review by Köhler et al. (2019) of the research on sustainability transition shows that the number of publications has grown exponentially in the first two decades of the twenty-first century from nearly zero to about 500 articles published in peer-reviewed scientific journals every year. The research covers a broad range of themes among which multi-scale transition and transformation processes across local, national and global levels have become an integrative form of sustainability research. The review does not ignore, but neither does it systematically discuss, the state of research with regard to non-violent and violent conflicts in the transition processes. The dispersed research described above is one reason for that, but another, related, reason is that much of the research on conflicts in the sustainability process is not done using the term "sustainability" and is carried out in other and broader fields such as environmental conflict research in political science and research on natural resource use and management (such as fisheries, water and land use management), peace and development, commons and common pool resource management and political ecology (ecological distribution conflicts), among others. To exclude this research from sustainability and transition research would be a fallacy of specialisation. Most of these conflicts arise from the interacting and interlinked fields of environmental policy and governance conflicts appearing in sustainable development or are of a specific nature such as conflicts about energy use or conflicts related to climate change and the migration conflicts that causes. A second problem with this review, like most other specialised research published in scientific journals, is that it is mainly communicated to and within the scientific community, with other specialised researchers—although many scientific journals describe their target audience as interdisciplinary, including practitioners and decision-makers. Large parts of sustainability and environmental research are not sufficiently and systematically communicated across the boundaries of science and practice, nor in interdisciplinary or transdisciplinary forms of knowledge communication, sharing and integration.

The organisation of academic research and education based on specialisation and the neglect of interdisciplinary and transdisciplinary research are not the only reasons for this insufficient boundary crossing in sustainability research; there are also misleading views and definitions of sustainable development and sustainability, for example the prejudices that conflicts are negative social phenomena that should be suppressed and prevented and have no positive function in driving development, social change and transformation processes. Such prejudices can be supported by the perception of conflicts as inherently violent and potentially escalatory—yet this is not the case with the majority of environmental and resource use conflicts, which can be addressed in routine processes of environmental policy and natural resource management.

Conflicts and conflict-generating crises of development and change are neglected in the discourses of global environmental change and sustainable development, which view the latter as a "win-win situation". Although sustainable development is implicitly conflict-related (masked by the rhetoric of intragenerational and intergenerational solidarity), no systematic efforts or solution processes and mechanisms have yet been established to address conflicts at different levels of the transformation processes. Conflicts remain part of sustainability diplomacy and other strategies to negotiate conflicts, for example through the established procedures of conflict mediation (Vasconcelos and Flavia Silva 2015). It is the complexity of the problems and of the sustainability process itself, rather than the conflicting interests of participants and inequalities, that is seen as the challenge to tackle in politically organised processes of governance. More elaborate strategies of transformation and continuous conflict mitigation are required to address the conflicting and vested interests that will arise on the road to sustainability.

Neglecting the social dynamics and complexity of global social and environmental change with development crises (including manifold conflicts between social groups, organisations, countries, national economies and actors and institutions in the international arena of sustainable development) causes major difficulties in the sustainability discourse and the connected economic and ecological discourses. When conflicts are not publically perceived and politically discussed, they find too little attention to become mitigated or mediated through organised governance processes, or tend to be seen as private or individual conflicts between persons. Conflicts experienced in the daily routines of the sustainability process as social, political, economic and ecological clashes over pollution or over access, distribution and redistribution of resources are analysed and discussed in limited and specialised research in economics, ecology and anthropology—as economic and ecological distribution conflicts or as economic cleavages between the Global North and South, or between the environmentalism of the rich and the poor—but they not dealt with systematically in national and international sustainable development policy. Conflict research as a topic dispersed across various disciplines—political science, psychology, sociology and economics—requires interdisciplinary cooperation and knowledge integration. Such integration can illustrate additional factors such as economic development and economically rooted conflicts that are missing in specialised research or theories of modern economy and society. In critical variants of political economy, institutional and ecological economic theories and theoretical concepts are more systematically used and applied for the analysis of environmental and sustainability problems, for example analyses of historical resource use crises and resolutions, historical transformations of economic and societal systems, different kinds of long waves and cycles of the modern economy and the dynamics of local, national and global processes interacting in economic development and resource use. In the sustainability process some of the latent dynamics of conflicts and the conflict theme have become visible since the mid-1990s in the discussion about the relations between economic globalisation and sustainable development (Box 7.4).

Box 7.4 Conflicting Processes: Globalisation and Sustainable Development

The Commission on Global Governance (1995) formulated in its final report a critical verdict of globalisation and the management of economic interdependence (the concluding chapter looked at problems that will arise in building more effective global governance institutions than the UN):

> The globalization process is in danger of widening the gap between rich and poor. A sophisticated, globalized, increasingly affluent world currently coexists with a marginalized global underclass.
>
> The pace of globalization of financial and other markets is outstripping the capacity of governments to provide the necessary framework of rules and cooperative arrangements. There are severe limits to national solutions to such failures within a globalized economy, yet the structures of global governance for pursuing international public policy objectives are underdeveloped.
>
> The time is now ripe—indeed overdue—to build a global forum that can provide leadership in economic, social, and environmental fields. This should be more representative than the Group of Seven or the Bretton Woods institutions, and more effective than the present UN system. We propose the establishment of an Economic Security Council (ESC) that would meet at high political level. It would have deliberative functions only; its influence will derive from the relevance and quality of its work and the significance of its membership.

The report was controversially discussed in the public policy processes. However, its clear conclusions and recommendations did not significantly influence the global policy discourse that followed. As in the Brundtland Report, and other reports of the UN about progress towards sustainable development, the diagnosis of problems and description of the situation is detailed and critical but the necessary action for building global environmental governance and sustainable development is described, until today, as not yet being underway. Assessing the problems to be dealt with the report described global governance as a continuing process in which many conflicting interests need to be settled—a realistic view of the connected processes of global governance and sustainable development.

Source: Commission on Global Governance 1995

Critical questions about globalisation continued after the report of the Commission on Global Governance, together with rapidly growing research on globalisation and its many faces and facets. The critical debate includes questions of ecology and sustainability, democracy and the political and cultural consequences of globalisation (Archibugi et al. 1998; Held and Koenig-Archibugi 2003; Lapka et al. 2007; Wolin 2008) in the broad fields of research on global governance and new normative orders.

7.3.2 Conflict Economics

The discussion of conflicts and controversies surrounding the collective use, distribution or sharing of natural resources provides an opportunity to discuss the use of economic methods and models of valuation of natural resources for their significance as means of conflict resolution. In the scarcely developed economics of conflict resolution (Bercovitch et al. 2009; Sandole et al. 2009; Croson and Bolton 2012; Kimbrough et al. 2017; Anderton and Carter 2019) resource use conflicts are studied from a disciplinary economic perspective, adopting the concepts, models and methods from conventional economics for the analysis of local and large-scale conflicts that appear continuously in the appropriation, production and processing of resources. In economics the conflict theme appears as a new theme insofar as the dominant view in mainstream economics is that conflicts are avoided by establishing optimal property rights and functioning markets. From an interdisciplinary and critical perspective there is not much new knowledge to be found in conflict economics—although it does correct some of the misleading assumptions of neoclassical economics in empirical research.

Conflict economics develops, similar to other sub-disciplinary specialisations of economics, especially environmental economics, with the use of the model-based research that is widespread in economics, and with a focus on the appropriation and use of natural resources. Conflicts change the view of the modern market economy and the processes of appropriation and allocation of resources, exchange and competition, usually considered as non-violent processes, peaceful action, where violence and conflict have been "civilised" through institutional rules and regulation within the economic system itself. With that the imperfect forms of economy in the earlier history of modern society, in the colonial economy, economic forms of robbery, plunder and booty, are considered as transformed into peaceful forms of the market economy that can be for the advantage of all; only violent political conflicts, civil war, war and terrorism seemed to create conflicts and violence that threaten the economic processes. This seems to be an illusionary view. The violent and non-violent conflicts in the economy do not come from outside, from imperfect policies or political conflicts. The conflicts emerge in the economic processes themselves, under conditions of increasing scarcity of natural resources, indicating the inherent conflict dynamics in the modern economy. The unleashing of resource use conflicts through exponential growth and the deregulation of markets in the globalisation process has created new conflicts, in an economy of decay, where power, violence and conflict return in the routines of economic processes. They could be kept out only temporarily, under conditions of resource abundance that did not require limitations on the appropriation and use of natural resources.

Regarding policy and regulation instruments for conflict resolution, economic conflict research has little to offer beyond what is found in existing knowledge about conflicts and the practices of conflict resolution, for example in environmental and sustainability policy. It does not offer much in terms of what would be necessary to explain the causes of conflicts theoretically, through the analysis of the complex systemic mechanisms in which economic action is framed and regulated. Economic action is not sufficiently explained by anthropological assumptions about rationality, or its lack, or by individual behaviour research. Policy instruments for conflict resolution include a broad class of economic and market-based instruments, negotiations, compensation payments for losses, subsidies and other tools that can be applied to many local resource use conflicts. But economic instruments alone do not resolve conflicts, nor prevent them by perfecting how economic institutions function or providing property rights and markets. For any local and limited conflict in resource use there is rarely one perfect method or one final solution, only temporary solutions that are created through combinations of different forms of policy instruments (see Chap. 8).

Property rights (see Chap. 4) are a complicated theme in economics, where property is not seen in its legal forms, but in its economic functions, as creating and maintaining access to resources. The functional logic of the market economy is not only that of exchange and buying and selling resources mediated through money and pricing. The constitutive processes that make the routines possible need to be analysed: what makes markets, buying and selling possible? What makes individuals and citizens, consumers and economic actors? How do the complex systemic structures and processes of which economic and resource use processes are a part affect economic functioning? In the modern economy the basic institution that enables markets to function is seen as the private ownership of resources and means of production. This is valorised in orthodox economics through the following reasoning: the appropriation of resources is made free of violence through perfect property rights that can be enforced without further costs to the owners. The abstract formulation of "perfect property rights" conceals the very assumptions that guide the reasoning behind and nature of property rights. Besley and Ghatak (2009) analysed the relationship between property rights and economic activity and surveyed some of the empirical and theoretical knowledge of the determinants of property rights. In similar forms questions about property rights have been studied in commons research (see Chap. 3), creating empirical knowledge that provides possibilities for differentiated reasoning, leaving aside general assumptions and ideas of a perfect form of property rights that guided the arguments of Hardin and others in the discourse of the commons.

With research in conflict economics develops a more critical view of appropriation. However, it is formulated in abstract models and idealised cases with single persons, for example in the power-based model of "guns vs. butter" that illustrates the functions of conflict: in reductionist form as a conflict between two players who have to decide between the reproduction of resources and the production of weapons with the only purpose being to appropriate the resources produced by the other player under conditions of competition (Hirshleifer 1991). The paradox of power that Hirshleifer constructed—that poorer and less powerful actors can often improve their position in a conflict relative to that of the strong and rich—should also help to explain political redistributions of income from the rich to the poor. But it has little explanatory power in the real and complex situations where redistribution should happen, for example in sustainable development.

The main and final idea that summarises economic thinking about conflicts and conflict resolution in economic and political processes, where the use and management of natural resources is at stake, is summarised by Mason and Muller (2007) in paradigmatic form as transforming environmental and natural resource use conflicts into cooperation of the conflicting parties. This is an example of a functional logic that could be specified as: whatever helps to transform conflict into cooperation is good for conflict resolution and re-establishing smooth processes of natural resource use in the economy. The irrationality of conflicts is thus seen as the different views and interests of the conflicting parties that do not understand the necessity of cooperation in economic processes.

7.3.3 Policy-Related Research on Environmental and Resource Use Conflicts

Policy-related research comes closer than conflict economics to discerning the complex reality of environmental and resource use conflict (Stepanova and Bruckmeier 2013). Environmental conflict research intensified after 1990 in the political sciences, with security and scarcity themes discussed by the Toronto school of conflict research (Homer-Dixon 1999; Swiss ENCOP-Project: Spillmann and Bächler 1995; Bächler 1999). In the debates about global environmental change that followed the connected processes of social and environmental change resulted in conflicts being intensified, resource conflicts being militarised and wars and civil wars being waged (Gleditsch 2004; Welzer 2008; Dyer 2011). The broadening of conflict research brought in new practices of conflict resolution, mediation and various forms of informal conflict resolution beyond those implemented by governmental institutions. The causes of conflicts, the social contexts of sustainability and the design of policies and strategies of conflict mitigation require interdisciplinary knowledge integration and theoretical analyses of social and ecological systems, their functioning and their coupling.

The different interests and aims of humans as resource users have normative meanings that affect conflicts from emergence to resolution. Environmental conflicts can be rooted in different worldviews, cultural values and social norms, in specific material interests, or in basic human needs such as food, shelter and health. In the political discourse and in environmental policy processes, conflicts appear in normative views and interpretations resulting from the interests, claims, aspirations, rights of access to and ownership of resources and the perceptions and valuations of conflicting parties, owners or citizens. Nobody can provide an "objective" view of the conflict. In conflict mitigation, the normative dimensions are visible in struggles over the definition, interpretation, aspiration, acceptance or rejection of claims and interests, and the contestation of formal or informal norms. To deal with the inherent normativity of conflicts the identification of worldviews, cultural values or clashes of values of the conflicting parties is not enough. A more critical question is: how is mutual recognition, cooperation and mitigation possible when conflicts are deeply rooted in the structures and processes of complex social and ecological systems? To find an answer, it is useful to turn to the knowledge, concepts and methodological problems of environmental conflict resolution from different disciplines. The discourse of social-ecological transformation provides opportunities to better integrate conflict research and conflict resolution processes in the broader sustainability process.

7.4 Relapse, Failure, Disasters: Unexpected Problems in the Transition to Sustainability

With the discussion of conflicts and crises another theme appears in the sustainability discourse: the failures of earlier policies and governance processes and of research (science failure), policy practice (policy failure) and economy (e.g. market failure). Of these different forms of failure only policy and market failure has received much attention. Science has seemed to have escaped controversy owing to its status as a producer of valid knowledge that cannot fail in theory, only insufficiently or incorrectly applied in practice. With the development of interdisciplinary environmental and sustainability research this traditional view of science collapses (reflected in the epistemological approaches of new knowledge production discussed in the preceding chapters). Science is now perceived more as a specific form of social action where failure in research and knowledge use is as widespread as it is in political and economic policy and practice. The failure in all three spheres can be seen as knowledge-related insofar as it begins with insufficient knowledge and inadequate forms of knowledge production and application when complex systems and processes are to be accounted for in sustainable development and in interdisciplinary knowledge integration. As the discourses of sustainability science and post-normal science have shown, science and governance proceed without safe knowledge, with tentative methods and procedures of knowledge integration, without exact data and without consensus on the long-term perspectives of change and transformative action. To deal with this epistemic situation requires changes in scientific knowledge production and in the transfer and application of knowledge in the sustainability process. Open, honest and public debates about the uncertainties in research and governance processes are a necessary part of the practices of knowledge use.

In addition to factors that block sustainable development such as conflicts or institutional complexities and inertia, there are other forms of silent, passive or manifest resistance to change and transformation that are not easily dealt with in policy and governance processes. They include different forms of resistance related to human, individual and collective behaviour (Table 7.1) that are known from social research. They need to addressed in different ways in the sustainability process, but particularly in environmental and sustainability education and in policy processes.

Table 7.1 Resistance to change

Causes	Definition	Interpretation	Sources
Values and attitudes	Values and norms that block environmental awareness; worldviews and views of humanity, for example dominant Western worldview, human exceptionalism paradigm; certain political and cultural ideologies; emotions (disinterest, fear of change); environmental scepticism	Cultural values and norms that cause resistance to behaviour change or engagement in environmental action are manifold and different; they do not appear in isolated form but are interwoven with the other forms of resistance described in this table	Social scientific research (sociology, political science, psychology): attitude research (e.g. value changes studied by Inglehart)
Interests and motivations	Vested interests as strong interests to maintain a system because of individual benefits, although the system works badly and brings disadvantages for all; motivations of egoism, greed or mistrust	Specific individual and collective interests and motivations are the main social explanations for human behaviour (intended or non-intended, conscious or unconscious) in forms of resistance to change	Psychological research
Social behaviour and action: dysfunctional routines	Habitus, habits, customs, routines and ways of life that prevent change; free-rider behaviour; certain forms of deviant behaviour	Socially institutionalised forms of behaviour that structure daily routines in the social lifeworld	Research in sociology and cultural anthropology
Knowledge: knowledge practices and knowledge cultures	Resistance caused by lack of awareness, knowledge and environmental education	Individual and collective knowledge practices that cause resistance to change are not simply a consequence of ignorance but of the selection of knowledge	Research on education and social communication in everyday life (ethnomethodological research)

(*continued*)

Table 7.1 (continued)

Causes	Definition	Interpretation	Sources
Power relations: power blocks, hegemonies	Asymmetrical power relations in national and international politics: traditional social, political and economic elites in power (hegemonic power relations through predominance of a social elite or group, or of a state over others)	Asymmetrical power relations exist at all levels of political and economic power (local, national and global), connected with social and economic divides; the most important problem for the sustainability process is the global divide between North and South, institutionalised in the modern economic world system since early modernity; its change requires long-term strategies and action additional to political programmes	Research in political science, interdisciplinary research (social and political ecology)

Sources: mentioned in the table; own compilation

7.4.1 Sustainability Transformation as Blocked Process

The sustainability process has faced immense challenges since the beginning, not only owing to the difficulties of conceptualising complex and long processes of change that reach into the unknown distant future. More important are the manifold obstacles in the present situation of sustainability transformation resulting from inertia and contradictory and maladaptive changes in societal, political and economic systems where actors and institutions are ill prepared for the challenges and changes that must be faced in transitioning to a sustainable future. To continue with the global sustainability process requires several initiatives of renewal: analysing the failure of earlier policies, clarifying the nature of the sustainability process as a social-ecological transformation of complex systems, analysing the social, political and economic obstacles to transformation and building transformative action groups at all levels of the process.

The idea and concept of social-ecological transformation developed from interdisciplinary discourses in social and political ecology, sustainability science and transformation research, not from the disciplinary knowledge bases that provide fragmented and specialised knowledge. In economics, for example, the complex problems of system change, transformation and the future are hardly reflected on and analysed. Great transformations, although studied in economic research of the history and development of human societies (Polanyi 1944; Polanyi et al. 1957), are not processes that can be organised and planned like the economic processes that are the themes of economic research. Societal transformations are only understood scientifically after they have happened: only then are the usual methods for knowledge generation through research applied. What is known in advance, about the problems and obstacles to a future transformation, is mainly knowledge about presently known problems and their potential solutions, whereas complex and newly emerging problems require continuous research and knowledge integratuion. The historical dis-simultaneity of industrialisation seems to make common strategies for economic development impossible; industrialisation started in countries of the Global South in the twenty-first century, by which time those in the Global North had begun the transition to a post-industrial society. Clashing forms of change (economic globalisation and sustainable development), social and economic inequality and accelerating processes of ecological change (global environmental change) make the processes of sustainability transformation confusing, evoking questions of the kind: when is it too late to mitigate climate change?

The knowledge available about the present state of society and nature cannot be neatly and completely translated into requirements, steps, procedures, means and ends of the further transformation. The present problems appear rather to be blocking the future and blocking the changes required to solve these problems: failures, delays, difficulties and closing windows of opportunity and time are the signposts of change that cannot be translated into planned action, or at least only into the logic of "trial and error", correcting errors through collective learning in a continually renewing process of transformation.

The present processes that block the transition to sustainability can be grouped into four forms: (a) systemic blocks resulting from the constitution of social, political and economic systems; (b) obstacles to collective action, agency and decision-making; (c) lack of knowledge and experience; (d) lack of consciousness and willingness to change routines of behaviour and institutionalised action. Of these factors the most discussed are obstacles to collective action, especially the lack of a unified social subject of transformation. The subject of transformation is not a specific social group or class, not a specific political organisation or government, not specialised scientific institutions and researchers, not economic institutions and enterprises, not social movements and not groups with joint worldviews and ideologies. All of these actors and institutions are necessary in the process, in different functions and for the building of transformative action groups, but the subject remains a fragmented social subject of collective action.

The twenty-first century appears as a century of conflicts caused by global social and environmental change: exponential growth of environmental disasters, growing conflicts and violence, terrorism, civil wars, resource wars, climate wars, uncontrolled urbanisation and migration, global mobility and refugees on the run from disasters. Opportunity structures for sustainability transformation cannot simply be created by coordinating temporal, spatial, social and knowledge-related conditions of collective action. The dilemmas of multi-scale sustainability governance require further reflection, analysis and joint discussion. How to prepare for what cannot be planned for, when actors in society and the economy are confronted with unpredictable events, surprises, turbulences and disasters, is the guiding question of the resilience discourse.

7.4.2 The Resilience Discourse

Resilient social, economic or social-ecological systems are characterised by the capacity to persist in spite of disturbance, disasters or catastrophes, for example by stabilising at a lower level of functioning or by changing the trajectory of development. Whereas such adaptive capacity appears as a strength of systems in preventing collapse, it is less clear how such capacities work in the sustainability process. It can be assumed that resilience is an adaptive capacity to cope with unexpected crises and turbulences that may also appear in the longer sustainability transformation. Coping successfully with disturbance would then enable the transformation process towards sustainability to continue. In this view there would be no problems to integrate the concepts of resilience and sustainability, connecting two temporal dynamics of short-term adaptation and long-term transformation that can be matched in a complex transformation process to maintain goal-attainment capacity and advance towards sustainability.

Another view of resilience emerges from ecological research: resilience as an adaptive capacity is replacing the more complex and complicated sustainability transformation. Resilience appears as the only dynamic of social-ecological systems that can be dealt with in environmental and sustainability governance, whereas sustainability appears too complex, long and complicated, insufficiently understood, and is overwhelming the capacity of agents and institutions. As a consequence, the idea of sustainability should be given up in ecological research; it may only be kept as a normative concept in politics, without scientific meaning, and no further research would be necessary (Benson and Craig 2014). This is an explicit reasoning which is not often given in ecological research. More often one finds a fuzzy reasoning, connecting resilience and sustainability and maintaining the idea of system change through sustainability transformation, but describing the change process in forms of resilience; for example, through the model of the adaptive cycle that is derived from ecosystems, where the development process is an iteration of four phases (Gunderson and Holling 2002). This conceptual model, although simplified and idealised, may have some significance in explaining changes in ecosystems that have another temporal dynamic than societal systems with the ability of sociocultural evolution, change and transformation. It seems that the historical dimensions and dynamics of societies are missing in the ecological interpretation of social-ecological systems. This implies, furthermore, ignoring large parts of social research about the long-term development, change and transformation of societal systems known from environmental history, historical ecology and social-ecological research.

The discourse of resilience developed from two original concepts in ecological research: that of "engineering resilience" (Pimm 1984), where an equilibrium or steady state of ecosystems is assumed and resilience is measured as the return time to such a state after disturbance or catastrophe, and that of "ecological resilience" (Holling 1973). Holling assumed and analysed ecosystem conditions far away from steady state, where disturbances result in shifting to another equilibrium state than the original (assuming that different states of equilibrium are possible). Resilience can then be measured as the magnitude of disturbance that a system can absorb before it shifts to another state. The two concepts do not exclude each other; they can refer to different types of systems with one (stability) or with several (multi-stability) stable states. What is controversial is whether the conditions of stability can only be theoretically constructed or derived from empirical research (Scheffer and Carpenter 2003).

Difficulties begin with the transfer of the resilience concept from ecosystems to social systems, and by constructing coupled social-ecological systems with the same concepts and in the same logic as ecosystems. This happened in the recent debate about coupled social-ecological systems, and also in Hollings's research, who applied the model of the adaptive cycle for ecosystems in describing the dynamics of social-ecological systems. With that came further controversies about the normative assumptions in defining the stable states of ecosystems (Scoones 1999; Manson 2008).

For the sustainability discourse and process the concept of resilience seems useful for differentiating between adaptive and transformative capacities of complex social and ecological systems, where transformation includes more and other forms of change than adaptation. Then resilience would appear as a reaction to disturbances, but not lumped together with other types of dynamics and further capacities of change. The reduction of processes of sociocultural evolution and development of human societies to the capacities of non-human agents in ecosystems is another problem when the resilience concept is applied in the forms and abstractions derived from ecosystem analyses. The minimum of clarification necessary for developing an interdisciplinary concept and theory of coupled social-ecological systems would be a clear differentiation of ecological and societal forms of resilience, such as would be effective in human societies and economies, where the model of the adaptive cycle cannot be applied in a meaningful way (it appears as a new form of mechanistic thinking in the analysis of societies), and such as would be effective in ecosystems, where the natural-scientific knowledge about physical, biological and ecological systems is that generating explanations. When the two forms of resilience are differentiated, it should be possible to maintain the concepts of resilience and sustainability as distinct in the sustainability discourse and process. However, it will be necessary to develop the analysis of crises, disturbance and disasters in other forms than have hitherto been proposed, studying such processes as opportunities to initiate changes other than that of adapting to or surviving catastrophes: as processes to support the more complex transformation to sustainability.

7.4.3 The Productive Potential of Failure and Crisis for Sustainability Transformation

Recent more systematic analyses offer potential routes to global sustainability. Sustainability is no longer seen as—only—paved through innovation, technologies, knowledge and capacities for development, but also through negative experiences such as failure, decline, conflicts and crises—and learning the lessons from these (see Box 7.5).

Box 7.5 The Productive Potential of Failure, Crisis and Decay

Newig et al. (2019) *discuss institutional dynamics hitherto ignored in the discussion of sustainability transformation*:

Current literature on sustainability governance and institutions is preoccupied with innovation, novelty, success and "best practice", but there is an emergent tendency to consider decline and failure as opportunities and leverage points to work towards and to achieve sustainability. Although failure, crisis and decay have been treated extensively, the link towards their productive potential has remained underdeveloped in the literature. Five archetypical pathways can be described, through which crisis, failure, deliberate destabilisation and active management of decline may facilitate sustainability transformation through adaptation, learning, providing windows of opportunity and informed choices regarding stability and change. The pathways that link aspects of failure to productive functions in the sense of sustainability (from literature in evolutionary economics, ecology and anthropology) include:

1. crises triggering institutional adaptations towards sustainability
2. systematic learning from failure and breakdown
3. the purposeful destabilisation of unsustainable institutions
4. making a virtue of inevitable decline
5. active and reflective decision-making in the face of decline instead of leaving it to chance

The common thread in these archetypical pathways is that all refer to institutional dysfunctionality, either taking dysfunctionality as an opportunity for productive change towards sustainability, or deliberately making normatively undesirable institutions dysfunctional, or using the opportunity to reflect and decide on the desirability of a declining institution.

Source: Newig et al. 2019

Newig et al. (2019) conclude from their discussion the necessity to connect the analytical concepts and the theoretical discussion of the productive potential of failure to empirical research about institutional change and governance, studying the functions of conflict, agency and of particular actors in institutional change processes. This coincides with the discussion of ecological distribution conflicts by Scheidel et al. (2018).

Ecological distribution conflicts as forces for sustainability are a specific form of disturbing events that may be studied for the possibility of converting them in forces for sustainable development (see Box 7.6). This is not a completely new idea, more a renewal of ideas discussed in earlier sociological debates about modern society and its dynamics of development. The discourse of critical theory, including various forms of system analyses and theories of modern capitalism after the critique of political economy by Marx and others, brought new ways of reasoning with crises, conflicts and instability. They are not only processes and events that threaten the existence and development of modern society, which makes them mechanisms of system transformation. They can also support capacity-building and the creation of agency for continued development, modernisation and growth such as, for example, in ideas like ecological modernisation. The sustainability discourse is still trapped in the "imperfect dialectics" of maintaining and transforming the modern economic system.

Box 7.6 Ecological Distribution Conflicts as Forces for Sustainability
Scheidel et al. (2018: 585) *argue against the view among scientists who feel that their voices are not heard: it might be that the scientists did not hear the voices of the activists fighting daily for sustainability.*

Can ecological distribution conflicts turn into forces for sustainability? Why, through whom, how and when can conflicts over the use of the environment take an active role in shaping transitions towards sustainability? A conceptual framework to map the linkages between different processes is required: (a) patterns of (unsustainable) social metabolism, (b) the emergence of ecological distribution conflicts, (c) the rise of environmental justice movements, and (d) their potential contributions for sustainability transitions. Ecological distribution conflicts can have an important role for sustainability; they show conflicting values over the environment as well as unsustainable resource uses affecting people and the planet. Environmental justice movements, born out of such conflicts, become key actors in political action against unsustainable resource use and they take sometimes radical actions to stop the conflicts. Through creative forms of mobilisation and action to reduce unsustainable resource use practices, people can turn from "victims" of environmental injustice into "warriors" for sustainability. But when will improvements in sustainability become lasting? This may happen, when the dynamics between the four processes are studied, resulting in a better understanding of ecological distribution conflicts, their effects and possibilities of solution in the sustainability process.

Driven by patterns of unsustainable social metabolism, ecological distribution conflicts often provoke the emergence of environmental justice movements.

(*continued*)

Box 7.6 (continued)

> Their collective actions to shed light on—and to transform—these resource uses damaging humans and the environment can contribute to transitions towards more sustainable futures in various ways .... From this perspective, conflicts bear a tremendous power of mobilizing social forces for change. (Scheidel et al. 2018: 595)

Environmental activists "are coming under increasing threat and repression. Violence against them has become systematic. Alliances for sustainability must, therefore, not only integrate in a fruitful way the work of academics and activists—for example through co-produced knowledge—but also seek growing institutional support for threatened grassroots activists. How such mechanisms of support and protection may look like in practice, remains to be explored. Relevance of developing such effective support is high as currently many of them are not only essential but also endangered actors for sustainability" (Scheidel et al. 2018: 595).

Source: Scheidel et al. 2018

The reversal of thinking about sustainable development with the discussion of the productive capacity of crises and of ecological distribution conflicts is part of a broader trend in the sustainability discourse and process. Reviewing and rethinking that what appeared in conventional views of sustainable development, which dominated in the discourse and process in the early phases of sustainability policies, helps to renew and strengthen the sustainability process. With the transformation debate it is possible to open a new discussion of sustainable development that makes better use of scientific and other knowledge through knowledge integration, and connects knowledge production and application processes more creatively.

7.5 Discussion and Conclusions

The experience after three decades of global sustainability policy and governance is articulated in the dilemma described in most global assessment processes by the UN (including the Millennium Ecosystem Assessment and others): *of the systemic changes necessary for a sustainable society and economy nearly nothing has been done and achieved so far. The time windows for initiating changes are closing; accelerating climate change undermines many of the change processes initiated through sustainable development policies*. The dramatic warnings are not efficient for a renewal of the sustainability process. Renewing the process requires a new conceptualisation of sustainability as social-ecological transformation and the analysis of problems and obstacles on the future road to sustainability. Conflict analysis and resolution is a necessary component of the renewal.

1. *Conflict resolution as pluri-dimensional process—knowledge, interests, power*: knowledge as a resource for conflict resolution in the sustainability process is more effectively used when it is "rationalised" through interdisciplinary and transdisciplinary knowledge integration. The integration makes visible the "blind spots" of disciplinary knowledge and neglected themes in the governance process, of which conflicts is but one, but the most important. In the processes of knowledge integration the conflicting normative assumptions and interests that guide researchers and other participants in the sustainability process need to be discussed. Interests in the aggregated form of joint interests for "our common future" (the title of the Brundtland Report) cannot be discussed in this abstract form in the governance process. Humanity as a collective subject cannot act jointly, its joint interests need to be articulated in the form of advocacy; someone needs to represent the interests of all humans. The interests that guide the sustainability process are such of specific groups, of influential actors and institutions that do not adequately represent the common interests of the present and future generations. Power-related questions, power asymmetries and vested interests of dominant groups need to be discussed in the future sustainability process to come closer to policies and governance processes that support the common interests of the Global North and South.
2. For the *creation of transformative agency as part of the renewal of the global sustainability process* it is necessary to take up the neglected themes of crises and obstacles and, furthermore, the cultural differences in the sustainability-related processes. To take into account neglected themes in the transformation process, the combination of the perspectives of system development and collective action in the governance process seems useful. System perspectives of transformation imply knowledge about the functioning and interaction of the complex adaptive systems of society, the economy and ecological systems in nature (questions of their reproduction, of the societal division of labour, of the social metabolism, of modes of production and accumulation regimes). Global environmental assessments such as the UN Millennium Ecosystem Assessment, global scenario analyses and transformation research all provide the knowledge relevant to the design and organisation of transformation processes at different spatial levels. Actor perspectives of transformation imply knowledge about the interests and behaviours of participants in collective action processes (social movements, networks, organ-

isations). The actors in sustainability governance are in a continuous learning process, which should be organised as such and not only as political process of cooperation; joint learning and cooperation can be supported though the creation of new rights and forms of environmental and global citizenship (Dower and Williams 2002). Creating transformative literacy, capacity and transformative action groups are the main forms of building and maintaining capacity and agency for the long transformation process. The social subject of sustainability transformation is more complex than the social subjects in earlier, national phases of modernisation with social classes and class movements. The transformation process comprises collective actors of different kinds (governmental institutions, political parties and organisations, business institutions, non-governmental actors like social movements and civil society actors, resource user groups and scientific institutions) that need to find concerted action. Environmental movements, only one group of transformative actors, are important for creating transformation capacity and mobilising people. To develop and connect the capacities of the participants and action groups, and the collective learning, strategies and action programmes for sustainability transformation, we need to experiment with different forms of action and actor constellations.

Summary of Chap. 7: Conflict, Relapse and Failure in the Sustainability Process

Conflict analysis and mitigation—realistic perspectives of sustainable development: assuming that environmental and resource use conflicts continue in the sustainability process, new development perspectives and improved procedures of conflict resolution need to be developed in the governance process if it is to become one of social-ecological transformation. Knowledge about global social and environmental change, conflict mitigation and sustainability transformation makes up the core of the renewal of governance. The scientific and practical learning in the transformation process is twofold: to learn how to integrate empirical research, theories and practical action, and to learn how to synthesise knowledge from social and natural-scientific research with practical and local knowledge. Learning from prior forms of policy and research, their misfits and their failures is necessary to prepare for more and intensifying conflicts as a consequence of global environmental change and the difficulties of conflict resolution. Beyond the transformation of conflicts to cooperation of the conflicting parties, to be able to solve the conflicts, more must be learned in conflict resolution to advance from the solution of single and specific conflicts to strategies where conflicts become driving forces of sustainability transformation.

Appendices: Further Information and Material

1. Questions and Individual Exercises

Learning Exercises

The information in Boxes 7.2, 7.3, 7.4, 7.5 and 7.6 (knowledge conflicts in the sustainability process, ecological distribution conflicts, the conflicting processes of economic globalisation and sustainable development, the productive potential of failure and conflicts in the sustainability transformation) can be used for further reflection and discussions in a group about sustainability and conflicts.

It is not possible with a single chapter to provide a systematic overview of the broad and diverse field of conflict research. You should choose one or several of the boxes and discuss them with the group to gain a first impression of what the difficulties are with regard conflict research for sustainable development. You can also watch a video on conflict research on the Internet to prepare for the discussion (to have an impression of the practice of conflict resolution). The only question you should try to answer is the one that runs through all conflict research: how broad and general should a conflict definition be (is it meaningful to rank under the term conflict such different types as intrapersonal and interpersonal conflicts, value conflicts, organisational goal conflicts, non-violent and violent conflicts, small and big conflicts, intergenerational conflicts, etc.)?

You should read further texts cited in the chapter when you work on conflict research: for example, when you plan or begin to write a thesis (see suggestions below).

2. Further Reading Suggested: Deepening, Thematically Specialised

Almered Olsson, G., & Gooch, P. (Eds.). (2019). *Natural Resources, Conflicts and Sustainable Development* (Earthscan Studies in Natural Resource Management). London: Routledge.

Croson, R., & Bolton, G.E. (2012). *The Oxford Handbook of Economic Conflict Resolution*. Oxford et al.: Oxford University Press.

Sandole, D.J.D., Byrne, S., Sandole-Staroste, I., & Senehi, J. (Eds.). (2009). *Handbook of Conflict Analysis and Resolution*. London and New York: Routledge.

Wärneryd, K. (2014). *The Economics of Conflict: Theory and Empirical Evidence*. Cambridge, MA: MIT Press.

Scientific journals (international, peer-reviewed) for the themes of this chapter: "International Journal of Peace and Conflict Resolution", "Journal of Conflict Resolution", "Environmental Policy and Management", "Environmental Management".

References

Allan, J. I. (Ed.). (2020). *The State of Global Environmental Governance 2019: Earth Negotiations Bulletin*. Manitoba: International Institute for Sustainable Development.

Almered Olsson, G., & Gooch, P. (Eds.). (2019). *Natural Resources, Conflicts and Sustainable Development* (Earthscan Studies in Natural Resource Management). London: Routledge.

Anderton, C. H., & Carter, J. R. (2019 [2009]). *Principles of Conflict Economics: The Political Economy of War, Terrorism, Genocide and Peace*. Cambridge et al.: Cambridge University Press.

Archibugi, D., Held, D., & Köhler, M. (Eds.). (1998). *Re-Imagining Political Community: Studies in Cosmopolitan Democracy*. Cambridge and Oxford: Polity Press.

Bächler, G. (1999). Environmental degradation and violent conflict. In M. Suliman (Ed.), *Ecology, politics and violent conflicts*. New York: Zed Books, 76–112.

Bannon, I., & Collier, P. (Eds.). (2003). *Natural Resources and Violent Conflict: Options and Actions*. Washington, DC: World Bank.

Bar-Tal, D. (Ed.). (2009). *Intergroup Conflicts and Their Resolution: A Social-Psychological Perspective*. New York and Hove, East Sussex: Taylor and Francis.

Bennett, J. W. (1976). *The Ecological Transition: Cultural Anthropology and Human Adaptation*. New York et al.: Pergamon Press.

Benson, M. H., & Craig, R. K. (2014). The End of Sustainability. *Society and Natural Resources, 27*, 777–782.

Bercovitch, J., Kremenyuk, V., & Zartman, W. (Eds.). (2009). *The SAGE Handbook of Conflict Resolution*. London et al.: SAGE Publications.

Besley, T., & Ghatak, M. (2009). Property Rights and Economic Development. In D. Rodrick & M. R. Rosenzweig (Eds.), *Handbook of Development Economics* (Vol. 5, pp. 4252–4595). Amsterdam: North Holland.

Bruckmeier, K. (2005). Interdisciplinary Conflict Analysis and Conflict Mitigation in Local Resource Management. *Ambio, 34*(2), 65–73.

Bruckmeier, K. (2015). *Social Conflicts in the 21st Century – the Shadows of Global Environmental Change*. Symposium of Debrecen University, Hungary, 2014. Metszetek 2015. Retrieved from http://metszetek.unideb.hu/.

Bruckmeier, K. (2016). *Social-Ecological Transformation: Reconnecting Society and Nature*. London: Palgrave Macmillan.

Bruckmeier, K. (2019). Environmental Conflicts: Towards Theoretical Analysis of Social-Ecological Systems. In G. Almered Olsson & P. Gooch (Eds.), *Natural Resources, Conflicts and Sustainable Development* (Earthscan Studies in Natural Resource Management). London: Routledge.

Burkett, P. (2006). *Marxism and Ecological Economics: Toward a Red and Green Political Economy*. Leiden: Brill.

Commission on Global Governance. (1995). *Global Governance: Our Global Neighborhood*. Report. Retrieved February 6, 2020, from https://web.archive.org/web/20000621111134/http://www.cgg.ch/contents.htm.

Corral, P., Irwin, A., Krishnan, N., Mahler, D. G., & Vishwanath, T. (2020). *Fragility and Conflict: On the Front Lines of the Fight Against Poverty*. Washington, DC: World Bank.

Croson, R., & Bolton, G. E. (2012). *The Oxford Handbook of Economic Conflict Resolution*. Oxford et al.: Oxford University Press.

Dower, N., & Williams, J. (Eds.). (2002). *Global Citizenship: A Critical Introduction*. New York: Routledge.

Dudouet, V. (2006). *Transitions from Violence to Peace. Revisiting Analysis and Intervention in Conflict Transformation*. Berghof Report Nr. 15. Berlin: Berghof Research Center for Constructive Conflict Management.

Dyer, G. (2011). *Climate Wars: The Fight for Survival as the World Overheats*. Oneworld Publications.

Elliott, D. (1997). *Energy, Society and Environment; Technology for a Sustainable Future*. London and New York: Routledge.

Fischer-Kowalski, M., & Haberl, H. (Eds.). (2007). *Socioecological Transitions and Global Change. Trajectories of Social Metabolism and Land Use*. Cheltenham, UK and Northampton, MA: Edward Elgar.

Gleditsch, P. (2004). Beyond scarcity vs. abundance: A policy research agenda for natural resources and conflicts. In *United Nations Environment Programme*. Nairobi.

Gunderson, L., & Holling, C. S. (Eds.). (2002). *Panarchy: Understanding Transformation in Human and Natural Systems*. Washington, DC: Island Press.

Habermas, J. (1991). *Erläuterungen zur Diskursethik*. Frankfurt am Main: Suhrkamp.

Hall, A. S., & Klitgaard, K. A. (2012). *Energy and the Wealth of Nations: Understanding the Biophysical Economy*. New York et al.: Springer.

Hannigan, J. (2006). *Environmental Sociology* (2nd ed.). London and New York: Routledge.

Held, D., & Koenig-Archibugi, D. (Eds.). (2003). *Taming Globalization: Frontiers of Governance*. Cambridge and Oxford: Polity Press.

Hirshleifer, J. (1991). The Paradox of Power. *Economics and Politics, 3*, 177–200.

Holling, C. S. (1973). Resilience and Stability of Ecological Systems. *Annual Review of Ecology and Systematics, 4*, 1–23.

Homer-Dixon, T. (1999). *Environment, Scarcity, and Violence*. Princeton: Princeton University Press.

Kasemir, B., Jäger, J., Jaeger, C. C., & Gardner, M. T. (Eds.). (2003). *Public Participation in Sustainability Science: A Handbook*. Cambridge: Cambridge University Press.

Kimbrough, E. O., Laughren, K., & Sheremeta, R. (2017). War and Conflict in Economics: Theories, Applications, and Recent Trends. *Journal of Economic Behavior & Organization*.

Klitgaard, K. A. (2013). The Failed Growth Economy, Stagnation, and Biophysical Limits to Growth. *International Journal of Social Science Research, 1*(1), 140–157.

Köhler, J., Geels, F. W., Kern, F., Markard, J., Onsongo, E., Wieczorek, A., Alkermade, F., et al. (2019). An Agenda for Sustainability Transitions Research: State of the Art and Future Directions. *Environmental Innovation and Societal Transitions, 31*, 1–32.

Langley, P., & Mellor, M. (2002). Economy, Sustainability and Sites of Transformative Space. *New Political Economy, 7*(1), 49–65.

Lapka, M., Rikoon, J. S., & Cudlínová, E. (Eds.). (2007). *Is Globalization Overpowering Democracy? The Challenge for Ecology, Economy and Culture*. Prague. Retrieved from www.dokoran.cz.

Lie, J. H. S. (2012). The Knowledge Battlefield of Protection. *African Security, 5*, 142–159.

Long, N., & Long, A. (Eds.). (1992). *Battlefields of Knowledge: The Interlocking of Theory and Practice in Social Research and Development*. London and New York: Routledge.

Manson, S. M. (2008). Does Scale Exist? An Epistemological Scale Continuum for Complex Human-Environment Systems. *Geoforum, 39*(2), 776–788.

Martinez-Alier, J. (1995). Political Ecology, Distributional Conflicts and Economic Incommensurability. *New Left Review, 211*, 70–88.

Martinez-Alier, J. (2003). Mining Conflicts, Environmental Justice and Valuation. In J. Agyeman, R. D. Bullard, & B. Evans (Eds.), *Just Sensibilities – Development in an Unequal World* (pp. 201–228). London: Earthscan.

Martinez-Alier, J., & Naron, S. (2004). Ecological Distribution Conflicts and Indicators of Sustainability. *International Journal of Political Economy, 34*(1), 13–30.

Martinez-Alier, J., & O'Connor, M. (1996). Ecological and Economic Distribution Conflicts. In R. Costanza, J. Martinez-Alier, & O. Segura (Eds.), *Getting Down to Earth: Practical Applications of Ecological Economics*. Washington, DC: Island Press.

Mason, S. A., & Muller, A. (2007). Transforming Environmental and Natural Resource Use Conflicts. In M. Cogoy & K. W. Steininger (Eds.), *The Economics of Global Environmental Change* (pp. 252–272). Cheltenham, UK: Edward Elgar.

Mason, A. S., Hagmann, T., Bichsel, C., Ludi, E., & Arsano, J. (2007). Linkages Between Sub-National and International Water Conflicts: The Eastern Nile Basin. In H. G. Brauch & C. Grin (Eds.), *Facing Global Environmental Change: Environmental, Human, Energy, Food, Health and Water Security Concepts* (Hexagon Series on Human and Environmental Security and Peace) (Vol. 4, pp. 325–334). Berlin: Springer.

Meadows, D., Meadows, D., Randers, J., & Behrens, W. W. III (1972). *The Limits to Growth: A Report to the Club of Rome's Project on the Predicament of Mankind*. New York: Universe Books.

Moore, J. W. (2014). The End of Cheap Nature, or: How I learned to Stop Worrying about 'the' Environment and Love the Crisis of Capitalism. In C. Suter & C. Chase-Dunn (Eds.), *Structures of the World Political Economy and the Future of Global Conflict and Cooperation* (pp. 285–314). Berlin: LIT.

Newig, J., Derwort, P., & Jager, N. W. (2019). Sustainability Through Institutional Failure and Decline? Archetypes of Productive Pathways. *Ecology and Society, 24*(1), 18. https://doi.org/10.5751/ES-10700-240118.

O'Hara, S. (2016). Production in Context: The Concept of Sustaining Production. In J. Farley & D. Malghan (Eds.), *Beyond Uneconomic Growth* (Festschrift in Honor of Herman Daly) (Vol. 2). Burlington, VT: UVM.

Pimm, S. L. (1984). The Complexity and Stability of Ecosystems. *Nature, 307*, 321–326.

Polanyi, K. (1944). *The Great Transformation*. New York: Rinehart.

Polanyi, K., Arensberg, K., & Pearson, H. (1957). *Trade and Market in the Early Empires*. New York: Free Press.

Rice, J. (2007). Ecological Unequal Exchange: Consumption, Equity and Unsustainable Structural Relationships within the Global Economy. *International Journal of Comparative Sociology, 48*(1), 43–72.

Sachs, J. D. (2005). *The End of Poverty: Economic Possibilities for Our Time*. New York: Penguin Books.

Sandole, D. J. D., Byrne, S., Sandole-Staroste, I., & Senehi, J. (Eds.). (2009). *Handbook of Conflict Analysis and Resolution*. London and New York: Routledge.

Scheffer, M., & Carpenter, S. R. (2003). Catastrohic Regime Shifts in Ecosystems: Linking Theory to Observation. *Trends in Ecology and Evolution, 18*(12), 648–656.

Scheidel, A., Temper, L., Demaria, F., & Martinez-Alier, J. (2018). Ecological Distribution Conflicts as Forces for Sustainability: An Overview and Conceptual Framework. *Sustainability Science, 13*, 585–598.

Schnaiberg, A. (1980). *The Environment: From Surplus to Scarcity*. New York: Oxford University Press.

Scoones, I. (1999). New Ecology and the Social Sciences: What Prospects for a Fruitful Engagement? *Annual Review of Anthropology, 28*(1), 479–507.

Southwick, C. H. (1996). *Global Ecology in Human Perspective*. New York and Oxford: Oxford University Press.

Spillmann, K. R. & Bächler, G. (Eds.), (1995). *Environmental crisis: Regional conflicts and ways of cooperation*. Bern and Zürich: Center for Security Studies and Conflict Research, Swiss Peace Foundation.

Stepanova, O., & Bruckmeier, K. (2013). Resource Use Conflicts and Urban–Rural Resource Use Dynamics in Swedish Coastal Landscapes. Comparison and Synthesis. *Journal of Environmental Policy and Planning, 15*(4), 467–492.

Varisco, A. E. (2010). A Study on the Inter-Relation Between Armed Conflict and Natural Resources and Its Implications for Conflict Resolution and Peacebuilding. *Journal of Peace and Development, 15*, 40–58.

Vasconcelos, L., & Flavia Silva, F. (Eds.). (2015). *MARGov Project*. Lisbon.

Wärneryd, K. (2014). *The Economics of Conflict: Theory and Empirical Evidence*. Cambridge, MA: MIT Press.

Welzer, H. (2008). *Klimakriege. Wofür im 21. Jahrhundert getötet wird*. Frankfurt am Main: S. Fischer Verlag.

Wolin, S. S. (2008). *Democracy Incorporated: Managed Democracy and the Specter of Inverted Totalitarianism*. Princeton and Oxford: Princeton University Press.

Part III

The Future: Sustainability Transformation

Science and Practice in the Sustainability Process

8

This chapter discusses the changing relations between science and practice with regard to knowledge creation for the transformation to sustainability. With the new forms of inter- and transdisciplinary knowledge creation and integration the disciplinary practices and the roles of the researchers change through the crossing of borders between disciplines and between science and practice (in the sense of management, policy, or governance). Economics and economic knowledge become here part of a broader discussion where the forms and conditions of the transfer of scientific knowledge to the practices in policy, economy and civil society need to be discussed critically, also with non-scientific and non-economic actors. This is missing in the conventional forms of knowledge transfer, in science communication, and in problem solving dominated by experts and scientists. Such unidirectional and hierarchical forms of communication are challenged through the new forms of bidirectional knowledge flows and knowledge sharing.

The science-practice relations and their changing forms are summarised in Fig. 8.1.

In *sustainability science*, conventional academic science is critically assessed and its methods found wanting, because they are inapplicable to both interdisciplinary research on nature-science interaction and knowledge relating to the sustainability process (Kates et al. 2001). This critique is part of the broader discussion about *new forms of knowledge production*, including inter- and transdisciplinary knowledge practices (Nowotny et al. 2001; Gibbons et al. 2010), where *knowledge is produced within knowledge application's social context.* Older knowledge cultures steeped in disciplinary specialisation are challenged through new, interdisciplinary practice-oriented forms of research and knowledge use, with the result that the processes of knowledge application become more complicated, even as the new forms of knowledge grow more transparent.

K. Bruckmeier, *Economics and Sustainability*,
https://doi.org/10.1007/978-3-030-56627-2_8

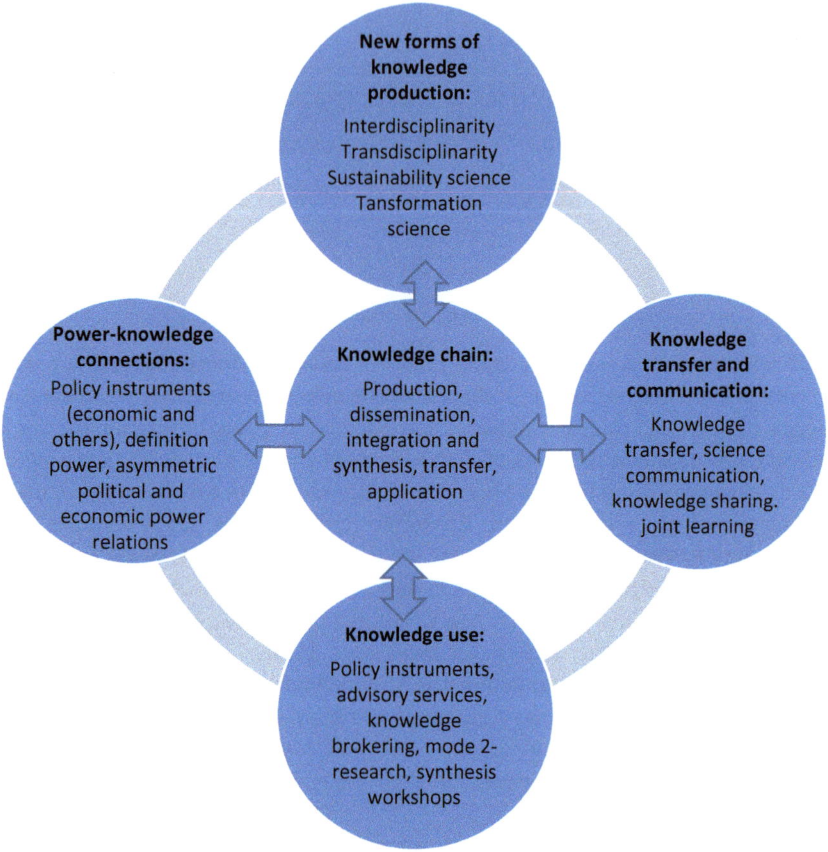

Fig. 8.1 Science-practice relations in the sustainability process

The difficulties of building knowledge bridges between science and practice for sustainable development are manifold (Bruckmeier 2009: 121ff). Consider: (a) Advanced specialisations in modern science separate research from the fields of application. (b) Theory-practice relations differ sharply in the natural sciences, the social sciences and the humanities. (c) The rapid, exponential growth of knowledge production in science and science's deepening specialisation make communication between specialised subjects more difficult and simultaneously more necessary to compensate the fragmentation of knowledge. (d) Knowledge transfer and application are complicated by competing research programs in other disciplines, especially in the social sciences, where conceptual boundaries are less clear than in the natural sciences.

Changes in scientific knowledge production and research approaches since the 1990s constitute reactions to the "explosion" of knowledge, as researchers must deal with vast amounts of knowledge, in order to bridge the gaps between science and practice. Deepening specialisation and differentiation in research subjects make the integration and synthesis of knowledge more difficult. Hence, knowledge integration, transfer and application require new methods to deal with the complexity of the systems and processes in nature and society. Inter- and transdisciplinary knowledge practices in the production, integration, and application of knowledge have, to this point, improved only slowly.

New methodologies and new forms of participatory research challenged conventional forms of research and their reliance on reducing complexity amid interacting systems via conceptual or mathematical models, theory connection and hypothesis testing. *Participatory research* and methods (Reason and Bradbury 2008; Chevalier and Buckles 2013) must, in short, develop together with the new knowledge production and become part of the change in science-practice relations. The roles of the participants in knowledge production and the forms and methods of knowledge integration and application need to be evaluated along epistemological lines. In environmental and sustainability research knowledge production requires working with uncertain results, inexact methods and heuristics, inter- and transdisciplinary knowledge integrations, and cooperation between scientists, practitioners and citizens. This scientific work functions as part of a transdisciplinary "mode two"-research method (Gibbons et al. 2010) that competes expressly with conventional academic research, or "mode one," and highlights the benefits of scientific cooperation with practitioners (Shinn 2002). Jamison (2002) described the changing forms of research during the twentieth century as the transition from disciplinary (academic) to multidisciplinary (bureaucratic) to transdisciplinary (commercial) research. This seems simplistic but captures important changes in the system of knowledge production and application. Newer environmental and interdisciplinary research provides examples for transdisciplinary research that reaches beyond commercial forms including the following: creating new capacities and skills (interdisciplinary communication, cooperation and knowledge integration, integration of science and governance), building networks and connections (through cooperation and participation of different actors and knowledge bearers), and creating extended processes of knowledge communication, exchange and mediation (communication between scientists, experts, decision-makers, resource users, political actors, movements, and citizen). Controversies about these new approaches's epistemological assumptions abound. Nevertheless, the new forms of knowledge production have gained a foothold in many fields of research, where they continue to develop.

8.1 Communication and Transfer of Knowledge in Environmental and Sustainability Research

Conventional practices of knowledge application have long been unidirectional, with knowledge flowing from science and research into organisations in the economy, in politics, education, health care and other services. In late modern society, by contrast, new knowledge forms, including inter- and transdisciplinary knowledge production, "mode two", "post-normal science", citizen science have expressly joined with social and environmental movements, and the environmental policy community to further sustainability discourses in a way that implements sociological theories of the risk society and promotes reflexive modernisation (Beck, Giddens).

Conventional forms of knowledge transfer from science to policy and decision-making are rooted in cultures of expertise, in which knowledge application is generally organised and channelled *through* technologies, as opposed to being diffused in a collective and critically self-reflective manner. Hence, the terms *knowledge management* and *knowledge transfer* describe forms of knowledge application in economic, political and professional practices in organisations, where the application is based on specialised knowledge produced specifically for professional application. Knowledge management includes practices of knowledge application in organisations with the help of information sciences, information systems, information technology and other technologies for purposes of improving organisational performance and innovation in competitive situations: that is in situations that are oriented toward traditional economic and business practices (Serenko et al. 2010; Serenko and Bontis 2017; Girard and Girard 2015). Knowledge transfer is similarly described as the organisational practice of disseminating knowledge to ensure its application for purposes of problem solving (OECD 1999; Argote and Ingram 2000; Perera et al. 2006). The terminology is not absolute, as actual practices produce overlap and change. Nevertheless, the underlying economic and organisational purposes are evident.

In environmental and sustainability research, however, forms of knowledge production and application expressly include newly-formed practices of transdisciplinary science, including ecological economics, sustainability science, social and political ecology, and emerging forms of transformation science. Amid the new knowledge practices, the knowledge flow from research to practice is becoming more varied, multidirectional, circular, and based on joint learning, with feedbacks between experience and practice and science and research becoming more intense and mutually sustaining. Moreover, new information technologies and the internet further support the intensification of knowledge exchanges and especially with reference to integration, review and assessment.

8.1.1 Critique and Change of Knowledge Transfer

The discourses about new knowledge production and sustainability criticise traditional forms of unidirectional knowledge transfer, on account of the latter's technocratic character, its tendency to distort through specialisation, its overemphasis on technology, and dependence on the alliance of expert power and political power. Interdisciplinary knowledge creation, integration, exchange and sharing require, however, more interactive and complex forms of communication. The existing differentiation between producers and users of knowledge is too simple to structure adequately the communication between science and practice. New forms of knowledge production, in contrast, including sustainability- and transformation-science, interact with more open, citizen-oriented and transparent forms of knowledge communication that cultivate bidirectional knowledge sharing and power equalisation.

The participants in the sustainability process, as in ecosystem-based management and natural resource use processes, work with several forms of knowledge, including scientific and practical knowledge from policy processes and natural resource management, as well as with traditional, local and indigenous knowledge. Thus, the knowledge transfer becomes a complex process of communication and translation between different knowledge forms. Given the difficulties that specialised research has in producing valid knowledge for complex forms of environmental and social change, new knowledge exchanges transformation-science for the construction of validity through interdisciplinary and discursive practices about knowledge's foundations and how to apply new knowledge equitably. From this perspective, competing knowledge from different theories or paradigms, as in economics and other social sciences, requires more formal discussion of knowledge's validity and of its production, application, influence on decision making and effect on collective action. In short, all forms of knowledge practices must become reflexive, reflecting always on their own procedures, methods and forms of communication.

For economic knowledge production and use that was discussed in the prior chapters, power is the traditional mechanism of knowledge transfer. It is not scientific knowledge's cogency alone that counts, but the power in political and economic processes that selects knowledge to fit specific interests and purposes. Thus, debates about growth, the green economy and TEEB are examples of how selective knowledge use serves the political and economic interests of powerful actors and institutions. It is not necessarily scientists and experts, but those who define power that frame discourses and define agendas in the political arena. Thus, only a coalition of powerful global players—political, economic and scientific institutions—will be able to define sustainable development as sustainable growth within a necessarily green economy, dismissing thereby contemporary doubts among scientists and politicians about the possibility of growth amid global limits on resources.

Power relations in science and knowledge production (Weiler 2011) comprise more than just cognitive apprenticeship in teaching and advisory systems among scientific practitioners. Power also courses through manifold forms of political, economic, cultural and scientific systems, forging a composite that is defined as knowledge by these different forms. In this respect, it is significant that Foucault broadened the concept of power from that of formalised relations of power and discipline in public policy and governmental action to include a universal subjective capacity, or a kind of strength that influences other persons by weaving together knowledge, discourse and communication. He analysed numerous examples of such disciplining power, including the power over individual lives or even over the lives of larger populations (Lynch 2011). Influence over behaviour within the bounds of science can take many forms, both violent and non-violent, since science comprises forms of social action and communication. Science is not, in this respect, the production of "pure knowledge". With regard to the society-nature relations that are often debated within the sustainability discourse, another form of power is relevant: collective human power over nature, which appears in phenomena as diverse as climate change, modification of ecosystems and genetically-engineered life, all of which are products of a specific knowledge-power nexus. Among the changes in power relations that are pertinent to the sustainability process are changes from hierarchies of power to networks of power—and this is obvious in public policy and governance, as the participation of non-governmental actors in the policy process has broadened the idea of political action. Along these lines, governance now implies new forms of legitimation of knowledge in the application processes.

The "triple helix" is another important form of knowledge production that is distinguished by new forms of cooperation between science (universities), politics (governments) and the economy (industrial enterprises) (Etzkowitz and Leydesdorff 2000; Shinn 2002). This approach is not simply an improved form of practice-oriented knowledge production, but also shifts back the filtering and selection processes, so that research is not defined purely by the coordinated interests of powerful actor coalitions that support primarily the commercialisation and commodification of knowledge production in exclusive pursuit of exploiting natural resources. Following this new practice, technology and innovation centres become important as mediating institutions that can alter commercialised knowledge production's trajectory.

The point is that in unidirectional knowledge transfer systems the interpretation, selection and modification of available knowledge—that is, when knowledge is applied—remains invisible. For that reason, the dominant emphasis on formal modelling within mainstream economics is insufficient for a proper analysis of knowledge's transfer and application. In effect, current methods are reductionist, limiting knowledge's ambit purely to cost-benefit analyses that are defined solely in economic terms. Nevertheless, economic knowledge about the functioning of economic institutions and processes is not relevant to practice, but applies exclusively to information about the monetary value of goods and resources. When practitioners think or act like economists and use economic knowledge in their analyses of management, policy and governance processes, the actual translation of knowledge from science to practice appears superfluous. Pearce (2002, see chapter six) formulated in precisely this sense the standard procedures for applying to economic analyses knowledge from environmental economics, neglecting all other forms of knowledge communication in the transfer process. The decision-making process in participatory resource management and co-management needs, however, to be much more complex, since economic knowledge must be balanced with not only ecological knowledge but also the practical knowledge of resource users. This kind of thing occurs within sustainability processes that emphasise bidirectional knowledge sharing.

Policy instruments (Box 8.1) or techniques of governance are a conventional form of transferring knowledge and exercising power in public policy. Knowledge transfer and application happens within in power-based forms, for example, through evaluation of policies and their success, or in decisions to change the economic processes of resource use. This can happen through economic policy instruments that imply coupled forms of political and economic action: taxes, subsidies, tradeable pollution certificates, tradeable quotas in fishery, or compensation payments for damages and losses. These economic instruments are characteristically adapted to the market economy and are also widely applied via neoliberal market deregulation. Other policy instruments, however, include persuasion and negotiation within knowledge communication in inter- and transdisciplinary knowledge practices. Policy instruments as techniques for knowledge transfer within the policy process reveal traditional forms of exercising political power in governmental action—that is, through political programmes and decisions, where power and knowledge are instrumental for regulating social action. Here, policy stimulates or sanctions the behaviour of social actors, resource users and economic agents.

Box 8.1 Policy Instruments as Knowledge Technologies

The conventional forms of knowledge transfer for complex processes of collective decision-making in policy and governance and practical application of scientific knowledge include knowledge standardisation through the construction of policy instruments and "science communication" for information of the public. In economics selective knowledge from policy research is adopted to construct policy instruments and action models for decision-making. The choice and combination of instruments are subjected to economic reasoning, which seeks exclusively cost-effective instruments and tools. The success of sustainability policy or governance depends, in this view, on combining policy instruments that are usually divided up in different classes:

1. *Political and law-based instruments*: political programmes, treaties, laws (governmental action) and control through police, military, law-and-justice systems
2. *Managerial and planning instruments*: public and political planning, management programmes for specific resources or ecosystems (wetlands, drylands, coastal zones, the seas)
3. *Scientific instruments*: forms of mandated and applied research, monitoring, review, assessment of policy programmes
4. *Economic instruments*: funding, economic incentives, subsidies, taxes, market-based instruments (for example, tradeable pollution certificates)
5. *Instruments for awareness building*: moral persuasion, consensus building, information campaigns, education, advisory systems

Most policy instruments are fashioned for national policies; international policy instruments are fewer (mainly treaties or agreements). International policy requires coordination of national and local political action, in addition to the combination of several instruments at different levels of policy. All policy instruments are, in effect, instruments for knowledge transfer and application, and with their use the knowledge process becomes complex (combining knowledge production, integration, transfer and application), although the knowledge process may still be conventionally unidirectional where knowledge flows from the authorised producer (science) to end users or target groups. The new forms of inter- and transdisciplinary knowledge production develop, however, more interactive, participatory, collective and learning-based forms of research and decision-making, with a special emphasis on bi-directional knowledge flows.

Sources: own compilation; Howlett 2012

Policy instruments are generally constructed for public policies and, as a result, are wholly instrumental in their view of policy and power relations. Yet, sustainability governance subjects to critical reflection precisely the power-knowledge nexus's role in shaping communication and knowledge transfer. Policy instruments reduce knowledge to a means for exercising power, in which compliance is enforced by rules that consider only knowledge transfer's effectiveness. The criteria of political and economic effectiveness (effectiveness for achieving the policy goals through instruments that understand cost effectiveness only in economic terms) are relevant only to conventional forms of political and economic action. Sustainability governance, however, assesses critically the adequacy and legitimacy of these criteria, holding up to scrutiny their ready justification of policy programmes and their tendency to ignore other criteria in environmental decisions. The social, spatial, and temporal limits of policy instruments and their inapplicability to the regulation of nature-society relations, becomes obvious at the point of application: they are power- and hierarchy-based knowledge technologies that use standardised processes of policy, management and planning and, for these reasons, need to be complemented by interactive forms of knowledge exchange, communication, cooperation, decision-making and learning.

Many different forms of intermediary institutions can facilitate a more robust transmission of knowledge from science to practice, including such advisory services as consulting institutions, knowledge brokers, and think tanks. All of these intensify exchange processes in a way that increases contacts between knowledge producers and users. With the new forms of communication and mediation within knowledge exchange, the transfer and application process becomes a complex system of organised knowledge flows that are sustained by multiple forms of communication, negotiation, mediation, sharing, feedback and joint learning. Unidirectional knowledge transfer becomes, in this respect, the shared, bidirectional and circular flow of knowledge.

Applied research, action research, and participant observation or participatory research, wherein scientists participate in the practices of action and knowledge application, cannot solve the problems of knowledge communication and sharing. These forms of research make the knowledge transfer more more complicated and imply the participation of scientists in knowledge application, as in the examples of medicine and psychotherapy. To legitimise and control the power relations in the various forms of interaction, communication, cooperation and negotiation about knowledge between scientists and practitioners requires open and transparent communication, ethical principles and guiding norms, in addition to epistemological and methodological rules for knowledge dissemination and application. This problem is exemplified in the life sciences. Contemporary biotechnological research routinely engages in genetic modification of plants, animals and humans, even though this practice's potential effects on humanity demand that stakeholder beyond scientists, experts and political powerholders be allowed to speak. The risks of such research need to be analysed in a way that elaborates new ethical rules and respects processes of democratic decision-making. Such open and critical reflection is possible only within inter- and transdisciplinary knowledge practices that emphasize sustainability in. research.

8.1.2 Complex Forms of Knowledge Exchange and Sharing

As a form of knowledge transfer knowledge sharing contrasts with bidirectional transfer between established binary groups. In the sustainability process three types of knowledge bearers cooperate: scientists and experts, practitioners and decision makers, and civil society actors. Knowledge sharing in unidirectional processes, as in the form of conventional teaching where knowledge flows from experts or teachers to disciples or laypeople, provides constitutes a selective approach that marginalizes questions of resource management and sustainability (Roux et al. 2006). Potential changes in the conventional approach emerge from science and technology studies, inter- and transdisciplinary knowledge production, and current research in environmental and sustainability sciences.

1. With the development of *science and technology studies* (Felt et al. 2017) science-practice relations are more differentiated, becoming circular in a way that allows us to study not only how science (research) and technology (innovations) influence society, policy and culture, but also the other way round. Scientific knowledge production is now seen more as a social process that can be analysed in terms of social action, not purely as a logical and methodologically structured process that follows accepted rules. In particular, the contingent nature of scientific knowledge's effects upon worldviews, paradigms, pre-analytic visions, and social norms and rules became an important topic in the wake of Kuhn (1962), which understood scientific revolutions not via progress in understanding, but in terms of paradigm changes.
2. *Interdisciplinary knowledge production* became an important theme since the 1970s with the intensification of environmental and sustainability research, as people have questioned existing boundaries between natural and social scientific disciplines; with the rise of transdisciplinarity developed, furthermore, a potential bridge between scientific and practical knowledge. This profound epistemological and social change in knowledge production augurs significant

changes for all traditional forms of knowledge transfer, but especially transfers within economics.

3. *Environmental research* also wrought significant changes of the views of knowledge production and application, especially within research about global environmental change, where intensive cooperation and knowledge synthesis is required. Traditional models and methods for verifying the scientific knowledge that they produce do not work for complex ecological systems and other processes that can be subject to rapid change. In this context in particular, scientific knowledge is understand as necessarily insecure and incomplete, with the result that is provides insufficient support for major decisions. Traditional forms of knowledge are, thus, deemed untrustworthy until they can be critically reviewed and discussed. This was paradigmatically formulated in the concept of *post-normal science* and in *sustainability science* (see Chap. 3), and reflected in the discourses of inter- and transdisciplinarity as new knowledge production with changing relations between science and practice.
4. In the process environmental policy and sustainability policy have also developed new forms of *public communication of science* that support public scientific literacy (Bucchi and Trench 2014), or, in the case of sustainability, a fully transformative literacy and capacity. The various forms of new knowledge production include significant changes in the relationship between science and practice. Environmental and sustainability research are paradigmatic examples: the weakness and insufficiency of scientific knowledge have been incorporated became part of the knowledge production and application processes, compensated through interdisciplinary cooperation and participatory research and decision-making. Scientific knowledge is negotiated in application processes and needs to pass through new procedures of selection and consensus building. Science no longer has a monopoly on valid knowledge. The consequences of these epistemic changes in the production and application of scientific knowledge have not discussed sufficiently and need to be taken into account in economic and environmental debates about the sustainability process.

The assumptions and premises of conventional economic research must be critiqued at the point of knowledge's application. (This happens more in other disciplines than in economics, public policy and governance.) Economics understands nature and the environment only with respect to scarcity, with the result that its theories and models of market-based exchange and economic valuation take no account of consequences, enabling thus the broader commodification of ecosystems and even of genetic resources. The normative assumptions guiding economic modelling and research were relatively immune to criticism as long as these were applied purely to natural resources that were traded as commodities, as in the cases of agriculture, fishing, logging and mining. The economic valuation and commodification of natural resources came under critical scrutiny when the underlying approaches were applied to other resources—that is, to ecosystems, their services, and genetic resources. It is not economic knowledge in general that has become controversial. Instead, it is mainstream economics's tendency to "economise ecology" that is now under review.

Scientific knowledge production's use in economics need to be connected and matched with different knowledge practices when it is applied in the sustainability policy and governance processes; sustainability cannot be reduced to or directed only by traditional principles in science and economics. Participants and practitioners must include groups that work with knowledge practices similar to those cultivated by scientists but that use scientific knowledge more systematically than, for example, managers and decision-makers. Other groups have different knowledge cultures, as in the cases of citizen groups and social and environmental movements. With different groups of actors cooperating within the sustainability process, different knowledge practices can be connected and mediated. Scientists become, thus, social actors with specific roles in the governance processes and. cannot, thus, define their roles only as producers of valid knowledge whose proper application is outside their portfolio; instead, they must become actors or stakeholders in processes that reach beyond research. Hence, environmental movements have served as actors who strongly criticise the dominance of economic thinking and practice from the perspective of the sustainability process (Box 8.2).

Box 8.2 Knowledge Practices of Environmental Movements

The knowledge practices of environmental movements are described by Jamison (2002) as "cognitive praxis" in three dimensions (consciousness is a part of all three). The development of environmental awareness happens at different levels, in different contexts, and. in different actor groups, where individual and collective awareness building are connected:

1. *Cosmological dimension*: (environmental action includes specific worldviews)
2. *Technical dimension*: ("small is beautiful": knowledge, awareness and learning related to technologies)
3. *Organizational dimension*: collective learning sites ("citizen science": includes organized processes of learning and knowledge use)

(*continued*)

Box 8.2 (continued)

Environmental movements differ in their application of knowledge, in spite of their ecological worldviews. Their practices and action, described by Jamison as citizen science, include scientifc, practical and normative knowledge in different combinations. The movements change their practices under the influence of internet-based information technologies. Jamison describes the knowledge culture that developed with the environmental movements as part of the new social movements in European countries in the second half of the twentieth century in form of "dimensions" with the common purpose of creating "knowledge for environmental action". This implies implicit or explicit critique of knowledge production and knowledge practices in academic science as inadequate for application in environmental and sustainability governance: the conditions for application, the selection of knowledge to apply, and the competition between different knowledge forms do not fit in the model of conventional knowledge transfer. The knowledge culture of environmental movements includes discussion and negotiation at all levels of the knowledge process from production to application.

Source: Jamison 2002

Jamison's analysis of knowledge practices of environmental movements and of changing forms of science and science-practice relations draws attention to further problems of application of scientific knowledge, including participative knowledge practices, normative components, and ethical reasoning in environmental governance processes.

8.1.3 Application of Knowledge from Environmental and Sustainability Research

The participatory knowledge cultures that develop within the sustainability process have already produced some new practices of knowledge production, integration and use, including joint learning wherein different knowledge forms are applied. Interdisciplinary assessments of social and ecological systems and their interaction have developed in research in global environmental change, where experts and scientists from many disciplines participate. Although this arrangement does not provide sufficient knowledge about the complex processes of change in the transformation to global sustainability, more knowledge is made applicable and continuous knowledge improvement becomes possible through learning from prior errors and mistakes. The new inter- and transdisciplinary knowledge practices developing in sustainability research, the differentiating forms of knowledge production, and the changing of principles can be described as follows (Bruckmeier 2009).

1. *Scientific knowledge production* has lost its reputation for producing safe knowledge: much of the scientific knowledge about the complex interacting systems in sustainable development is uncertain, inexact, insufficient, and devalued rapidly through changes with the systems studied and the rise of new knowledge production, as reflected in sustainability science and the epistemology of post-normal science.
2. *Knowledge production in the policy process* happens mainly through monitoring, review and evaluation of conventional forms of expert and power-based knowledge practices that are based on selective use of scientific knowledge. The review processes and their methods require critical reflection and improvement through more open and interdisciplinary review practices.
3. *Knowledge production in civil society and by social movements* is oriented toward the transformation of routines and behaviour in resource use and consumption in everyday life. Normative and cultural knowledge components in the social lifeworld need to be reflected, discussed and legitimised in the sustainability process when they are combined with scientific and policy-related knowledge production.

The significance of power relations in knowledge application: The creation of a policy-internal knowledge process in routines of monitoring, evaluation and assessment that is more open to reflexive and interactive forms of knowledge communication and joint learning with the aim of accelerating the creation of a new knowledge culture for the sustainability transformation. The established system of monitoring and evaluation cuts off, however, the knowledge flow between research and practice by establishing specialised intermediary institutions that monopolise knowledge's application and control. The system limits, thus, the discussion about knowledge integration and application to forms that support the interests of the actors in political and economic power positions. The experience of decision-makers and their fear that the policy process cannot deal with the manifold (and often contradictory) competing forms of knowledge offered by different disciplines and fields of research gains support and justification through the establishment of new routines for policy review and evaluation. Counter-arguments hold that the policy process would collapse if stakeholders brings conflicting interests and alternative knowledge practices to the process. In this sense, only controlled and limited forms of participation can be practiced.

Knowledge filtering, selection and review mean that the journey from science to policy and into application is simply part of a closed and obscure process, because asymmetrical power relations among the actors in the governance process are determinative. Knowledge does not become power by itself or through an inherent quality of scientific knowledge, but through the selection of certain knowledge that overrules others in political process; in the case of global sustainability policy this is the knowledge from orthodox economics. The power relations inherent to public policies make the use of knowledge a question of power throughout the whole knowledge process—that is, from production to dissemination and, finally, to application. "Definition power" is too simple a term for understanding the complexity of-power-knowledge interaction; but it illustrates something about the power struggles that occur within political knowledge practices. In the sustainability process a more systematic analysis of knowledge practices and a more open process of knowledge transfer and communication will be necessary to deal with the difficult questions of asymmetric power relations, knowledge selection, different interests and vested interests. To overcome the blocking of the sustainability transformation changes in power relations, empowerment of actors, new forms of knowledge use and control of decision-making will be necessary. Re-conceptualising the practices of production and application of scientific knowledge in the sustainability process includes a theoretical component (to analyse the relations between science, politics, economy and civil society), and a practical component, (to reorganise the use and application of scientific knowledge, economic knowledge and knowledge of other kinds); both components are necessary for creating collective agency and strategies for sustainability transformation.

The role of ethics in environmental decision-making: The complicated interactions between knowledge and power in the policy and governance processes become further complicated through the necessity of ethical reflection and justification as part of the environmental decisions about the interests of both nature and future generations. A classical ethical principle discussed in environmental ethics relevant for this purpose is that of "following nature" (Box 8.3).

Box 8.3 Ethics in Environmental Decision-Making

A classical theme in environmental ethics: Can and shall we follow nature in environmental politics and action?

Rolston (1999): *Ecocentric ethics* have as their guiding idea "nature first". There are different meanings of "naturam sequi" ("following nature" or "nature knows best"):

1. *Absolute meaning*: humans are not able to act against the laws of nature—morally irrelevant variant (but historically relevant in early modernity and especially in Enlightenment thinking, for example in physico-theology: see Chap. 2)
2. *Meaning of action as artefact*: "all action is unnatural"—we cannot follow nature through action—morally irrelevant variant
3. *Relative meaning*: (more or less to follow nature), on four variants:
   (a) *homeostatic meaning*: nature as total network of creation ("retinity") which needs to be taken into account in responsible action for the environment and for future generations
   (b) *imitative meaning*: trying to live naturally
   (c) *axiological meaning*: identifying nature's values
   (d) *pedagogical meaning*: orientation in life (ways of life or practice of life) through experience of nature

Rolston's reasoning reveals the difficulties in arguing for an ecocentric ethic, although the problem is intensively discussed in environmental ethics and has motivated many environmental movements and activists. An environmental virtue ethic that focuses on human excellence and fulfilment in nature (Cafaro 2001) is less discussed in environmental ethics. All arguments for an ecocentric ethic can be refuted and disproved through other approaches. Hence, no final and convincing argument has been made for an intrinsic value of nature.

Environmental ethics is a controversial discourse that competes with different worldviews and principles of moral behaviour. To make ethical values and principles applicable for environmental politics and action cannot simply be done through decisions for one or another ethical approach. Ethical criteria develop through continuous discussion, in a discourse, where the consequences of different forms of ethical reasoning clash and need to be negotiated. In modern discourse ethics as that of Habermas (1995) the principles of discursive reasoning in public and political ethical discussions are elaborated for discourses about democracy and legislation. But discourse ethics is not limited to these spheres and is also applicable in environmental ethics.

Sources: mentioned in the text

The discussion of ethical questions in the sustainability discourse is based on the concepts of *solidarity* with regard to humans (responsible action for the well-being of all now living and future generations) and *retinity* with regard to nature (humans are part of the network of creation in nature and are responsible for maintaining this network; the idea originally developed in Christian ethics: Krämer and Vellguth 2017). Both ideas can be interpreted as similar in their implications, as they highlight the responsibility for maintaining life, nature and the natural resource base in all forms of human action, especially as formulated with the notion of sustainable development for the practices of environmental and sustainability governance. Regarding the economic valuation and commodification of nature and natural resources, arguments against unlimited economic appropriation and use of nature can be developed from different ethical approaches, including *ecocentric ethics* and the idea of an intrinsic value of nature as discussed by Rolston (Box 8.3) and the *virtue ethic* as discussed by Cafaro (2001), where the alternative to a purely economic view of nature is to see economic activities as supporting human welfare and well-being, rather than as powering an engine of endless growth, acquisition and consumption (Cafaro 2001: 13).

The consequences of the principle of responsibility for maintaining nature and the conditions of life on the earth are apparent in natural resource use. In the final analysis the earth's functioning as total ecological system must be maintained and the life of both humans and other species protected. The conditions for this maintenance of the functioning of the earth can be formulated, to a certain degree, with respect to ecological indicators and in terms of positive knowledge. But in practices of resource use, to observe the conditions and the limits of resources requires complementary ethical principles such as that of intra- and intergenerational solidarity. The empirical information about limits to growth cannot replace ethical principles: as soon as empirical information is applied in practices of resource use, it requires ethical and normative principles and rules for the use of natural resources and their sharing between resource users. Ethical reasoning is not just an abstract or theoretical framework for environmental research and action; it specifies the rules and principles for a code of conduct that governs humans as individuals, as members of modern society, and as community of all humans. The different approaches in environmental and sustainability ethics require intensive communication of ethical principles and norms for environmental and sustainability governance in the sustainability discourse. The process of translation and application of ethics in different fields of practice, including the economy, is essential for the ethics of sustainability as discussed by Kibert et al. (2011).

Environmental and sustainability ethics cannot be practiced in the forms they are discussed in philosophy and science; they are mainly learned in the practice of governance and resource management. The ethics cannot be formulated in simple codes of moral conduct like the ten commandments of Christian religion. Ethical principles need to be translated in various forms: in rules for private and public life and for the evaluation of individual and collective action, in professional practices, in case-based reasoning (casuistry), in discourses in sustainability politics and governance, in the practices of business, in normative and legal rule systems like the universal human rights. There are no institutions for the enforcement and sanctioning of ethical rules, except for some professional ethics (for example, for the medical professions). Although ethical norms are often universal, addressed to all humans, their acceptance depends on individual belief. The application of ethics is not well structured and specified through rules, methods and procedures or reasoning, in spite of an intensifying debate in applied ethics (Singer 1986; Chadwick and Schroeder 2001; Cohen and Wellman 2005).

8.2 Economics and Ecology: Constructing Science-Practice Links in Resource Use

Since the earliest discussions of how economics and ecology can learn from each other (Barbier 1993), conventional economic knowledge about the use of exhaustible resources has been criticised as ineffective or misleading when it is applied in natural resource management or environmental regulation. showed in the Ecological discussion has shown, for example, that the Hartwick-Solow rule (the depletion of natural capital can be compensated through investments in reproducible, non-exhaustible capital, which enables a constant consumption over time) is unrealistic in its assumptions. That there is sufficient substitutability between natural and reproducible capital and that all economic values are known and reflected in market prices are unrealistic assumptions. The economic value of natural capital cannot be calculated because of the insufficient knowledge about its value (Barbier 1993: 3f). The uncertainty of economic knowledge in the valuation of natural capital and of ecological knowledge about the ecological functions of natural capital, such as tropical rainforests, make economic valuation of natural capital unrealistic and lead to incorrect conclusions with respect to the efficiency and sustainability of resource use.

The problem of the contradictory framing of economics in the growth and non-growth paradigms is discussed by Kläy et al. (2015: 73): "Science for sustainable development is … confronted with a fundamental contradiction arising from this double normative framing of science policy: can scientists really live up to their role of contributing to sustainable development, while at the same time helping societies achieve only greater economic growth, at the expense of equity and the environment?" A new science for sustainable development begins with a critique of the social and environmental impacts of growth and competition on the global market. In the new modes of knowledge production accountability and value orientation are seen as fundamental for not only science but also sustainability science (Miller et al. 2014). The following can be concluded: the successful transformation to sustainability requires a transdisciplinary transformation of science. This requires the building of new arenas in the transformation process where transformation action groups can develop, cooperate and learn.

Economic principles cannot be implemented without observing additional criteria, norms and restrictions: restrictions in political and legal forms become necessary to prevent overuse of resources and environmental damages. In situations of using knowledge from different disciplines and observing multiple criteria in decision-making about resource use, the intentions of economic valuation of natural capital and commodification of natural resources become a problematic: when and where should the commodification of nature be limited or excluded? The overuse of resources cannot be prevented through economic self-regulation, for example, through market-based principles of resource use and management, and ecological restrictions through political and legal regulation became necessary. When the sustainability process has become more effective and efficient as a social-ecological transformation process, it will still be necessary to continue such policies and to introduce further restrictions of resource use to prevent the overuse of resources or limit environmental damages:

1. *Ecological regulation and management of resource use, material and energy flows.* Examples for such debates include local and global environmental governance, ecosystem based management, adaptive management, regulation of natural resource use in single economic sectors, especially the primary production in agriculture, fishery, forestry, or the management of energy systems and their ecological transformation (see Chap. 4).
2. *Social, symbolic and knowledge-based regulation management of resource use.* These forms are complements to ecological management, supporting cooperation and participation of stakeholders in resource management, defining principles of collective resource use, defining and limiting the rights of resource owners, ethical rules and principles of resource use and sharing, environmental information and education, broadening of environmental rights, ecological citizenship, an new normative orders (ideas that come close to the debates in politics and political science about new forms of democracy, cosmopolitanism, a new world order).

3. *Restriction of economic principles and action through public policies.* Such restrictions exist in form of protection of nature and species through partial or total ban of resource use in certain areas or at certain times. In all societies such restrictions are found today in forms of protected areas (national parks, biosphere reserves, nature protection programmes, marine protected areas, and others). Also, under the present economic globalisation and deregulated markets the number of such areas is rapidly increasing. They seem the last resort because of the ineffectiveness of many voluntary and non-prohibitive regulations.

Although there was some improvements at regional and national levels, environmental policy and regulation before the beginning of the global sustainability policy in the 1990s remained largely ineffective. The global overuse and pollution of resources is increasing in the twenty-first century according to new ecological indicators. In this situation sustainable development programmes are under pressure to improve and to find more efficient concepts, approaches and political instruments. Such efforts can be seen, for example, in the search and application for new measures as nature-based solutions. Ecologists and environmental movements and other policy fields, especially climate policy, continued to argue that there is not much time to change the course of development, resource use practices and environmental pollution. Witness these many variants: "action is urgently needed", "time windows are closing", and "it is already too late". None of this provides yet ideas and arguments to build transformation strategies. When ideas and strategies for social-ecological transformation of the economy are lacking or ignored in policy and practice, the reasoning for sustainable development tends to fall back on doomsday prophecies and environmental scepticism. That "*philosophical, psychological, social and political transformation is needed for humanity to develop sustainable ways of life*" (Rayner 2010: 112, using ecological knowledge about energy flows in ecosystems to formulate principles of sustainability as critique of economic principles and practices) is not easily translated into strategies for sustainability transformation that require something more systemic than mere behaviour changes. With the new forms of inter- and transdisciplinary knowledge production in sustainability and transformative science, one can expect more knowledge to build new forms of action and new transformative action groups for the economy's transformation.

Re-thinking the knowledge economy—material goods and information processes. The knowledge economy is understood as the production of goods and creation of services by use of knowledge (Houghton and Sheehan 2000; Rooney et al. 2005; Western 2012). This inexact description coincides with the popular terms of the knowledge or information society, where knowledge, human capital, skilled labour, information technology and the digital economy are seen as determinants of a development beyond the industrial society. The economic trends and changes are connected with the forms of knowledge generation described as new inter- and transdisciplinary knowledge practices. New knowledge practices with more intensive use and communication of knowledge are essential parts of the sustainability transformation and of building transformative action groups. Strategies of social-ecological transformation require a new "epistemology of complexity" (Bruckmeier 2009: 3f, 6f) for the analysis of social-ecological systems, their transformation, and for the development of forms of transformative action and action groups. Transformation as (a) knowledge seeking process needs to be connected with transaction as (b) creating collective agency (transformation action groups) and (c) transformation as creating power for transforming asymmetrical power relations in politics and the economy. The new forms of digital and knowledge economy can support the sustainability transformation, making the complex processes of knowledge management and application technically feasible, but they are also part of the growth-based economy. Therefore, the building of transformative knowledge and transformative agency is necessary to dissolve asymmetric power relations in the global policy and economy that help to maintain a growth-based economy.

8.3 Re-Connecting Science and Practice: Sustainability Science

The sustainability discourse, which developed from the new practices of inter- and transdisciplinary knowledge creation, introduced forms of reconnecting science, policy, resource use practices, and economy with both each other and nature. Paradigmatic examples are sustainability science, human, social and political ecology. Ives et al. (2018) describe five general links to connect humans and nature in material, experiential, cognitive, emotional and philosophical forms, similar to Rayner (2010), which was mentioned above (Sect. 8.2). The term philosophy as it is used for understanding normative connections through worldviews and views of humans tends to become problematic here, that is in the discussion of society-nature relations, because it does not differentiate adequately between scientific and knowledge-based forms of philosophical thinking and philosophy as the trivial management philosophy or philosophy of worldviews, where all normative and value-oriented reasoning in the social lifeworld appears philosophical. Useful as such conceptual models can be for specific purposes, they mislead through their assumption that there is one connective relation which comprises all specific connections. In the framework of Ives et al. it is the assumption that societal connections with nature can be constructed through connections of people with nature that deceived, because it ignores that societal and ecological systems are not simply systems of people and species, but comprise a plurality of functions and structures that cannot be reconstructed as purely material or energetic connections. In societal systems these include the different functions and levels of social systems in culture, science, education, politics, and economy. In nature these include the complex organisation and interdependence of different forms and levels of physical and biological systems, ecosystems, ecological cycles, and the earth system. For the proper connection of nature and society, material and energy flows are important, but they. cannot be analysed without accounting for human and societal changes of nature, their knowledge and value bases.

Sustainability science—interactions between nature and society. Following the foundational statement of Kates et al. (2001), sustainability science can be described as problem driven, interdisciplinary, and practice-oriented research within the thematic scope of sustainability that sees preserving the life-support systems of planet Earth as the essence of sustainable development, and that understands sustainability science as identifying the fundamental relations between nature and society (Box 8.4).

Box 8.4 Sustainability Science

Kates et al. (2001): sustainability science seeks to understand the fundamental character of interactions between nature and society. Such an understanding must encompass the interaction of global processes with the ecological and social characteristics of particular places and sectors. The regional quality of much of what sustainability science is trying to explain implies: sustainability research needs to integrate the key processes of sustainable development across the scales from local to global. It requires fundamental advances in our ability to address issues as the behaviour of complex self-organizing systems and the responses of the nature-society system to multiple and interacting stresses. Combining different ways of knowing and learning will permit social actors to work in concert, even in situations of uncertainty and limited information.

Components of a research strategy: (1) span the range of spatial scales between such diverse phenomena as economic globalization and local farming practices; (2) account for both the temporal inertia and urgency of processes like ozone depletion; (3) deal with functional complexity (evident in recent analyses of environmental degradation resulting from multiple

(*continued*)

Box 8.4 (continued)

stresses); (4) recognize the wide range of outlooks regarding what makes knowledge usable within science and society. Pertinent actions are not ordered linearly in the familiar sequence of scientific inquiry, where action lies outside the research domain.

Core questions:

How can the dynamic interactions between nature and society—including lags and inertia—be better incorporated into emerging models and conceptualizations that integrate the Earth system, human development, and sustainability?

How are long-term trends in environment and development, including consumption and population, reshaping nature-society interactions in ways relevant to sustainability?

What determines the vulnerability or resilience of the nature-society system in particular kinds of places and for particular types of ecosystems and human livelihoods?

Can scientifically meaningful "limits" or "boundaries" be defined that provide effective warning of conditions beyond which the nature-society systems incur a significantly increased risk of serious degradation?

What systems of incentive structures (including markets, rules, norms, and scientific information) can most effectively improve social capacity to guide interactions between nature and society toward more sustainable trajectories?

How can today's operational systems for monitoring and reporting on environmental and social conditions be integrated or extended to provide more useful guidance for efforts to navigate a transition toward sustainability?

How can today's relatively independent activities of research planning, monitoring, assessment, and decision support be better integrated into systems for adaptive management and societal learning?

Source: Kates et al. 2001

Sustainability science as it is described in this mainstream variant has been criticized, in spite of its efforts to construct a framework for institutionalising interdisciplinary knowledge production about interactions between nature and society, for having neglected of social-scientific knowledge that does not fit with in the construction of sustainability science from a natural-scientific perspective, logic and knowledge base. The more specific and critical analyses of societal systems and structures to be considered in sustainability science include the specific nature of modern society and the economic world system as causes for maladaptive social and environmental change, including the mechanism of economic growth, inequality in North-South relations, property relations, the capitalist mode of production and its societal metabolism (see Glossary). Brown (2012) elaborated such critique to emphasise the necessity of studying material consumption and the structures of the consumer society, the functions of technologies and their support of unsustainable social practices, the macroeconomic theories assuming growth as necessary for human welfare and wellbeing.

What remains to be done in the sustainability discourse and process, is the development of *pluralistic knowledge practices and forms of knowledge transfer and integration*. Such knowledge practices help to deal with a variety of problems: the interaction and integration of knowledge from economics, ecology and policy science; the critique of conventional models of knowledge transfer and science communication; the development of transdisciplinary models of participation, cooperation and joint learning; the integration of empirical, theoretical and normative knowledge for sustainable development; the critical analysis of technocratic and power-based epistemic coalitions of science, policy and business, as for example in the "triple helix of knowledge production", that exist also in the sustainability process; critical and comparative analyses of the competing concepts and models for transitions to sustainability (green economy and green growth; cyclical concepts and models of the economy; high-tech and digital economy; models of the steady state, of limits to growth and degrowth; social-ecological transformation). These themes figure prominently in the sustainability process's research and governance agenda.

8.4 After Sustainability Science: Transformation Science

Since about 2010 sustainability research has adopted the concept of transformation as the key for conceptualising and designing the transition to sustainability via societal change. The initiatives for transformation research emerged from different fields of research and environmental action (Box 8.5).

Box 8.5 Transformation Research

More than sustainability science, the newly developing transformation science uses and produces social scientific knowledge. Integrating both approaches can strengthen the research and collective action for sustainable development. Transformation research has been supported through several initiatives:

1. *Early and limited forms of conceptualising transformative research and specifying it in research programmes came from the USA'S National Science Foundation*, where it was defined as "research that has the capacity to revolutionize existing fields, create new sub-fields, cause paradigm shifts, support discovery, and lead to radically new technologies" (Arden and Bement 2007). A similar concept from the European Research Council is "frontier research", which is also used by other institutions.
2. *Transformation research was specified through the formulation of transformative conerstones by the International Social Science Council* (2012) that should be observed in all social-scientific research about global change and global governance: (1) Historical and contextual complexities (analysing the full complexity of global change to understand the political economy of climate change and other global processes). (2) Consequences (identifying and mapping the threats and impacts from the global change processes studied). (3) Conditions and visions for change (what drives change processes: how change happens, at what levels and scales, in what directions). (4) Interpretation and subjective sense making (personal and collective values, beliefs, interests and worldviews, needs that guide the reactions of people to global change). (5) Responsibilities (how global and inter-generational solidarity and justice can be created, including ethical approaches). (6) Governance and decision making (how political decisions can be made in situations of uncertainty; determinants of success or failure of political agreements).
3. *The German Scientific Council for Global Change developed a concept of transformative research in the broader perspectives of transdisciplinary science* (WBGU 2011) as research supporting the transformation processes towards sustainability. From these ideas developed a transformative science for the study of global change processes, especially for climate change (Schneidewind and Singer-Brodowski 2013) Transformative science includes the building of transformative literacy and capacity (overview on transformative literacy in higher education:

(*continued*)

Box 8.5 (continued)

Leal Filho 2015). Knowledge relevant for transformation processes and for transformative literacy is gained from the integration and balancing of knowledge about technological, economic, cultural and institutional forms and dimensions of transformation. Transformative science has also developed a critical view of technological change and innovation that emphasizes this change's negative social, economic and environmental consequences. Economic knowledge needs to be embedded in the functional relationships of modern society and its systemic constitution, to understand how the economic, political and other systems interact and how they change in transformation. Schneidewind et al. (2016) specify the functions of transformative science as observing, describing, initiating and catalysing societal transformation processes with the aims of improving their understanding and increasing the societal capacity to reflect on them.

4. Asara et al. (2015) *discussed sustainable degrowth from the perspective of social-ecological transformation, connecting sustainability science and transformation research.* They concluded that the mismatch between a growing scientific field of sustainability science and sustainable social–ecological change can be explained by different factors: insufficient scientific engagement with stakeholders, anachronistic academic institutions and incentives, lack of meta-studies making transdisciplinary sustainability research available to scholars and practitioners, and a missing link between knowledge production and action. To support transformation requires fundamental reforms in the academy, comparative studies that make knowledge about sustainability accessible and applicable, and a new social contract between scientists and society in which scientists participate in the co-production of knowledge for action with other stakeholders. If these changes are not connected with a broader critique of the modern economy, as that offered by degrowth and other transformation approaches, sustainability science can hardly support the social-ecological transformation required to solve the global environmental crisis. Uncovering the ideology and practice of economic growth (connected to capitalism) as the ultimate driver of unsustainability may help sustainability science to develop and to become more influential.

Sources: mentioned in the text

Transformative approaches go beyond maintaining the socio-ecological system. They aim to change the fundamental attributes of the entire system, such as the economic mode of production, political institutions, ideologies, societal norms, everyday life, ecology, and combined socio-ecological assemblages. Transformations involve non-linear processes, because they deal with dynamic multidimensional and complex systems and understand social innovation as a key driving force of such processes. They involve multiple scales and system levels, from the local to the regional, national and international levels, and functional levels such as the markets, states and civil society (Asara et al. 2015). Transformational literacy, capacity, agency are necessary conditions for sustainability transformation that includes changes of the imperial ways of life in Western countries and of the systemic structures (power relations, market and growth-based economy), integrating actor- and system-perspectives of sustainability transformation (see Chap. 7).

8.5 The Future as Knowledge Problem for Science and for Sustainability Governance

Sustainable development as social, economic and ecological transformation needs to deal with conflicts and dilemmas in collective action and resource use. The intergenerational and inter-systemic perspectives in the transition to sustainability require transformative knowledge practices that stretch in action into both the near and distant future. Transformative action groups and the creation of agency—that is, of capacities to change social institutions—require the creation of new social subjects for sustainability transformation, environmental movements and other transformation actors. The transformation cannot be mainly driven by the political and business elites that have contradictory motives and interests with regard to sustainable development.

8.5.1 Dilemmas in the Sustainability Process

From the sustainability discourse's inception a dilemma in natural resource use and economic action more broadly has been understood as the contradictory relationship between short-term and long term conditions of economic and societal development (Western 2001):

Human success in short-run development, in the organisation of the necessary conditions for survival and the fulfilling of basic needs, was achieved long ago, through "ecosystem simplification", as human societies produced agro-ecosystems, simplified food webs, and landscape homogenisation. Modern society has augmented this through the use of high nutrient and energy inputs into agricultural production. The modification of ecosystems that made possible short-term success in creating conditions for welfare and well-being have undermined, however, the long-term future of human society and its survival. Scientific knowledge, especially from ecology, that balances the short and long-term conditions of human survival is currently far from sufficient, but the requisite knowledge is growing and new modes of knowledge production are creating better conditions for integrating knowledge from different disciplines. More than simply accumulating knowledge through research, contemporary approaches are becoming aware of environmental risks and are developing the individual and collective willingness to act in a way that explores the limits of knowledge and action.

Ostrom (2007) discussed these problems with regard to practices of natural resource use. Her work shows, that the organisation of present resource use can limit the possibilities to maintain the resource use in the long run; it is necessary to specify the limits of resource use att all levels from the local to the global. Her article and the others in the same issue criticise the assumption that simple, predictive models of social-ecological systems can be made, and that universal solutions or panaceas derived from these can solve problems of overuse or destruction of resources. Moving beyond panaceas requires the study of complex, multivariable, nonlinear, cross-scale, and changing systems, in which many variables affect the patterns of interactions and outcomes that are observed in empirical studies of resource use practices. "*A step toward developing a diagnostic method is taken by organizing these variables in a nested, multitier framework. The framework enables scholars to organize analyses of how attributes of (i) a resource system (e.g., fishery, lake, grazing area), (ii) the re-source units generated by that system (e.g., fish, water, fodder), (iii) the users of that system, and (iv) the governance system jointly affect and are indirectly affected by interactions and resulting outcomes achieved at a particular time and place. The framework also enables us to organize how these attributes may affect and be affected by larger socioeconomic, political, and ecological settings in which they are embedded, as well as smaller ones*" (Ostrom 2007: 1581).

The sustainability dilemma to consider within resource use practices both the short-term and long-term consequences of action translates into a more differentiated view of the future. The dominant view, especially of the long-term future, is that the future is unknown and unpredictable, which seems to exclude talking in terms of knowledge about the future. But the idea of safe knowledge and predictions of the future are misleading in their reflections on the future and possibility of change and transformation. Other forms of knowledge can be used to deal with the future. The relation between knowledge and ignorance is not only for the future time but also for the present. Long ago, this problem was discussed by an economist, who noted that the progress of science "*while increasing the stock of knowl-*

edge, increases also the area of conscious ignorance" (Marshall 1923: 671). This point is underscored by recent epistemological debates and also the newly developing field of sociology of ignorance. It seems necessary to act for sustainable development with another view of the future that imagines it as a time when we, the presently living people, are dead. Thus, thinking beyond traditional time horizons in economics will allow us to develop inter-generational solidarity, new principles in resource sharing, greater integration of management of ecosystems and social systems, and transformative capacity (see Chap. 9).

8.5.2 Temporal Dimensions of Sustainable Development

Understanding the sustainability discourse as a form of collective thinking and action that makes possible an integrated future for society, economy, ecology, and ways of life, requires us to conceptualise the future in a way beyond the physical construction of time as linear process, symbolized in the metaphor of the time arrow. A socially transformed and constructed time, understands time as a component of the lifeworld in the sociological sense of the term (Schütz 1932). This approach opens new perspectives on social construction and the transformation of time in social action in the lifeworld. The forms of social action in the lifeworld are usually routine processes in the present time, whereas sustainable development requires from the actors to become aware about the future, how a wanted future can be created through social action, beginning with the question, when and how does the future that is imagined in the sustainability concepts begin?

The future of sustainability reaches beyond the time horizon of political action and its conceptualisation in terms of goals, their achievement, the implementation of policy programmes that are limited through the normative framing and the constraints of the policy process, which cultivate a short-term perspective. Significantly, sustainable development and its long-term perspective, adopted with the concept of transformation as specific form of change, are also not adequately conceived in the ecological models of the adaptive cycle, where the future is seen as an iterative, repetitive process running through several phases and as an extension of the cyclical time model of ecology that is bound to seasonality and the rhythms of biological growth processes. Also, the risk concept in its conventional form as projecting the probability of future consequences of specific forms of economic and other action is bound to the short-term perspective of social action and is mainly conceptualised in terms of negative and unwanted consequences. To go beyond the limited perspective of the future imagined as created through the same processes and forms of action as today, or as reproduction process of social and ecological systems, it is necessary to conceptualise the future in terms of change, changing thus the pathways of development and creating new pathways. Breaking through the path dependency of development cannot be done in everyday routines of social action. The modern economic world system's complex processes, and especially its economic, political, and cultural processes of reproduction need to be accounted for in sustainability transformation.

The minimum conceptual requirement in constructing the temporal dimensions of sustainable development as a transformation towards another economy and society is the additional and longer structuring of a process that can be maintained over many decades, through unforeseen changes and events, and that does not end when the present conditions and ideas of sustainable development change or lose support (Table 8.1). The process needs to be renewed repeatedly, and one such required renewal is the present re-conceptualisation of social-ecological transformation (see Chap. 9).

Table 8.1 Trends and megatrends

Name	Definition	Interpretation	Sources
Trends of change	Trends that can be derived or extrapolated from processes of development and change that happen presently in society (politics, economy) and in ecological systems	The long waves ("Kondratiev cycles") in modern industrial economies describe a dominant economic trend that can be derived from the study of the economic history of industrialisation and its change; the cycles can be connected with sustainability transformation as factors influencing the sustainability process	Research about cultural, social, political, economic, environmental change
Megatrends	The simple definition as large-scale and long-term trends is not sufficient; the trends need to be specified in terms of social and ecological forms of change, so that they include the selective combinations of these components (cultural, economic, political etc)	Megatrends in forms of global social and environmental change (globalisation, urbanisation, climate change, biodiversity reduction) are dominant forms of megatrends influencing the sustainability process directly. They require integration of different policy fields and processes	Future research

(*continued*)

Table 8.1 (continued)

Name	Definition	Interpretation	Sources
Unexpected events and incalculable changes that influence trends and megatrends	Climate-related disturbances and disasters; conflicts and disturbances caused through the instability of the national and global political systems and institutions; disturbance through instabilities of the global economy	Resilience research: unexpected changes and catastrophes and capacities to cope with them provide knowledge about ecological and social-ecological resilience as part of the sustainability process	Resilience research
Temporal structures of change and transformation	The generalised forms of time perception in past, present and future cannot be changed, but must be differentiated in social and collective action and agency; forms of differentiation and combination of temporal perspectives (as short and long duration, near and distant future, cyclical and non-cyclical processes) are relevant for the regulation of processes in the sustainability transformation	Important in the sustainability discourse and process is temporal structuring through action strategies that are based on scenarios	Future research

Sources: mentioned in the table

8.6 Discussion and Conclusions

The changing relationship between science and practice that paves the way to a future sustainable society and economy can be described as a series of paradigm changes in science and policy—that is, in the generation, use and application of knowledge in science and in practice. One problem that has arisen in discourses about inter- and transdisciplinary knowledge production is that one must understand the changing relationships between knowledge in its different forms and ignorance. Intense debates, during the second half of the twentieth century, about new knowledge production are part of the changes in modern society and have yielded the development of environmental movements and other social movements that are critical of modern society and its forms of life, including environmental policies at national and international levels. Overall, the criticism is characterized by a growing awareness of the need to limit growth, of the increasing scarcity of natural resources for industrial society, of unequal development and access to resources between the Global North and South, and of global social and environmental changes, including climate change, biodiversity loss, land use change, exponential growth of the global population, economic globalisation, technological innovations of the information and knowledge society. Sustainable development can, thus, be seen as part of a. larger attempt to combat the negative aspects of development.

To achieve this global transformation requires the following: rebuilding solidarity as a global societal project of sharing resources within and between generations; developing and combining different models and pathways for sustainability transition; formulating new ideas about a future global economy that does not destroy the environment; new forms of a "mixed economy" that includes local economy, market economy, reciprocity, sharing and redistributive economy; finally, ideas for re-embedding and differentiating the economy and building sustainable economic systems.

Regulating coupled social-ecological systems, creating intergenerational solidarity, stimulating processes of collective learning, knowledge transfer and knowledge governance, and building of transformation action groups are core processes in the sustainability transformation that begins with the forms of inter- and transdisciplinary knowledge integration and application in new knowledge cultures and practices. The renewing of sustainable development must focus on the new knowledge debate on the triangle of economics, ecology and policy science, with an emphasis on the following question: how to integrate economic knowledge with knowledge about global environmental change and global environmental governance? Further efforts will be required to deal with the knowledge and method problems in the epistemic triangle of empirical, theoretical and normative knowledge (ethics, values and worldviews) for sustainable development. Insofar as the sustainability process offers the chance to address the deficits in inter- and transdisciplinary knowledge integration and application, this cannot be done with the methods that are created in academic science, in economics, ecology or political science, which focus only on research.

Summary of Chap. 8: Science and Practice in the Transition to Sustainability

Changing relations between science, politics and civil society in the research and in the practice of sustainable development are part of the broader processes of strengthening civil society and new knowledge production which began in the early 1990s, or about the same time as the sustainability discourse itself. With these changes come critical reflections about knowledge transfer, dissemination, sharing and application, described with the two terms of "mode two" or "transdisciplinarity" that influence sustainability research. In the epistemological discourse and the research about science and technology studies, inter- and transdisciplinary forms of knowledge production, post-normal science, and sustainability science new forms of knowledge transfer and communication develop by utilising scientific knowledge more systematically and actively (in discursive knowledge creation), by emphasizing socially efficient production (through joint learning), and by embracing practical effectiveness (achieving the goals of collective action).

Such knowledge practices are necessary for renewing the process of sustainable development that is currently stuck in the routines of managing complex system processes without adequate knowledge, tools and strategies for managing an overburdened goal system, as is visible in the development goals and targets formulated by the UN. The future sustainability process can, as social-ecological transformation of economy and society, correct the failures of the past and create both more realistic and a more effective forms of goal attainment in a long-term process, in which the global problems to be dealt with can be deliberated more effectively and solved more consequently. The problems include the modification and functional disturbance of ecosystems, the overuse and limits of natural resources, the unequal development and exchange, climate change, and other forms of environmental change. The adoption of new knowledge practices and the building of transformative action groups in the sustainability process is a component of the joint learning of scientists, practitioners and activists in the further transformation process towards sustainability. It began half a century ago with the new environmental discourse in science and policy making, and the spreading of interdisciplinary practices in environmental research. It is now developing to a broader process of social change and transformation of economy and society.

Appendices: Further Information and Material

1. Questions and Individual Exercises

Learning Exercise 1: Example of a Traditional Concept and Model for Connecting Science and Practice

Download from the Internet the model "The Bridge" in the short text by Poteet, M. L., Zugec, L., & Craig Wallace, J. (2016). The Bridge: Connecting Science and Practice. *The Industrial Psychologist, 53*(4). Retrieved from https://www.siop.org/Research-Publications/TIP/TIP-Back-Issues/2016/April/ArtMID/20280/ArticleID/788/.

Discuss this model critically with regard to its strengths and weaknesses. Is the description sufficient to provide concrete rules and procedures for science-practice integration?

Learning Exercise 2: A Complex Model of Science-Practice Relation—Systemic Interaction

To understand the complex forms of science-practice relation in the new forms of knowledge production described and discussed in this chapter, it is suggested to read the following article about sustainability research:

Kläy, A., Zimmermann, A. B., & Schneider, F. (2015). Rethinking Science for Sustainable Development: Reflexive Interaction for a Paradigm Change. *Futures, 65*, 72–85.

Discuss critically (in a group) how the suggested changes for developing a transformative science for sustainability can be realised practically. What are the obstacles? Which questions and problems need to be addressed that are not discussed by the authors?

Learning Exercise 3: Transition and Transformation

Read and discuss a short article to explore the differences between the concepts of transition and transformation:

The concepts transition and transformation are both used in this book, differentiating vaguely between "transition" as a term designating more simple changes and less complex systems, not necessarily the large scale system transformations as that of the global economy or modern society discussed in the sustainability process and specified as "social-ecological transformation". The concepts are not strictly standardised in definition and application. To see, how they are applied in sustainability research and practice—how they differ and how they can be connected—read the following short article and consider different perspectives on non-linear societal change:

Hölscher, K., Wittmayer, J. M., & Loorbach, D. (2018). Transition Versus Transformation: What's the Difference? *Environmental Innovation and Societal Transitions, 27*, 1–3.

It is suggested that you read the article and discuss individually or in group the differences between the terms, as well as advantages and disadvantages of both.

Learning Exercise 4: Responsibilities of the Present Generation Towards Future Generations

Download from the internet the "Declaration on the Responsibilities of the Present Generations Towards Future Generations" from the General Conference of the United Nations Educational, Scientific and Cultural Organization (UNESCO) from November 12, 1997.

Discuss critically the twelve articles of the declaration with regard to their realisation and implementation through policies and law. How (and how far) can the principles be translated into concrete rules for social action and collective behaviour?

2. Further Reading Suggested: Deepening, Thematically Specialised

Bucchi, M., & Trench, B. (Eds.). (2014). *Handbook of Public Communication of Science and Technology* (2nd ed.). London and New York: Routledge.

Felt, U., Fouché, R., Miller, C. A., & Smith-Doerr, L. (Eds.). (2017). *The Handbook of Science and Technology Studies* (4th ed.). Cambridge, MA: MIT Press.

Kibert, C. J., Monroe, M. C., Peterson, A. L., Plate, R. R., & Thiele, L. P. (2011). *Working Towards Sustainability: Ethical Decision-Making in a Technological World.* Hoboken, NJ et al.: Wiley.

Rooney, D., Hearn, G., & Ninan, A. (2005). *Handbook of the Knowledge Economy*. Cheltenham, UK and Northampton, MA: Edward Elgar.

Scientific journals (international, peer reviewed):

"Global Change", "Earth System Governance", "Ecology and Society", "Education for Sustainable Development", "Environmental Values", "Review of Environmental Economics and Policy", "International Journal of Sustainability in Higher Education", "Journal of Education for Sustainable Development", "The Journal of Sustainability Education", "Social Studies of Science", "Science", "Technology & Human Values", "Science, Technology and Society", "Science as Culture". "Sustainability Science"

References

Arden, L., & Bement, A. L. (2007). *Transformative Research: The Artistry and Alchemy of the 21st Century*. USA: National Science Foundation. Retrieved March 10, 2020, from https://www.nsf.gov/news/speeches/bement/07/alb070104_texas.jsp.

Argote, L., & Ingram, P. (2000). Knowledge Transfer: A Basis for Competitive Advantage in Firms. *Organizational Behavior and Human Decision Processes, 82*(1), 150–169.

Asara, V., Otero, I., Demaria, F., & Corbera, E. (2015). Socially Sustainable Degrowth as a Social Ecological Transformation: Repoliticizing Sustainability. *Sustainability Science*. https://doi.org/10.1007/s11625-015-0321-9.

Barbier, E. (Ed.). (1993). *Economics and Ecology: New Frontiers and Sustainable Development*. London et al.: Chapman & Hall.

Brown, H. S. (2012). Sustainability Science Needs to Include Sustainable Consumption. *Environment: Science and Policy for Sustainable Development, 54*(1), 20–25.

Bruckmeier, K. (2009). Sustainability Between Necessity, Contingency and Impossibility. *Sustainability, 1*. https://doi.org/10.3390/su10x000x.

Bucchi, M., & Trench, B. (Eds.). (2014). *Handbook of Public Communication of Science and Technology* (2nd ed.). London and New York: Routledge.

Cafaro, P. (2001). Thoreau, Leopold, and Carson: Toward an Environmental Virtue Ethics. *Environmental Ethics, 23*(1), 3–17.

Chadwick, R., & Schroeder, D. (Eds.). (2001). *Applied Ethics: Critical Concepts in Philosophy. Critical Concepts in Philosophy*. London: Routledge.

Chevalier, J. M., & Buckles, D. J. (2013). *Participatory Action Research: Theory and Methods for Engaged Inquiry*. London and New York: Routledge.

Cohen, A. I., & Wellman, C. H. (Eds.). (2005). *Contemporary Debates in Applied Ethics*. Malden MA: Blackwell.

Etzkowitz, H., & Leydesdorff, L. (2000). The Dynamics of Innovation: From National Systems and 'Mode 2' to a Triple Helix of University-Industry-Government Relations. *Research Policy, 29*, 109–123.

Felt, U., Fouché, R., Miller, C. A., & Smith-Doerr, L. (Eds.). (2017). *The Handbook of Science and Technology Studies* (4th ed.). Cambridge, MA: MIT Press.

Gibbons, M., Limoges, C., Nowotny, H., Schwartzman, S., Scott, P., & Trow, M. (2010 [1994]). *The New Production of Knowledge: The Dynamics of Science and Research in Contemporary Societies*. London et al.: Sage.

Girard, J. P., & Girard, J. (2015). Defining Knowledge Management: Toward an Applied Compendium. *Online Journal of Applied Knowledge Management, 3*(1), 14.

Habermas, J. (1995). *Between Facts and Norms*. Cambridge, MA: MIT Press.

Hölscher, K., Wittmayer, J. M., & Loorbach, D. (2018). Transition Versus Transformation: What's the Difference? *Environmental Innovation and Societal Transitions, 27*, 1–3.

Houghton, J., & Sheehan, P. (2000). *A Primer on the Knowledge Economy*. CSES Working Paper No. 18. Melbourne: Victoria University of Technology, Centre for Strategic Economic Studies.

Howlett, M. (2012). *Designing Public Policies: Principles and Instruments* (2nd ed.). New York and London: Routledge.

International Social Science Council (ISSC). (2012). *Transformative Cornerstones of Social Science Research for Global Change*. Paris. Retrieved from www.worldsocialscience.org.

Ives, C. D., Abson, D. J., von Wehrden, H., Dorninger, C., Klaniecki, K., & Fischer, J. (2018). Reconnecting with Nature for Sustainability. *Sustainability Science, 13*, 1389–1397.

Jamison, A. (2002). *The Making of Green Knowledge: Environmental Politics and Cultural Transformation*. Cambridge: Cambridge University Press.

Kates, R. W., Clark, W. C., Corell, R., Hall, J. M., Jaeger, C. C., Lowe, I., et al. (2001). Sustainability Science. *Science, 292*, 641–642.

Kibert, C. J., Monroe, M. C., Peterson, A. L., Plate, R. R., & Thiele, L. P. (2011). *Working Towards Sustainability: Ethical Decision-Making in a Technological World*. Hoboken, NJ et al.: Wiley.

Kläy, A., Zimmermann, A. B., & Schneider, F. (2015). Rethinking Science for Sustainable Development: Reflexive Interaction for a Paradigm Change. *Futures, 65*, 72–85.

Krämer, K., & Vellguth, K. (Eds.). (2017). *Creation: Living Together in Our Common House* (One World Theology) (Vol. 11). Freiburg: Herder.

Kuhn, T. (1962). *The Structure of Scientific Revolutions*. Chicago: University of Chicago Press.

Leal Filho, W. (Ed.). (2015). *Transformative Approaches to Sustainable Development at Universities: Working Across Disciplines*. Cham et al. Springer.

Lynch, R. A. (2011). Foucault's Theory of Power. In D. Taylor (Ed.), *Michel Foucault: Key Concepts* (pp. 13–26). Acumen Publishing Ltd.

Marshall, A. (1923). *Industry and Trade*. London: Macmillan.

Miller, T. R., Wiek, A., Ansong, D., Robinson, J., Olsson, L., Kriebel, D., et al. (2014). The Future of Sustainability Science: A Solutions-Oriented Research Agenda. *Sustainable Science, 9*, 239–246.

Nowotny, H., Scott, P., & Gibbons, M. (2001). *Re-Thinking Science, Knowledge and the Public in an Age of Uncertainty*. Cambridge: Cambridge University Press.

OECD. (1999). *Managing National Innovation Systems*. Paris: OECD Publications Service.

Ostrom, E. (2007). A Diagnostic Framework for Going Beyond Panaceas. *PNAS, 104*(39), 15181–15187.

Pearce, D. (2002). An Intellectual History of Environmental Economics. Annual Review of *Energy and the Environment, 25*, 57–81.

Perera, A. H., Buse, L. J., & Crow, T. R. (Eds.). (2006). *Forest Landscape Ecology, Transferring Knowledge to Practice*. New York: Springer.

Poteet, M. L., Zugec, L., & Craig Wallace, J. (2016). The Bridge: Connecting Science and Practice. *The Industrial Psychologist, 53*(4). Retrieved February 10, 2020, from https://www.siop.org/Research-Publications/TIP/TIP-Back-Issues/2016/April/ArtMID/20280/ArticleID/788/.

Rayner, A. D. M. (2010). Inclusionality and Sustainability – Attuning with the Currency of Natural Energy Flow and How This Contrasts with Abstract Economic Rationality. *Environmental Economics, 1*(1), 102–112.

Reason, P., & Bradbury, H. (Eds.). (2008). *The Sage Handbook of Action Research: Participative Inquiry and Practice*. London et al.: Sage.

Rolston, H. (1999). Nature and Culture in Environmental Ethics. In K. Brinkmann (Ed.), *Ethics: The Proceedings of the Twentieth World Congress of Philosophy* (Vol. 1, pp. 151–158). Bowling Green, OH: Philosophy Documentation Center.

Rooney, D., Hearn, G., & Ninan, A. (2005). *Handbook of the Knowledge Economy*. Cheltenham, UK and Northampton, MA: Edward Elgar.

Roux, D. J., Rogers, K. H., Biggs, H. C., Ashton, P. J., & Sergeant, A. (2006). Bridging the Science–Management Divide: Moving from Unidirectional Knowledge Transfer to Knowledge Interfacing and Sharing. *Ecology and Society, 11*(1), 4. [Online]. Retrieved from http://www.ecologyandsociety.org/vol11/iss1/art4/.

Schneidewind, U., & Singer-Brodowski, M. (2013). *Transformative Wissenschaft. Klimawandel im deutschen Wissenschafts- und Hochschulsystem*. Metropolis: Marburg.

Schneidewind, U., Singer-Brodowski, M., & Augenstein, K. (2016). Transformative Science for Sustainability Transitions. In H. G. Brauch (Ed.), *Handbook on Sustainability Transition and Sustainable Peace* (pp. 123–136). Cham: Springer.

Schütz, A. (1932). *Der sinnhafte Aufbau der sozialen Welt: eine Einleitung in die verstehende Soziologie*. Wien: J. Springer.

Serenko, A., & Bontis, N. (2017). Global Ranking of Knowledge Management and Intellectual Capital Academic Journals: 2017 Update. *Journal of Knowledge Management, 21*(3), 675–692.

Serenko, A., Bontis, N., Booker, L., Sadeddin, K., & Hardie, T. (2010). A Scientometric Analysis of Knowledge Management and Intellectual Capital Academic Literature (1994–2008). *Journal of Knowledge Management, 14*(1), 13–23.

Shinn, T. (2002). The Triple Helix and New Production of Knowledge: Prepackaged Thinking on Science and Technology. *Social Studies of Science, 32*(4), 599–614.

Singer, P. (Ed.). (1986). *Applied Ethics*. Oxford: Oxford University Press.

WBGU. (2011). *World in Transition: A Social Contract for Sustainability*. Berlin: WBGU - German Advisory Council on Global Change.

Weiler, H. N. (2011). Knowledge and Power: The New Politics of Higher Education. *Journal of Educational Planning and Administration, XXV*(3), 205–221.

Western, K. I. (Ed.). (2012). *Foundations of the Knowledge Economy: Innovation, Learning and Clusters*. Cheltenham, UK and Northampton, MA: Edward Elgar.

Western, D. (2001). Human-Modified Ecosystems and Future Evolution. *PNAS, 98*(10), 5458–5465.

Re-Thinking Temporal Perspectives of Sustainability Transformation

9

The main challenges to furthering the process of sustainable development include: the design, structuring and process organisation of sustainability governance under the guiding idea of transformation; the development of collective learning and transformative action groups; the cultivation of continued interdisciplinary knowledge production and integration that produces gradual improvement in policies, governance, and the state of the environment. The critical challenge in the complex sustainability transformation is the temporal structuring of the process in phases, so that there is continual progress toward a future state of sustainability. Well-structured planning and resource management approaches explicitly connect past development practices with the present situation, while formulating specific goals and timetables for the coming years. It is not sufficient simply to insist that the sustainability transformation reach over the time horizon into the near future. New ideas are necessary in this respect, as we also need to reflect on what may happen in the distant future and to consider the effect of both expected and unexpected changes on the possibility of achieving a sustainable economy and society. Such a future economy and society was discussed in the previous chapters. It is not a mere vision or a simple desideratum, but a future built on three concrete conditions for sustainability: post-industrialism, post-growthism, and post-capitalism. A sustainable society cannot be achieved, in short, without a fundamental transformation of the global economy.

Figure 9.1 shows the complex temporal structuring of the sustainability transformation as a long-term process.

There is a growing awareness that sustainable development cannot be organised as a political process only, in which the emphasis is on changing individsuals's behaviour and consumption patterns. Instead, the process must ramify through all

K. Bruckmeier, *Economics and Sustainability*,
https://doi.org/10.1007/978-3-030-56627-2_9

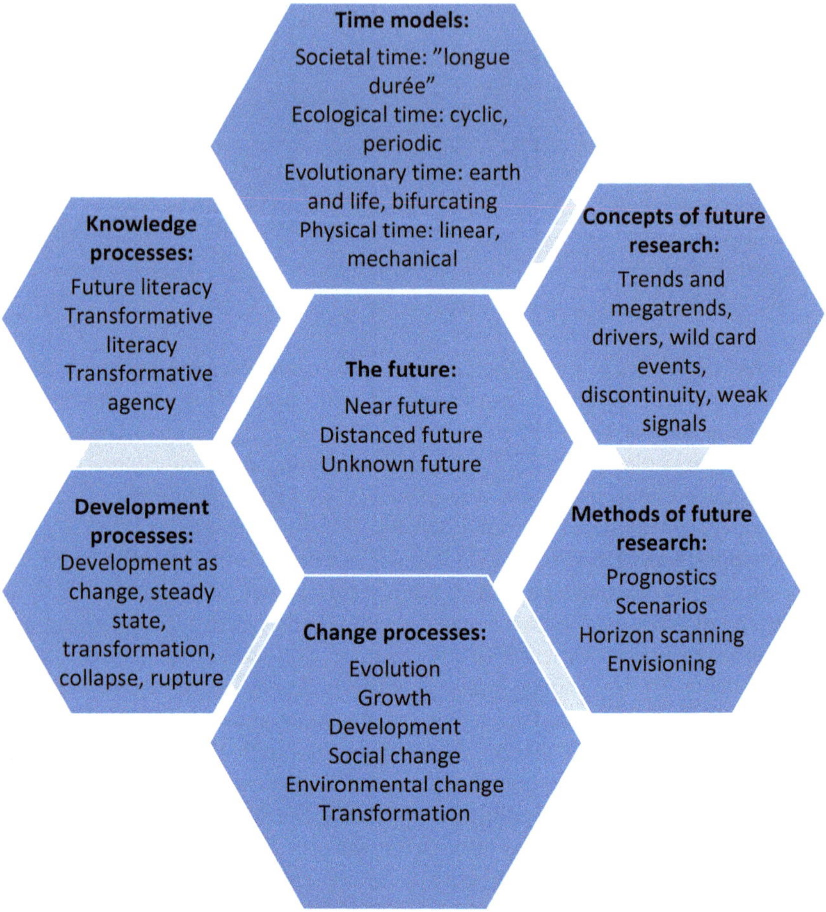

Fig. 9.1 Temporal perspectives of sustainability transformation

systems and spheres of modern society and across the networks of ecological systems on the earth in which our society and economy are embedded. This implies that sustainability governance must analyse and clarify different kinds of processes: (a) processes that can be politically managed and regulated, (b) processes that can be influenced only partly or indirectly through political and collective action, and (c) processes that cannot be influenced in forms of management, policy, planning and governance (mainly processes in ecosystems). To deal with the complexity inherent to coupled social-ecological systems, the process of sustainability transformation needs to be structured by more than just the sustainable development goals and policy programmes that currently proliferate. Contemporary developments in theories of social-ecological transformation in critical variants of ecology and economics (see Chap. 6, 7, and 8) remain widely ignored in the policy programmes for sustainable development, which emphasise little more than setting limited goals to be achieved. Although programmes and policy documents have adopted meanwhile as a new keyword the idea of transformation, it remains unclear whether transformation has been reduced to another term to describe fragmented changes in different technical, economic, social or ecological fields. No systematic analyses of the interaction between social and ecological systems inform current programmes. Nor are there any analyses of the systemic constitution of the global economy, of the limits to growth and the consequences of growth, and of the North-South divide and how it can be overcome. Moreover, there are not sufficient analyses about coordination and integration of national policies for sustainable development, about global redistribution of resources, about cooperation of actors and building of transformation action groups, and about conflict resolution on the way to sustainability. In short, no long-term perspectives and phased action programmes have been formulated, and no concepts and approaches for knowledge integration and application in the future are currently available.

The deficits in policy and governance strategies and processes point to a big problem: sustainable development as it is currently institutionalised and implemented in public policy cannot be called a strategy or programme to support the transformation of the societal and economic systems towards sustainability. The global power elite and its epistemic community control public policy and, thus, limit the programme of sustainability to concern for continued growth. In its periodical reviews of sustainable development programmes the UN laments that the necessary processes are still not on the way, and the most alarmed calls from the new journal "Environmental Sustainability" (Box 9.1) hold that the sustainability process has hit a dead end. These warnings are useless, if they are not followed by concrete suggestions for initiating transformation processes in a global economic system that emphasizes economic growth. UN strategies for sustainable development constitute nothing more than a list of seventeen development goals, none of which support transformation. For example, the research published in the journal "Environmental Sustainability", where mainly local examples of change are studied, emphasizes biotechnology as a potential source for change. Neither the UN's development goals nor the application of biotechnology will make the current system environmentally sustainable. None of the critiques challenge the systemic mechanisms of economic growth, the global market economy, the overconsumption of natural resources and the pollution of the environment that are causing global environmental change. The current dominance of economic and technological measures in the pursuit of sustainability policies leads to a paradoxical situation in which policy makers seek " to solve the problems with the means that caused them".

Box 9.1 The Irrecoverable Loss of Environmental Sustainability

Arora (2019)—*Degradation of the earth system in the last fifty years*:

The causes—population growth and the industrial economy: "The industrial, agricultural and the infrastructural revolution have resulted in over exploitation of resources and pollution of every nook and corner of the planet. The technologies which were developed to adorn and ease our routines" ... are "threatening the survival and ... made it very clear that no human science can replace 'nature's perfect systems' which have been carving the environment and ecosystems of earth to balance it in the zone of habitability." (Arora 2019: 2)

The rapid and continuing deterioration of the state of the global environment since 1970: "Most of the striking changes have appeared in last 50 years or so. … humans have destroyed about 83% of wild mammals and half the species of plants till date. On the whole, humans have consumed 30% of the known resources resulting into (sic!) scarcer ecosystem services for future generations. If these trends continue, the Earth will soon be experiencing mass extinctions and we will be left with an even more degraded planet". (Arora 2019: 1)

The present level of carbon dioxide in the athmosphere "has exceeded 411 parts per million (ppm)", it is "much higher as compared to about 323 ppm about fifty years ago, resulting in major environmental issues such as global warming and climate change". (Arora 2019: 1)

Dimensions of environmental pollution: "about 500 million tons of heavy metals, toxic sludge and hazardous solvents were estimated to be released in global water supply in 2007 making it unsafe to consume. Plastic pollution is also a big nuisance caused by humans on Earth. The stats show that annual

(*continued*)

Box 9.1 (continued)

production of plastics during 1970s was about 50 million metric tons and it has increased to over 348 million metric tons at present." Regarding biodiversity loss "humans have eradicated 60% of the Earth's wildlife in less than 50 years. About 20% of Amazon forests are lost in the last half century. A recent study revealed that of total global tree cover loss between 2001 and 2015, 27% depreciation came from commodity driven deforestation i.e. conversion of forests permanently in order to expand commodities such as meat, minerals, oils and gas." ... "the population of freshwater animals has plummeted by 75% since 1970s. Reports say that the damage done is so rapid that even if we end it now, it will take centuries to replenish the natural world" (Arora 2019: 1f)

The solutions suggested—technical fix: "Last 50 years have changed the demography of the planet Earth in a negative way resulting in its deterioration. Goals for next 50 years should be to carry out the green upliftment of the Earth so as to bring it back to normalcy and natural form as it was about a century ago. The aim of the journal "Environmental Sustainability" is exactly to promote the greener technologies and biotechnological interventions so as to heal the planet Earth back to normalcy and sustain it for the survival and flourishing of all the life forms." (Arora 2019: 2f)

The unsustainable activities and technologies should be replaced with viable measures: "employing green and biological methods, technologies and inputs such as the use of biofertilizers and biopesticides as substitutes to their chemical counterparts, biofuels in place of petroleum products, bioremediation instead of traditional remediation methods, bioplastics and biofilters to minimize pollution, biotechnological advances for food and energy security, study of metagenome for better knowledge of diversity and working of micro-ecosystems at the molecular and biochemical levels." (Arora 2019: 2)

Source: Arora 2019

The deficits in current sustainability policy cannot be explained by lack of knowledge or insufficient research. Abundant knowledge is available from the research in environmental sustainability research, and the knowledge is rapidly growing. Large parts of this scientific knowledge are, however, not used in the policy and governance processes, but are only communicated and discussed among specialised researchers, with the result that the knowledge remains within the science system. A dysfunctional system of knowledge transfer and communication (see Chap. 8) leaves much useful knowledge unused. The limited capacity for using available knowledge, the lack of safe knowledge about many social and ecological processes, the limits of general scientific knowledge, and the ignorance about the future—all of these conditions shape the sustainability process in ways that must be reflected upon, analysed, and discussed more systematically.

Uncertainty of knowledge: The uncertainty of the scientific knowledge produced within environmental and sustainability research, the many competing forms of knowledge, and the general ignorance of many aspects of complex and interactive social and ecological systems and processes, must be addressed in the knowledge application processes. The relations between knowledge and ignorance in environmental and sustainability research have long been discussed (Keil and Stieß 2007; Voß et al. 2006). It has been concluded: *what is known about the problems to solve through sustainable development is as important for policy and action as that what is not known.* To deal with the uncertainties in scientific knowledge for the sustainability process is, however not simply a question of developing a method for uncertainty analysis, as Schaubroeck et al. (2020) discuss. Required is a broader understanding of the relationship between knowledge and ignorance, and between science and practice, to deal with uncertainty in the policy processes (Walker et al. 2013).

For example, proposals from Cash et al. (2003) for linking knowledge and action on behalf of sustainability remain limited to mobilizing the connections between "science and technology" as a means to transfer knowledge from science to practice. Their ideas aim to reduce the problems gaining credibility and legitimacy for new knowledge practices through such methods as case studies, the translation of knowledge in action programmes, the mediation and managing of boundaries between expertise and decision-making. Thus, they concentrate on the communication and knowledge exchange between scientists or experts and decision-makers or practitioners. They overlook, however, the need to develop inter- and trans-disciplinary sustainability research forms that opens broader perspectives and accelerate knowledge communication and application beyond accept bounds.

Uncertainty in sustainability transformation refers to changes in the forms of research, the relations between science and policy, and the forms of knowledge application in sustainability-related action. Transformation cannot be designed as conventional policy, planning or regulation, that is, as processes in the sphere of public policy that require only decision-making. Transformation includes extended action, collective learning, integration and networking of different spheres of political and non-political action, *and the application of theoretical knowledge about social-ecological systems that has so far been largely unused.* To overcome uncertainty requires, thus, new forms of review, evaluation, assessment, and knowledge integration that make once external perspectives core components of sustainability transformation. This situation resembles, to some degree, the processes of reflexive modernisation analysed in the sociological theories of Beck and Giddens (Beck 1992; Beck et al. 1994), where the neglected risks and dangers of the early industrial modernisation are corrected by actors and institutions who learn from past failures. Unlike the sustainability process, however, these sociological theories exclude analyses of the interaction between social and ecological systems or nature and society, of the systemic mechanisms of modern economy and society, and of capitalist emphases on accumulation. As a result, sociology does not grasp the complex reality of interacting social and ecological systems.

9.1 The Structuring of Time in Social and Ecological Processes

One theme often neglected by most social-scientific and economic research into modern society and its problems is the temporal structuring of processes of development, change and transformation. To deal with temporal structures requires that one reflect on the future. In general, however, most practising scientists reject the notion that the future can be a matter for science. They see the future as simply unknown. Consider, for example, the economist John Maynard Keynes ironic quip that, "in the long run we are all dead". Consider also Karl Popper's five arguments against the possibility of predicting the future of a society, which he included in the preface to his book "The Poverty of Historicism" (1957). They are: (a) The course of human history is influenced by the growth of knowledge; (b) Scientific methods do not allow for predicting the future development of knowledge; (c) Therefore, also the future course of human development cannot be predicted; (d) This implies the impossibility of a theoretical science of history; (e) Both historicist methods and historicism can be rejected. He saw it as necessary to add that none of this excludes the possibility of social prognoses. Seen from this perspective, a science of the future seems to become a theoretical science of history, a form of speculation, or of ideology, with visionary and utopian ideas. Within this view of history and the future two forms of reasoning seem problematic: (1) formal logic and logical inference are insufficient, when it comes to argument; (2) prognoses and predictions are not the only way to reflect on the future. The relations between knowledge and ignorance require further discussion and clarification, as also does the concept of the future and its connections to human action.

In complex processes multiple time-space connections frame collective action. These connections and the many forms of social structuring of time and space are, however, insufficiently reflected in current policy and governance processes for sustainable development. Temporal structuring of the future and of future-oriented action appear simplified with regard to the different structures of time that are found in social and ecological systems in modern society. For the dominant time regime—the "'vulgar time' of the clockmakers—the measured, linear time of past, present and future", as Bourdieu calls it—we "can break with this point of view by reconstructing the point of view of the acting agent, of practice … revealing that practice is not in time but makes time (human time, as opposed to biological or astronomical time)" (Calvelli 2015: 36). The times of actors and their manifold subjective perceptions of time are not the only forms of complex temporal structures in modern society, as there are structures of cyclic, seasonal, repetitive, linear, punctuated, bifurcated, physical, biological, social and historical, subjective and objective times. Some of these are connected in the temporal model of historical analysis propounded by Braudel (see Chap. 2).

The temporal structures that dominate in public policies, in planning, and in economic analyses stretch only in the near future. Hence, with them "the future cannot begin" (Luhmann 1976). Instead, they are oriented to the present time and situation and merely extend these forward via incremental changes and peer forward only toward a limited time horizon and assumer, moreover, that long-term developments in the relationship between social and ecological systems can be understood via incremental improvements. What Luhmann and others see as the main idea of the future in modern society, namely that it differs from the past, means that one can now include non-realised possibilities of development and change in one's view. The future's greater complexity becomes visible in the sustainability process, where the future is understood as a complex system of change via the overlaying of different time regimes onto the systems that are affected by sustainable development. The difficulties inherent in taking action for the future, which is required for the sustainability transformation, demand an improved structuring and priority-setting for action taken in different phases of the sustainability process. It is not simply a process of achieving politically formulated goals. Instead, the process is more complex, as actors from other social spheres are involved, and knowledge based on both short-term and the long-term perspectives can be integrated.

Short-term perspectives on policies for sustainable development comprise adaptation processes. Adaptation is the mode of action in adaptive management or governance, resilience-oriented action, and in the planning of the near future. Forms of cooperation and collective action are oriented toward such action where the near future appears as continuous and the distant future is reduced to visions. The scaling of governance and transformation processes is, hence, reduced to a present-bound time modus, which means that the matching of action at multiple temporal, spatial and ecological scales is simplified or ignored, or, better, is subjected to time regimes that seek to maintain a social system under conditions of controlled change.

Long-term perspectives on policies for sustainable development comprise long historical processes in both society ("longue durée") and nature (evolution). They are processes about which knowledge exists only in the form of history. Sustainable development and social-ecological transformation are, by contrast, long processes that stretch into the distant future and are unknown in their course and progress. It is, therefore, difficult to conceptualise them within policy and governance processes. The transformation cannot be sufficiently structured in forms of social and collective action. Only the near future is foreseeable and can be planned for in terms of action. The distant future, for its part, appears under the forms of uncertainty and ignorance, and as an unknown, impossible to structure in terms of time and action. Time horizons in economic and ecological processes, as they are conceptualised in economics and ecology, are variants of short-term perspectives for which the future is merely adaptive change or repetition. In ecological research about ecosystems, however, resilience and sustainable development the temporal mode of governance is not one of transformation and system change, but another variant of short-term perspectives, that is, ecosystem development, resilience and sustainable development, are conceptualised in the short-term mode of adaptive cycles, with time understood as a cyclic process of adaptation and continuity. By contrast, developing long-term perspectives on collective action in economy and society, and in environmental and sustainability governance, includes, by contrast, such new processes of knowledge creation, dissemination and application as future literacy, transformative literacy, transformative science, and transformative agency. One method from future studies that allows us to think and structure the future in terms of action beyond the short time horizons is the use of scenarios. The regular deployment of such scenarios can play an important role in the sustainability process. Scenarios often work with archetypes of an imagined future that are relevant for the economic future of society, including continued growth, collapse, steady state, and transformation (Amer et al. 2013: 25).

9.1.1 Changing Views of Knowledge in the Sustainability Process

Changing views of knowledge and ignorance, and their roles in and relations to policy and governance processes are discussed less than other knowledge-related questions. In the modern society, knowledge production has become the monopoly of science. This field's influence on all parts of society and daily life has grown, as no other forms of knowledge have proved able to compete. The production of scientific knowledge is growing exponentially, as measured in terms of publications and other indicators. Nevertheless, much less is known about (or even discussed, for that matter) the application and applicability of scientific knowledge. Scientific knowledge seems self-evidently to be made to transfer technologies that will improve the conditions of life. This can no longer be the case. From the moment that the sustainability discourse began, a broader discussion emerged about scientific knowledge's limits, its applicability, and its relation to other forms of knowledge. Indeed, forgotten forms of knowledge (from which scientific knowledge developed) have been recovered, including practical and local knowledge. A crisis of knowledge production and application has resulted, to a degree, in the changing of knowledge cultures and practices within science and politics, and, moreover, in a reassessment of knowledge's limits and of the relations between knowledge and ignorance.

The conventional modern view is that scientific knowledge production is unlimited and that it will result in ever more knowledge being available for social development, reducing thus ignorance over time and transforming all "black boxes" into, in effect, "white boxes" in a way quite similar to how the earth's continents were "discovered" in previous centuries. This view does not take into account, however, that science no longer produces only safe and valid knowledge but has reached its limits with regard to the complex global change processes. As a result, science is rapidly becoming a fount of uncertain knowledge.

The relationship between knowledge and ignorance is central to sustainable development. The traditional view, that more knowledge leads to less ignorance, no longer seems valid. In fact, growing scientific knowledge may bring an increase in ignorance. In environmental research, for example, the application of knowledge reveals ignorance, as is betokened in the creation of technologies that modify nature and destroy ecosystem, such as had occurred in fishing, forestry, agriculture and industry. Disturbed ecosystems (Sousa 1984) become more complicated and more difficult to understand, creating thus more problems through additional catastrophic changes, such as the loss of species (Box 9.2).

Box 9.2 Ignorance

Routledge Encyclopedia of Philosophy (2020)—*Summary of the article "ignorance": the examples show, that problems of ignorance and socially produced ignorance are relevant to environmental and sustainability discourses*

Epistemology has largely neglected ignorance, focusing on knowledge and what is necessary for knowledge, (for example, epistemic justification). Between 2000 and 2020 the tendency has changed, especially in debates at the intersection of ethics and epistemology. Ignorance is no longer simply seen as absence of knowledge that does not deserve significant philosophical attention. However, it is difficult to define ignorance exactly; even if it is seen as absence of knowledge, ignorance requires philosophical attention.

Ignorance is an important concept in four philosophical areas:

1. *Epistemology*: the core questions are what the nature of ignorance is, what varieties of ignorance exist, and whether ignorance has any epistemic value. Several issues in the philosophy of religion (such as negative theology, which says that we are inevitably ignorant of the divine) also touch on ignorance.
2. *Ethics and morality*: ignorance's ethical dimensions appear in certain moral virtues, such as modesty; ignorance can be an ethical excuse; the fact that ignorance itself sometimes excuses one for one's ignorance leads to the so-called tracing problem.
3. *Criminal and civil law and philosophy of law*: ignorance can provide a legal exculpation. For example, if one is ignorant of a law, one may not have committed a crime in the first place. Ignorance can also be a legal excuse. One did commit a crime, but one is not guilty on account of ignorance. But ignorance can also count as inculpation. One is culpable for committing a crime because of one's ignorance.
4. *Topics in social philosophy and the philosophy of science*: agnotology is the study of ignorance or doubt that is intentionally induced or maintained by the government or influential business companies (example: the ignorance claimed by the tobacco industry about the effects of smoking). Debates about minority groups show different forms of socially produced ignorance. Debates about privacy refer to the obligation of other people to remain ignorant about certain facts about one's personal life. Philosophers of technology argue that societies should aim at collective ignorance regarding particular technologies (for example, weapons of mass destruction) because of their enormous potential for social harm.

(*continued*)

Box 9.2 (continued)

Ignorance is easily discussed in the sustainability discourse, but many strategies for sustainable development ignore its salience for discourses about the reasons, causes, consequences and potential solutions required for the sustainability transformation. Not only political institutions and governments ignore potential strategies (degrowth, for example) or reduce solutions (through technical fixes). Ignorance is as widespread in science and economics. For example, some scientists are ignorant of much of the social scientific knowledge that is relevant for sustainability in natural scientific and ecological research. Traditional forms of disciplinary specialisation in academic research can, thus, be seen as cultivating ignorance through their very practices of knowledge formation. Although inter- and trans-disciplinary research is widely practiced in environmental and sustainability research, knowledge selection and exclusion continues in research in political action and governance. In considering the causes of ignorance, one must differentiate between "real ignorance," or lack of knowledge (for example, the climate system is still poorly understood) and "socially produced ignorance," or, what one does not want to know, which includes all forms of selective and biased knowledge that are used in science and politics.

Source: Routledge Encyclopedia of Philosophy 2020

Ignorance and uncertainty can be broken down into in various forms (Allen 2000). Nevertheless, logical reasoning and differentiation are not sufficient to deal with the effects of ignorance and uncertain knowledge in the practices of decision-making and collective action for sustainable development. We need not assess ignorance and uncertainty either qualitatively or quantitatively. Instead, we must discuss them critically within the sustainability process, noting always that uncertainty and ignorance have different consequences for the different actors involved. To deal with the problems of mediating between knowledge, uncertainty and ignorance requires specific forms of social communication and learning in both science and governance. An important way to enable such learning is to develop within sustainability research transdisciplinary knowledge production, communication and integration. Transdisciplinarity, understood here as knowledge production in the contexts of application and interpretation of scientific and non-scientific knowledge, is essentially a dialogue between science and society, between researchers, decision-makers, and civil society actors, about the forms and conditions of knowledge application for action. Ignorance and uncertainty that cannot be dissolved methodologically through evaluation and assessment needs to be dealt with dialogically, with the aim of creating temporary consensus for collective action.

9.1.2 Views of the Future: Knowledge and Ignorance

In spite of intensive discussion of future changes in economy and society through global think tanks, such as the "Strategic Foresight Group" or the "Tellus Institute for a Great Transition", the methodologies and the possibilities for projecting long development trends are limited, and the connections between policy analysis and future research remain weak (van Dorsser et al. 2018: 76). Before the distant future can become part of the sustainability discourse, it will be necessary to differentiate the concept of the future. How and how far, with which methods, can the future beyond the time horizons of planning be taken into account in sustainable development? This question is connected to the identification of design options and key competencies for sustainable development (Rieckmann 2011), which include such capacities as anticipative thinking, dealing with complexity, and imagining unforeseen problems (Box 9.3).

Box 9.3 An Economics of the Future
Some of the cognitive problems in the economics of the future reflected by Hagens (2020: 13f) *are relevant for the sustainability process*:

1. *The necessity to come away from the predominant view of climate change in public, policy and large parts of science implies*: climate change is not the only systemic risk to be dealt with in the foreseeable future of modern society, reducing all other negative changes to minor or less important ones.
2. *The realistic alternative for the sustainability process, is*: to see climate change as part of larger and interconnecting maladaptive changes and dysfunctions in all parts of society and economy that include and can only be solved together, through a system transformation of society and economy.
3. *The transformation implies*: to prepare in science and governance for the possibility of increasing social, economic and environmental risks, continuing disturbance and partial collapses of the global economy triggered through increasing scarcity of resources; "collapse" is, as the

(*continued*)

Box 9.3 (continued)

author comments, a misleading term, because it signals final ends and complete destruction, whereas the fall of states, empires or civilisations in human history did not imply the end of society and development, only transitions to other forms of societies and development.

4. *To deal with the foreseeable disturbances in a realistic way requires*: to prepare for changes towards an economic system developing in accordance with the biophysical processes that direct and determine natural resource use, no longer and only through economic growth in monetary terms: "The real challenge will begin, when economic growth ends." (p. 13)
5. The foreseeable process of sustainability transformation requires continuous discussion of what can be kept from the present system, what will be changed, and how to change it; not much planning will happen in the changes that will be influenced from other forces than human will and environmental governance: from physical, biological and ecological processes, changes and disturbances.

Source: Hagens 2020

Projecting the foreseeable future during this century as Hagens did, requires more systematic discussion of how the future can be dealt with in the scientific and political discourses. To create a framework for policy related future studies, van Dorsser et al. (2018: 78ff) one must connect the taxonomy of different types of futures by Voros (2003) with different levels of uncertain, plausible and unknown futures that are used in policy analysis (Thissen and Walker 2013). The types of futures (in terms of thought futures) can be sub-divided into *potential futures* (everything beyond the present moment), *possible futures* (what can happen), *plausible futures* (what could happen, based on our knowledge of physical and social processes), *probable futures* (what is likely to happen, extrapolating current trends), *projected futures* (the most probable of futures), *preferable futures* (what should happen, according to our normative views). Extreme types of futures are *preposterous futures* (judged as impossible) and *predicted futures* (foreseen by someone).

The distinctions between types reveal how difficult it is to form a unitary concept of the future. The list above allows only rough estimations of different future states of systems in probability-related forms, although these are not necessarily calculated quantitatively. The authors cited above see this combined typology as a step toward a fuller integration of policy research and future studies. However, the framework's long-term perspective is limited to about thirty years, a time horizon in which the future appears already as unknown or deeply uncertain (van Dorsser et al. 2018: 78). For sustainable development to be conceptualised as fully transformative longer timeframes are necessary, and the question for all types of constructed futures is, where does the knowledge come from on which the construction of a certain type of future is based? Scenarios, for the purpose of generating knowledge about possible future states of society or economy can be connected with many research methods from future studies and other disciplines. They, can be used in research and policy making, as well as for improving the connections between research and practice (Ramirez et al. 2015; van Dorsser et al. 2018).

Another form of thinking about the future and how to achieve wanted futures of sustainability is to direct the capacities and capabilities of the actors involved toward achievement of sustainability. The approach of "education for sustainable development", during the UN-decade of education for sustainable development 2005–2014, is an example for anchoring sustainable development in all institutional forms. In this respect, education should include key competencies of all citizen (prospective thinking, dealing with complexity, understanding one's own situation and role in the global society) so that they can participate actively in the sustainability process (Rieckmann 2011). This is an important part of transformation strategies, although its focus is on individual competencies and capacities, rather than organised, institutional, collective and systemic capacities for transformation.

Knowledge for the sustainability transformation needs to be translated into knowledge for action. This can be done mainly with global scenarios. It is not a matter of using technical knowledge to restore ecosystems, to clean the environment, or to correct non-anticipated consequences of earlier technologies in resource use. More important for achieving sustainability is knowledge about the negative effects of present economic production, consumption and resource use, and to consider how this knowledge can be used to bring about change. This requires connecting empirical knowledge about overuse of resources and destruction of the environment with social visions or scenarios about the future society as a sustainable society, in which a plural economy with markets, sharing, solidarity, and redistribution is achieved. The global scenarios discussed today are, however, only informed by empirical knowledge and trends that are known from present environmental, social and economic research and have, thus, little contact with theoretical knowledge about modern society and economy. On only finds theoretical scenarios about the future society far afield of the sustainability debate, that is, within social-scientific theory discussions about the transformation of modern capitalism. A classical variant that accompanied the debates about transformation and the future society is Wallerstein's world system theory. This approach has been complemented by a theory of world ecology and the "capitalocene" (Moore 2016) and the recent debates about "another great transformation" and a social-ecological transformation described below (Table 9.1).

Table 9.1 Different types of transformation affecting the sustainability process

Type of transformation	Definition	Interpretation	Sources
1. Technological transformation: sociotechnical transitions	Sociotechnical transitions (from one socio-technical system to another) where the economy and technology change in slow processes of innovation are part of the sustainability process (for example, the transition from fossil to renewable energy)	Similarities with ecological modernisation, where green and biotechnologies are seen as creating social change and ecological adaptation; the whole sustainability process cannot be conceptualised as sociotechnical transition or coupled socio-technical transitions; difficulties in socio-technical transitions are similar to that in the total sustainability process: rigidity of economic, political, cultural structures, norms and systems	Journal "Environmental Innovation and Societal Transitions"
2. Social-economic transformation: globalisation	Globalisation as deregulation and dis-embedding of markets (neoliberalism): a global process of social change, with the consequence of growing resource flows; globalisation in this form is in conflict with sustainable development	New variants of neoliberal globalisation: market-based regulation, commodification and valorisation of nature (TRIPS, TEEB)	de Senarclens and Kazancigil 2007; Robinson 2009

(*continued*)

Table 9.1 (continued)

Type of transformation	Definition	Interpretation	Sources
3. Social-economic transformation: "post-industrial economy of singularities" (Reckwitz)	The post-industrial society is not only one with a dominant service sector: cultural changes since the 1980s are transforming the industrial economy of mass production to a new cultural economy of singularities (new cultural goods, no standardised production; consumption and lifestyles become more singular and creative ("creative industries": Davies); the new economy is described as "creative economy" and "experience economy"	The post-industrial economy described in this way is "the future that is already present": it came with post-Fordist accumulation regimes and with the digital economy or the knowledge society; it does not describe the total economic and social reality of the globalising economy, but the dynamic part of it that drives other changes; singular lifestyles and individualisation (Beck) are part of the changes; these changes remain within the capitalist economy or "cognitive capitalism" (Lorey and Neundlinger), based on economic growth and globalisation	Reckwitz 2020; Howkins 2001; Lorey and Neundlinger 2012
4. Social-economic transformation: post-capitalism	Post-capitalist economy and society is not a coherent set of ideas but selective views of that what makes the economy no longer capitalistic: knowledge and information-based economy, internet, weak property rights, the neoliberal project is supposed to collapse (Mason)	Technological innovations, knowledge and participatory economy (Hahnel), similar changes as described as post-industrial society	Drucker 1993; Hahnel 2012; Mason 2015; Srnicek and Williams 2015

Sources: mentioned in the table

The variants one and two in Table 9.1 mark the spectrum of system-immanent transformation ideas in side which the term transformation is somewhat misleading: they are not transformations in the strict sense of the term, where the whole economy or society is transformed into a new system, but refer only to partial transformations with the aim to create further development and growth (although the transformations discussed for parts of the economy—the digital economy, for example, are strong transformations). The changes discussed remain within the growth and innovation paradigm of the industrial economy, describing another form or new era of industrial society. The variants three and four of the post-industrial and the post-capitalist society can be better connected with ideas about transformation to sustainability. They do not show the future society in total, but reveal only the paths that lead to it and the prospective phases of development according to which a new society can merge with the globalised and digital economy.

9.2 Reconceptualising Time in the Sustainability Process: Interaction of the Past, the Present, and the Future

Following Miller (2018), if one wishes to discuss the future with an eye toward achieving sustainability, then it seems best to look, first, at how the historical sciences have conceptualised development, change and transformation of societies over time. Braudel's conceptual models of time (see above, and Chap. 2) are a first step toward rethinking time's position within social and economic development processes. In addition, phenomenological sociologist have discussed temporal structures and their connection to the lifeworld (Schutz 1967; Schutz and Luckmann 1973), although they concentrate less on understanding future time than on analysing past or present time. Anticipative thinking and prospective action are, however, a component of social action according to the sociological analyses. To connect the historical analysis of societal development with the analysis of potential development in the future is the second step. This step is more speculative. So, the use of scientific knowledge becomes difficult here, especially since the future is not only unknown but also imagined. Yet, simply to impose a "veil of ignorance" on the future, reduces the discussion to nothing but subjective visions and wanted futures. In this respect, scenario construction and analysis are important, because they envision anticipated futures that sustain the creation of knowledge about future changes.

The view of the future in modern society differs from that of earlier times in its assumption that the future is constituted by change, development, progress and improvement. This view is closely connected with industrialisation, which incorporated an evolutionary view of development, in which industrial society was not only the only way to improve of society, but also the source of endless improvement (Spencer). An older conception of the future (exemplified by Machiavelli) is rooted in medieval thinking, as a result of which no developmental dynamic is at work. Instead, the future appears as a repetition of the past. As Heilbroner put it: "*whoever wishes to foresee the future must consult the past; for human events ever resemble those of preceding times. This arises from the fact that they are produced by men who ever have been, and ever will be, animated by the same passions, and thus they necessarily have the same results*" (Heilbroner 1968: 21).

The complexity in coupled and networked social and ecological systems around the globe exceeds the capacities of traditional natural resource management, ecosystem based management, or adaptive management, since these all deal with single ecosystems, social systems, or local areas. True spatial and temporal complexity can never be analysed through one coherent approach or project. The process of sustainability transformation will appear more like patchwork, at least initially, but can be improved through better coordination with knowledge about the spatial and temporal complexity of social-ecological systems. The knowledge of different forms of temporal structuring in social and ecological systems is a precondition for designing and formulating action strategies and interventions whose potential side-effects and unintended consequences need to be assessed, too. The complexity is, in fact, dissolved through piecemeal governance that takes action at different spatial levels and at different points in time. For example, governance often regulates resource use and resource protection for a specific type of natural resource, even though the consequences for entire ecosystems still need to be assessed. Indeed, transformative activities differ for specific types of action. Climate change adaptation differs from biodiversity management, from management of consumption behaviour, and from regulation of growth-based economic production.

The construction of scenarios with possible futures of the global economy and natural resource management is an important method for sustainability governance, because it reconceptualises the relationship between the past, the present and the future in a way that supports the sustainability transformation. The scenarios can be divided into useful binaries: wanted and unwanted futures; futures that more easy or more difficult to achieve; futures that result from the continuity of the present economic and societal systems versus futures resulting from transformation, rupture and changes of development paths. Scenarios allow us, furthermore, to calculate future resource use independently of specific forms and rates of production and consumption. This can, for example, help with discussing the conditions of future food production and of food security for a growing global population. That global food security requires industrialised, high-tech agriculture, with continually increasing yields through use of agrochemicals and synthetic fertilizers, or genetic modification of plants and animals, is meanwhile doubted. Many global food scenarios are available—and these open the possibility of developing and combining different food systems that are sustainable (Erb et al. 2009; Hubert et al. 2010; Institute for the Future 2011; Millennium Institute 2013; Barilla Center for Food and Nutrition 2016; World Economic Forum 2017; van Wuren et al. 2019). The scenario by Erb et al. (2009) shows that the global population can be fed from wholly organic production, or from conventional production, or from agricultural systems with several components. In the global scenario debate (Box 9.4) several scenario variants have developed that are relevant for decisions about sustainability.

Box 9.4 The Global Scenario Debate

> Perhaps the key theme in the story of the 21st century will be how humanity addresses multiple threats to the stability of the planetary social-ecological system. Over the past 10,000 years of the Holocene Era, Earth's self-regulating mechanisms have kept climatic and biogeochemical processes within a narrow range, providing relatively stable conditions—a safe operating space—for civilization to develop and thrive ... The increase of population and economic activity since the Industrial Revolution has ushered the coupled human-environment system into the Anthropocene, a new geological era where humanity plays a dominant role in driving planetary change ... Further intensification of anthropogenic stress could impel the Earth system out of its safe operating space with dire consequences for society and ecosystems, potentially undermining opportunities for humanity to thrive on Earth in the future. (Gerst et al. 2014: 124; see also Raskin et al. 2010)

The present transformation can be described in changing forms of risk and vulnerability and the changes in forms of resilience and sustainability:

Risk: a term originating from social sciences/economics—the probability of negative consequences of actions/decisions; the new types of risks relevant for sustainable development are the big "systemic risks" discussed in the theory of risk society (Beck 1992) and the risks through human-made global environmental change.

(*continued*)

Box 9.4 (continued)

Vulnerability: a term for the exposure of humans, their communities and societies, and other species to natural or man-made risks, dangers and disasters

Resilience: the adaptive capacity of ecosystems and social systems to survive catastrophes (with the possibility to achieve a qualitatively different state)

Sustainability: capacity of maintaining coupled social and ecological systems in the long run, without disturbing their social and ecological contexts

The scenarios presented by Raskin et al. (2010) and Gerst et al. (2014) present views of the future as they are discussed in the broader sustainability discourse and process. These are: forms of positive future states that are in line with sustainability (for example "adapting mosaic") and negative or unsustainable future states (for example "fortress world"); sufficient changes to achieve sustainability (for example "great transformation") and insufficient changes (for example, "business as usual). The global scenario debate shows, furthermore, how scenarios for sustainable development can be improved methodologically by combining (1) scenario analysis with (2) information about planetary boundaries of resource use and (3) the targets of human development.

Sources: mentioned in the text

Extant global scenarios show that different transition paths are possible within the safety margins of the limits to growth and planetary boundaries. They show which scenario variants yield better results in achieving sustainability and which yield fewer. The "great transition" scenario, a combination of transformative socio-cultural and technological change, is a realistic scenario variant that can be adapted to changing conditions in the process of global sustainability transformation. The global scenarios do not reveal the systemic nature of the future sustainable society economy and society; rather, they reveal how we can imagine the future in simplified forms of social practices for production, consumption, politics and the economy. However, they also introduce more elements of theoretical knowledge from theories of modern society, its relations with nature, and from analyses of complex social-ecological systems. Thus, they deal with the complexity of the transformation process. The scenarios do not provide a theoretical, sociological or economic description of a new mode of production and societal metabolism. (This is done in the theories discussed in the sustainability discourse.) Instead, the scenarios mark the beginning of the process of the sustainability transformation, when the knowledge about a new economy and society of the future is still limited. Knowledge and practical experience are growing in the continuing process and need only to be better captured through various strategies of knowledge work. It is not necessary to predict or to imagine the future society in its final state, or in an ideal or utopian form, because the transformation processes can improve continuously as the global environmental and social state improve.

The knowledge relevant for the structuring of the sustainability transformation processes is not necessarily that of specialised disciplines, of environmental history, environmental economy, environmental sociology and environmental policy analyses, although knowledge from this research will be useful in one or another form. It will not be used, however, in the forms in which research originally produces it, namely as fragmented knowledge about single facets of a single process, but in the forms created through inter- and transdisciplinary knowledge practices. Beyond the gathering and production of knowledge from research, the knowledge created under conditions of transformative action will be applied to knowledge for social-ecological transformation. For the success of sustainable development it is important to manage manifold knowledge processes and, above all, to interconnect them within "knowledge-action systems" (see Munoz-Erickson 2014: knowledge-action systems for sustainable urban development).

In discussing knowledge from historical science, it will be necessary to learn from earlier social-ecological transformation processes in human history, considering why and how they happened and bringing to bear the perspectives of social ecology and cultural anthropology. There are, however, further possibilities for learning and capturing knowledge from historical analyses of the development of modern society that focus on the political and economic processes discussed in the first and second parts of this book, including especially economic growth, political modernisation, and changing forms of natural resource use. Three examples of such integrative knowledge work are summarised below. They not examples of historicism in the sense condemned by Popper, since they allow for other conclusions about the science of history and knowledge of the future. History does not simply appear as a non-theoretical science and collection of "dead facts". Instead, as the sustainability discourse shows, it provides knowledge about the origin of social and environmental problems that need to be solved in future.

The past and the future in social-scientific and ecological analyses of problems:

1. Vogt (1948), who was one of the founders of North American environmentalism and was as important as Leopold (1949) with his ethic of the land and Carson (1991) with her ethic of the sea, opened with his book "Road to Survival" with a Malthusian discourse about population growth. He discussed connections between global population growth, economic growth, and the foreseeable destruction of the resource base for future generations, presenting therein a dystopian future, of war, hunger, misery and societal collapse. On account of this, the population researcher and feminist author Betsy Hartmann called him a founder of "apocalyptic environmentalism". Living amid Western affluence and mass consumption Vogt began to argue ecologically against economic growth and wealth. In his view, affluence was not an achievement, but modern society's biggest problem. He described the industrial society and its pursuit of growth and consumption as "the Great Illusion" and the "fallacy of industrialization" (Vogt 1948: 57ff). Robertson (2012) analysed critically the history of the Malthusian concerns in North American environmentalism. But none of this history can be denounced as exemplifying ignorance. Instead, is must be understood as a selective form of reasoning that was based on Vogt's natural-scientific and demographic interests. Vogt assumed that after a few more decades of the maladaptive and destructive development, the process would be beyond science and politics's control. Hence, the time factor cannot be eliminated (Vogt 1948: 151f, 265). Vogt's suggestions for "ecological health" included renewable resources for producing wealth on a sustained-yield basis and adjusting our demand to the supply by reducing resource use and population growth. He underestimated the depth of the demographic transition. Nevertheless, but the future that he foresaw has come to pass—only not yet in form of big catastrophes. In his last and short chapter about the "History of Our Future" he describes the ecological trap in more detail, always seeing chances to avoid global bankruptcy and collapse through immediate action. What he describes as "to regain ecological freedom" (Vogt 1948: 284) is an adequate description for the changes to be achieved by sustainable development. He describes the process as learning to know "our dependence from the earth and the riches with which it sustains us", which is in his words "the long road back" (Vogt 1948: 286, 288) to sustainable resource use within the limits of the earth's capacity. The history of modern society appears here as a false programming of economic development based on the illusions of industrialisation as humankind's cornucopia.
2. Heilbroner (1968) brought the discussion about the future of modern society to the point where the discourses of limits to growth and sustainability could follow up these arguments. The optimistic philosophy and orientation towards the future that came with the industrial society did not come from the unaided abil-

ity of humans to make their destiny, but from historic forces that promised to improve humanity's future. He included among historical forces the dynamic potential of technology and the democratic aspiration, that developed "*within a self-sustaining mechanism for economic growth, which first opened the future to optimistic expectation; and later it was the very blindness, the determinism of these forces which fortified men's faith in them*" (Heilbroner 1968: 34). This deterministic view of development, technology and progress binds the perception of development to a silent acceptance of history's forces and, moreover, to specific perceptions of historical forces. The rapid disenchantment with the ability of self-government came through the forms in which industrial society was realised, that is, in the capitalist mode of production—"*it was capitalism which pre-eminently represented the automatic achievement of progress. And whereas it was soon clear, that capitalism had unquestionably changed economic conditions, it was far from evident that it had improved them*" (Heilbroner 1968: 38). The disenchantment came with the classical political economists, Malthus, Ricardo and Mill, and still more with Marx critique of political economy resulting in pessimism about the future of modern society and economy in the form of capitalist and industrial society. The future of modern society does not look less grim in Heilbroner's analysis than in that of Vogt, but the former derived his arguments from a more coherent social-scientific and economic analysis of modern society, where the values, worldviews, and paradigms appear as ignored sources for shaping and driving a modern economic growth machine.

3. Tilly (1997) described the "Roads from Past to Future" from a more sociological perspective, although he also discussed population processes and resource use, scientific development, in addition to change, political modernisation and democracy, changes of class structure and migration in the historical epoch of modernity. Like Vogt, Tilly thinks the future mainly in terms of the past but does not describe the future in terms of action required for transformation, discussing. only changing views of time and future with the conceptual model of "*the past as a sequence of futures*" (as Stinchcombe writes in his review essay, in Tilly 1997: 387ff). Thereupon follow the subjective views of the actors in the processes of societal change, with humans as intentional and calculating animals building history out of their pictures of the future. Hence, there is no sense "*in being rational about anything but the future, because one can't do anything about the past*". The future is the only state that can be influenced by human action, but when it comes to the mental models of the future that humans construct, our "intelligence about the future" is always wrong. This is the first thing to become aware of: Our (wishful) thinking about the future is, in Tilly's view,

strongly influenced by power. The subjectively wanted, proclaimed objectives could almost never be achieved in the short run, and they became dependent on anticipation, cumulation, and coordination of efforts with that of many other actors. Both parliamentary democracy and its decision-making institutions have developed a mechanism for building the future by mobilising resources through power struggles. The variations of the mechanisms can be divided into five aspects: lengthening the time into the future; increasing knowledge about the future; the conditions for future action; structures that provide guarantees for the future, including laws, contracts, and the continuity of institutions. Complications come with social crises and conflicts. Tilly's analysis of political rationality and its institutional forms contrasts with simple rational choice models in economics.

Heilbroner (1968) analyses the modern economic system more systematically and from an interdisciplinary perspective. His work can be seen as a theoretical knowledge source for the sustainability discourse, as it analyses the systemic components and the worldviews that guide action in the economy and in public policy. The ways into the future of a sustainable economy or society depend on many factors that influence political, economic and ecological processes. The roads from past to future which Tilly (1997) investigates are complex and long-term processes of industrialisation, state formation, and social development, all of which are "defined by changing configurations of social interaction" (Tilly 1997: 1). He has studied the changes on the way to the future in micro-processes of collective action and has developed a theory of political rationality (Stinchcombe in Tilly 1997: 406). The context of collective action, the social and societal systems through which collective action is framed, requires complementary analyses of the complex, large-scale systems that influence, support or hinder the collective actors in the transformation to sustainability.

9.3 The Ways to the Future: Sustainable Development as Transformation

Horizon scanning (van Rij 2010) is a method used to identify different types of threats and opportunities. It can be used as first step in the formulation of scenarios for sustainability transformation, where the short-term and the long-term perspectives of change and transformation are combined. In the foreseeable future, for example, international relations includes complex multi-scale interactions between states and other international organisations (Archibugi and Held 1995; Held 2010).

With globalisation states have differentiated their functions, becoming part of the global network of political and economic institutions and systems. Acting in the global policy processes, nation states became multi-functional and multi-scale states through decentralisation (strengthening of forms and functions of the local state), and internationalisation of the state (development of more and more international regimes under the management and control of states and governments). However, in the goal setting for sustainable development, in public policies and policy agendas, the time that is necessary for the long transformation is not reflected, which leaves the misleading impression that the pace of change depends solely on political actors's willingness to act of, thus ignoring the broader knowledge we have gained about the transformation of complex systems. The temporal horizon of the "Agenda 2030" is, for that reason, unrealistic when it some to the achievement of the sustainable development goals, and it is still more unrealistic for a sustainability transformation, since the stated goals say nothing about the necessity of systemic change in the global economy.

9.3.1 The "Longue Durée": Conceptualising Social-Ecological Transformation

How far in the distant future is the completion of a new sustainable society and economy? With the presently available knowledge about societal transformations it can be estimated that the transformation time for the completion of institutional changes, country by country, in both national and international change processes, will be much longer than the time frame of the "Agenda 2030". Also, with respect to the EU reform project, to bring the economies of the member states to a zero-level of greenhouse gas emissions by 2050, assuming that this plan succeeds, is still not a breakthrough to global sustainability. By this century's midpoint, the global economy will have sailed into further turbulence, thanks to the inevitable slowing of global economic growth. Environmental pollution in the newly industrialising countries will, thus, have reached extreme levels. The slowdown of growth will, in turn, impede further global industrialisation and will, moreover, be blocked global limits in natural resources. Limited progress in sustainability transformation in some countries, in the end, will not change the global trends.

It is unlikely that global sustainability can be achieved during this century. The processes will continue into the distant future and reach far beyond the time horizons of current planning and policy programming. The problem of resource scarcity will, moreover, create additional difficulties among not only individual countries but also the Global North and South. Progressing climate change will produce worse living conditions and diminished food production in many parts of the world. Global migration will increase, and defensive attempts by affluent societies to build a "fortress world", as the EU and in the US are doing, will become less effective over time. The collective elaboration of long-term perspectives for global action under such unfavourable circumstances will require combinations of many different forms of knowledge creation and management, including scenario analyses, more interactive forms of learning and creation of agency, the development of transformative literacy, transformative science, transformative action groups, and governance under conditions of unsecure knowledge.

Socio-metabolic profiles of resource use and development of the countries of the world (Krausmann et al. 2008; Schaffartzik et al. 2014) highlight the possibilities of socio-ecological transformation. They give potentially useful information for critical countries or country groups according to their development paths. They also help to identify necessary changes in development paths during the long transition to sustainability. The risk of local or regional economic and ecological collapses is especially high in newly industrialising countries. The empirical data show:

1. Not all industrial countries have very high levels of resource use. But three critical clusters have risky development paths. The highest level of per capita consumption is in industrial countries with low population density, especially in North America and Australia.
2. A lower level of per capita consumption, but a high level of total resource use is found in the densely populated industrial countries of Europe.
3. The developing countries with high population density (China, India, South Asia, Central America) have the largest share of global population of poor people (66%). Although much closer to the metabolic profiles of agricultural societies and with low consumption per capita, these countries consume large parts of the global resources through their large population size. When these countries industrialize rapidly, which began to happen in the 1990s, the resource and regeneration capacities of the earth system will be overwhelmed. In China and India, both big and densely populated countries, the crises of development and regional collapse seem possible even with the continuing industrialisation.

Combining the information about the countries with global scenarios discussed above (Sect. 9.2) leads to the following conclusion: *the countries that need to change their development paths most are those with high resource use levels, relatively (per capita) or absolutely (total resource use).*

4. For the industrialised countries in Europe and North America, the main requirements of sustainable development are to reduce their high levels of resource use by saving energy and materials, reducing economic growth. They could also invest in "green technologies", thereby, mitigate climate change through the expected reduction in CO_2-emissions.
5. For the newly industrialising countries (especially China and India) the requirements of sustainable development will more difficult to meet. These include reduction of population growth and changing the industrialisation strategies through partial industrialisation, reducing pollution, and supporting more rural development, agriculture and food production. For saving of energy and material use it is necessary to experiment with locally and regionally adapted development forms. For the success of sustainable development in these countries and the others in the Global South all that is necessary is what the Brundtland report demanded: better cooperation and sharing of resources globally; the reduction of unequal development and inequality of resource use between the North and the South.

The analyses of Krausmann et al. (2008) and Schaffartzik et al. (2014) were among the first to apply the analysis of societal metabolism to the comparison of countries with the express goal of assessing the difficulties that the transformation to sustainability presents. The analyses can be developed and updated with the continuing research on societal metabolism that becomes an important knowledge source for the future process of sustainability transformation. The socio-metabolic profiles show more clearly the difficulties of transforming national economic systems than do other economic or ecological indicators.

9.3.2 Future Literacy: Anticipating the Future as Process of Knowledge Creation

Future literacy as a capacity to re-think the future under perspectives of change and transformation is a form of transformative, action-bound literacy: to understand knowledge and to apply it for discussing possible futures. This requires, for example, being able to anticipate rebound effects, or to adapt to changes in ecological systems and resource availability (Schneidewind 2013: 82). *Transformative literacy* is oriented toward balancing technological, economic, institutional (governance-related) and cultural dimensions of knowledge application. Knowledge creation for prospective and transformative action utilizes methods for the analysis of complex processes of global social and environmental change as they studied within sustainability science and other form of transdisciplinary knowledge production, including semi-quantitative models, participatory research, or "real laboratories" that experiment with change.

Anticipating and creating the future in terms of action that produces desired futures is not based exclusively on scenarios. Further tools and methods for future studies aiming at creating individual and collective agency are being discussed. Miller (2018) proposes a conceptual framework for future(s) literacy that allows for the continuous development of anticipation as the capacity to "use" the future in both the practice of collective action and the collective thinking of the future. It develops, thus, new forms of mapping, manifesting and formulating possible paths of the future. These forms include future literacy laboratories, knowledge creation through collective thinking and action, learning from innovation policies, or developing new methods for measuring wellbeing and for empowering different actor groups. With future literacy the knowledge creation about the future has gone far beyond heuristics for creative thinking about problem solving that are characteristic of traditional management science and planning methods. As has been noted, traditional planning philosophy combines "wicked problems and clumsy solutions" (Hartmann 2012), because it struggles with the complexity, uncertainty and normativity of planning. Ideas such as "seven generation planning" address problems of natural resource use and landscape management by emphasizing community-based planning, reacting to climate change, and maintenance of health and welfare. Good examples are the website of the "US Climate resilience Toolkit", and other planning companies and institutions. These approaches to planning deal with local problems, but are embroiled, nevertheless, in similar difficulties with the maintenance of wealth and transformation of growth as discussed in the controversies between mainstream and ecological economics. (However, in local venues it may be easier to formulate strategies of sustainable development than it has proved at national level).

By applying futures literacy and transformative literacy to societal processes of long-term change it will be possible to translate understanding into action strategies. These forms of change-oriented literacy are ways of learning and acquiring knowledge about the unknown future via continuous processes of knowledge creating, learning and gaining experience. Such methods of creative thinking about the future are relevant for dealing with the sustainability paradox of building a new economy and society without knowing how to do it. Thus, we must emphasize improving knowledge continuously and learning and gaining experience collectively in the socio-metabolic transition to sustainability (Haberl et al. 2011). The metaphor "navigation of social-ecological systems" towards the future (Berkes et al. 2003) is closely connected with resilience thinking and research dealing with managing and planning for the unplannable (disturbances and catastrophes), but it can become part of the overarching sustainability process. The toolbox of future studies includes, beyond scenario analysis, more methods (Popa et al. 2015; Ahvenharju et al. 2018; van Dorsser et al. 2018) to be used for sustainable development as the transformation process connects the near and the distant future.

9.4 The Distant Future

In the present sustainability discourse the time horizon of global scenarios is the end of the twenty first century or less. About the distant future nothing is known. Many potential and unforeseeable changes may change the development path of modern society—that is, changes outside the sustainability process. The first step to deal with this situation is horizon scanning for the near future through which one can identify a series of different potential futures and future change events which can, for example, be used in scenarios (van Rij 2010). If this is done, the conceptual differentiation between a series of potential factors influencing the future development of an economy or society can be conceptualised (Saritas and Smith 2011). Possible factors include: foreseeable *short term changes*, *drivers of change*, *trends* (including potential trends through innovations and branching trends through branching trends), and *megatrends* (over many generations, centuries and millennia, such as climate change), *unexpected and low-probability events* (future surprises, shocks or "wild cards", disasters), *continuities and discontinuities*, *weak signals* (warnings or first indications of change in future, still too incomplete to estimate their impact). When the factors are classified in this way, the periodical updating of horizon scanning generates improved knowledge about trends and megatrends. With this differentiation it becomes possible to create a basic classification of potential changes influencing further development, in both the short and the long run. The potential changes in the long run—megatrends or weak signals—become more concrete and assessable in the course of time. With periodically revised planning, horizon scanning and foresight, scenario construction, and phase models of transformation, the distant future can be connected with the near future.

With respect to the sustainability process as something that stretches to the distant future, the key factors can be divided into five groups (the following examples come from "big picture survey" among future experts described by Saritas and Smith 2011):

1. *Drivers of change*: environmental policies and changes of policies and regulatory practices, resource use practices, scientific and technical innovations, economic globalisation, new forms of governance, scarcity of resources, slowdown of economic growth, international cooperation and networking;
2. *trends*: climate change, increasing conflicts, use of renewable energy sources, development of mobile information technologies, increased mobility and migration;
3. *wild card events*: natural disasters, shifting of the gulf stream, tipping points in climate change, terrorist attacks, pandemic diseases, nuclear accidents and wars, accelerating climate change;
4. *discontinuities*: scarcity and limits of resources, transformation of energy systems, new communication technologies, lifestyle changes, advances in medicine, changes of political systems;
5. *weak signals*: consequences of climate change, recurrent wars and conflicts, exploitation of the human genome, artificial intelligence, rising individualism and nationalism.

Beyond strategies to think about the future and programmes in public policy and governance, the future sustainability process requires the building of action groups for social-ecological transformation. So far, the discussion about transformative action groups has not brought sufficient results. The sustainability transformation is a process without a single social subject as in earlier transformations of economy and society, where new social classes emerged as transformative collective actors. The coordinating and guiding political institutions of the UN are weak when it comes to transformation processes. The governance process, the further social processes of collective action, and the processes in the global economy include many social groups, organisations and institutions of different kinds that are not coordinated for the achievement of sustainability. The international classes that develop with the economic globalisation process are bound through their common interests to the maintenance of the global growth-based economy. The idea of humankind as a collective subject or actor can be realised only in new forms global environmental governance. The education for sustainable development as it was initiated by the UN can help to spread knowledge and awareness globally, but even with that humankind does not yet become a collective actor. The building of transformative action groups at different levels of the sustainability process, and the experimenting with different coalitions of groups will continue in the transformation process.

9.5 The Great Ignorance: Unknown Ways to Solidarity

Sustainable development as a long process tends in all policy and governance processes to be reduced to the short-term future to be managed through policy programmes and regimes. Sometimes, also in the economic discourse, it tends to be reduced to a future that is already coming through the development of new technologies and innovations that substitute for finite resources and maintain the growth-based market processes. This confidence in the creative forces of economy and technology supports misleading views of sustainable development, including especially the belief that nothing needs to be done to achieve sustainability. If sustainability comes through the continuing economic development processes, then. institutional changes in economy and society are unnecessary. Sustainable development is, in this respect, seen as in resilience, that is, as a transformation strategy that is reduced to preparing for and managing unforeseeable disasters. Other ideas that lack transformation strategies include normative visions and constructions of the future with ethical principles that are not translated into transformation processes, thus allowing only partial, fragmented and limited processes of change in economic and resource use practices.

Korten (2009) offers one example for how economic ideas can transform the growth-based economy through reforms of the financial and banking sector and of the monetary system, more generally. He locates the driving forces of the growth economy in the financial sector and its speculative institutions, which create the "phantom wealth" of money unrelated to the creation of real value or utility, and he contributes thereby to discussions of degrowth and its connection to environmental sustainability and sustainable resource use. But the necessary analyses of the complex system of the modern economy and the interactions between social and ecological systems are lacking. The idea of reform of the financial sector remains weak without such analyses and tends, moreover, to emphasize small economic changes and reforms with which the sustainability transformation cannot be achieved. Another example of weak and soft strategies of change that are not sufficient for sustainable development is the idea of new contracts for social and economic order from the Lisbon Club.

The main requirements of sustainable development for building a new society and economy include changes in the societal metabolism of the industrial society that can, in turn, end global pollution, environmental change, and overuse of natural resources. For this purpose various projects are necessary: (a) the creation of transformative action groups and transformative agency to enable the sustainability transformation; (b) building a new system of global environmental governance for regulating social, economic and ecological change; (c) building of new institutions of the global economy that achieve through a sharing economy what seems impossible in the growth-based and deregulated market economy, namely the redistribution of resources, and the invention of new forms of production and resource use that learn from industrial society's failures. Once the requirements of transformative agency and global solidarity are seen as the core of the societal project of social-ecological transformation, new knowledge and experience will have to be created in science, the economy, politics, and civil society. As we have said, the transformation begins when the actors do not know how to build a sustainable economy and solidary society. Therefore, continuous processes of joint learning are necessary on the way towards sustainability. In the course of this collective learning it will be necessary to deal with unsecure knowledge, limited knowledge, and the relations between knowledge and ignorance.

The unknown part of the social-ecological transformation, the project of societal change that has not yet begun, is the *building of new solidarities for a sustainable society*. The notion of solidarity gives the misleading impression that it will only be a process of learning how to cooperate peacefully and to share the limited resources that are available locally. The rhetoric of sustainable development ignores what is required for the building of a solidary society, namely the international redistribution and sharing of resources, and the dissolution of unequal development and the North South divide in the global economy. As long as these core processes of transformation are not in the offing, sustainability cannot be achieved. The decisive part of global transformation is blocked through illusionary ideas and strategies. But this can be avoided through further economic growth and industrialisation in the context of economic globalisation. Further growth and industrialisation are presently the dominant processes, supported by the influential global actors in a situation where global limits of natural resources are achieved and alternatives to growth need to be sought. However, for growth's renewal as sustainable growth, the alternatives of degrowth and knowledge are still lacking. This is a major impediment to the sustainability transformation that requires not only contradictory activities but experimenting with both strategies in order to be overcome. The asymmetrical power relations in the global policy processes and the global economy serve as obstacles in the sustainability process need to be changed through the building of complex transformation strategies (as summarised above in three conditions) that can be specified with regard to counteracting and supporting processes of sustainability transformation: economic globalisation, trans-political governance processes, and building new economic institutions in a "mixed economy" where the growth based market system are balanced, embedded and controlled through other forms of economy.

1. *Globalisation and sustainable development*: Globalisation is sometimes seen as supporting sustainable development, but it generates the economy of unleashed growth and social and environmental destruction against which sustainable development is fighting. Not returning to earlier historical constellations in which

the nation state was dominant can align the conflicting processes. Nevertheless, the transformation of globalisation itself is necessary in order to make it compatible with the social-ecological transformation toward global sustainability. The development of new institutions of global environmental and economic governance, which support the building of cosmopolitan democracy, are, thus, necessary conditions for the strengthening if the sustainability process.

The publication by the United Nations University of a discussion about regulating globalisation through critical approaches to global governance (de Senarclens and Kazancigil 2007) shows the dilemma behind regulating globalisation. There are wide gaps between aims and reality, between the erosion of democracy and the process of globalisation. Notably, the chief idea presented for future global governance—to build regulatory institutions that "tame" globalisation—collapses before the difficulties that arise in the process. The conclusion of the editors is that globalisation is uncontrollable (de Senarclens and Kazancigil 2007: 275) and, hence, the future of globalization is uncertain. This is an echo of conventional economic and political discussions, where crises and disasters are seen as casual events, unforeseeable and incalculable, that is, as "black swans" (Taleb 2007). As long as global processes are seen to be too complex for understanding and regulation, cognitive barriers against sustainable development and sustainability transformation will perdure. The chapter by Coicaud in this same publication reveals the gap between the social reality of globalisation as a crisis-prone process and wishful thinking about the virtues of global governance. The author highlights the many changes required, including the creation of global justice, the strengthening of the human rights and global governance (as foreseen in the Charter of the United Nations 1945 to support social progress and better standards of life), the balancing of power at international levels, and the strengthening of developing countries in global governance, the harmonisation of relations between national and international norms of governance, and the application of democratic pressure (Coicaud in de Senarclens and Kazancigil 2007: 249ff). The broadening of the discourse and its interdisciplinary opening in the concept and project of Earth System Governance brought some progress in seeking alternatives to the present institutional forms of global governance. In this form global governance becomes part of the global sustainability transformation. The critical analysis of the interactions between globalisation (Robinson 2009) and sustainability needs to be developed in future sustainability research.

2. *The creation of transformative agency for global sustainability*: this is not merely a matter of reforming environmental policies and promoting wider participation of non-governmental actors and organisations in knowledge production and gov-

ernance. The process goes beyond the public policy sphere and includes changes in civil society, changes culture and economy, and changes in institutions in all spheres and at all levels of modern society. New social and environmental movements are part of this broader process of improving collective action, especially at trans-national levels. This debate has been ongoing (Edelman 2001), but the theme advances slowly. What Edelman first saw, namely the emergence of new movements that use the internet and other forms for renewing transformational movements and trans-national activism, has in the meantime advanced through the development of new internet-based social movements. Nevertheless, these movements's capacity to drive the global transformation processes is limited. The search for new forms of transformative action groups in which social and environmental movements are participating has, thus, not advanced sufficiently. The new transformation research creates new opportunities. This implies a broadening of our perspective, to see new forms of action, and to include new actors who can participate actively in the sustainability process, which can no longer be limited to social and environmental movements. It can no longer be a policy process, but must include actors in various social spheres, especially actors in civil society, which are not necessarily subject to the guidance of government policies.

3. *Rebuilding and newbuilding of economic institutions*: this can only be done by supporting a sustainability transformation that re-embeds the economy in society and nature (Adaman et al. 2003) and works with ideas from critical institutional economics. This is, of course, another project with the long sustainability process that complements other strategies for creating transformative agency and also builds new forms of solidarity. In this context, critical interdisciplinary economic and ecological theories and approaches such ideas can also be discussed. If the economy is to become environmentally sustainable, it needs to be transformed by developing new social and political controls (Adaman et al. 2003: 357)—and this is one precondition of the transformation process. This process does not end, however, as social re-embedding, but it triggers an additional transformation of the economy that includes the ones already discussed in political economy, world system theory, social ecology, and in the heterodox forms of economics. Building on new non-market based institutions of the economy, this approach will enable the transition towards a post-growth economy by building new socio-metabolic regimes, new accumulation regimes, new regulatory mechanisms for the economy (included among which will be self-regulation), new political regulation, and new regulation through the civil society. Only a genuine transformation processes confronts the real problems and the difficulties of building a new economy for the sustainable society.

9.6 Discussion and Conclusions

The future sustainable society (a post-industrial, post-capitalist and postgrowth-society) is not a globally homogeneous system. It includes different forms of economy, production, exchange, distribution of resources, consumption and ways of life. Each is only similar to the other with regard to the ecological criteria of not polluting the environment and maintaining the natural resource base for future generations, and creating more equal and better conditions of life for both the people now living and for future generations. For these purposes economic growth cannot be the sole measure of a long-term perspective, because the approaching global limits of resources require transforming growth into processes of sharing and redistribution. Different cultures and lifestyles and also a higher global population seem possible in such a future sustainable society. Global population growth, the dystopian scenario of the Malthusian and Neo-Malthusian discourses about limits to growth, are not the main problems the future will confront. Demographic transition will likely result in reduced birth rates, so global population growth may continue only for a limited time. More than discussing population growth it will be necessary to deal with unequal development, wealth and resource use, and unfettered. economic growth (growth that is not necessarily connected with population growth). Finding alternatives to the commodification of nature and to the market-based forms of growth that cause extreme forms of inequality and concentrations of wealth cannot be limited to experiments with green growth and the green economy, in which again the economic interests of the powerful global players, the UN, the OECD and the EU are dominant. New forms of resource sharing and redistribution of the global resources are hardly developing or remain only in experimental forms of sharing in the boundaries of the newly developing digital economy.

Practically seen it will be necessary in the future transformation processes to deal with conflicts, disturbances and disasters, because these will happen more often, as a consequence of the growing scarcity of natural resources and of the further deterioration of the environment and of the ecosystem's functioning. In terms of a "Promethean revolution", the transition to new energy regimes in economy and society will transform the long historical process of forever intensifying resource use by producing a de-intensification that is rooted in the secure knowledge of natural limits to growth and planetary boundaries.

How to create transformative agency and how to build new social, economic, ecological solidarities in the sustainability governance process are the big questions in a future sustainability process; they are the unknown factors of transformation. With regard to the building of transformative action groups, the knowledge creation and the discussion are advancing with recent research in transformative science about transformative literacy, capacity and action. Transformative action is socially learned in projects and experiments at local, community, and urban levels mainly. Here action groups are limited in size, which means that the means of transformation is easier to foresee, and the requirements of change can be communicated better. Then the decisive question becomes, how to connect and network local and global processes to deal with the social and ecological complexity in the transformation process?

Summary of Chap. 9: Sustainability and the Distant Future

The sustainability goals of the UN do not include everything that is necessary for the social-ecological transformation of economy and society into post-industrial and post-capitalist forms. The development goals provide a bare minimum of ideas about how complex the transformation process will be, and they give the misleading impression that they can be achieved in few decades. To work against the illusions of a short, smooth, conflict-free sustainability process, in which all gain, will be the first requirement in the continuing process.

To build a continuously improving process of sustainability transformation requires additional and new strategies, governance programmes and finding pathways of transformation, in which (1) the policy processes supporting sustainable development and (2) the processes beyond politics, on other social spheres and in the civil society, are connected. The policies with which sustainability governance overlaps include the regulation of global environmental change (climate change, biodiversity reduction, land use change, and urbanisation), with earth system governance as overarching policy following the same goals as sustainability governance.

The processes in society and nature that should be influenced, modified or converted through sustainable development are economic globalisation and global environmental change as processes that are generated through the interaction of societal and ecological systems in modern society. The synergy effects that can be achieved through the integration of policies and further processes of social and ecological change beyond policies are gained in terms of knowledge and agency for regulating the social-ecological transformation to global sustainability. The further and complicated process of change implies something that has been widely ignored in the sustainability policy and governance up to now, namely the needs to foresee conflicts and disturbances, tot deal with asymmetric power relations in the global arena, and to build a solidarity-based economy and society. The social-ecological transformation that includes all sustainability-related processes will require a much longer time frame, with a differentiation of the process in several phases.

Appendices: Further Information and Material

1. Questions and Individual Exercises

The chapter described and discussed attempts to deal with the future and how it can be influenced through knowledge and action strategies for the long process of global sustainability transformation. Much of this discussion is connected with the development and application of methods for creating knowledge about the future.

Read about the methods for future studies and for future literacy in one or several of the following articles and discuss them in groups. The guiding questions for the discussions are: Which methods help to create new knowledge that can be applied in the transformation processes in sustainability policy and governance? Which methods create visions, utopias and speculative forms of thinking about the future, for which it is difficult to describe their functions in the sustainability process?

Amer, M., Daim, T. A., & Jetter, A. (2013). A Review of Scenario Planning. *Futures, 46*, 23–40.

Ahvenharju, S., Minkkinen, M., & Lalot, F. (2018). The Five Dimensions of Futures Consciousness. *Futures, 103*, 1–13.

Popa, F., Guillermin, M., & Dedeurwaerdere, T. (2015). A Pragmatist Approach to Transdisciplinarity in Sustainability Research: From Complex Systems Theory to Reflexive Science. *Futures, 65*, 45–56.

van Dorsser, C., Walker, W. E., Taneja, P., & Marchau, V. A. W. J. (2018). Improving the Link Between the Futures Field and Policymaking. *Futures, 104*, 75–84.

van Wuren, D. P., Bijl, D. L., Bogaart, P., Stehfest, E., Biemans, H., Dekker, S., et al. (2019). Integrated Scenarios to Support Analysis of the Food-Water-Energy Nexus. *Nature Sustainability, 2*: 1132–1141.

2. Further Reading Suggested: Deepening, Thematically Specialised

Miller, R. (Ed.). (2018). *Transforming the Future: Anticipation in the 21st Century*. London: Routledge.

Voß, J.-P., Bauknecht, D., & Kemp, R. (Eds.). (2006). *Precaution, Foresight and Sustainability: Reflection and Reflexivity in the Governance of Science and Technology*. Cheltenham UK: Edward Elgar.

Scientific journals (international, peer reviewed):

"Ecology and Society", "Economics of Disasters and Climate Change", "Earth System Governance", "Environmental Sustainability", "European Journal of Futures Research", "Futures", "Global Environmental Change", "Global Environmental Governance", "International Journal of Social Ecology and Sustainable Development", "International Journal of Disaster Risk Science", "International Journal of Sustainable Development and World Ecology", "Journal of Futures Studies", "Journal of World-Systems Research", "Review of Environmental Economics and Policy"

References

Adaman, F., Devine, P., & Ozkaynak, B. (2003). Reinstituting the Economic Process: (Re) embedding the Economy in Society and Nature. *International Review of Sociology, 13*(2), 357–374.

Allen, P. M. (2000). Knowledge, Ignorance, Learning. *Emergence, 2*(4), 78–103.

Amer, M., Daim, T. A., & Jetter, A. (2013). A Review of Scenario Planning. *Futures, 46*, 23–40.

Arora, N.K: (2019). Earth: 50 years challenge. *Environmental Sustainability, 2*(1),1–3.

Archibugi, D., & Held, D. (Eds.). (1995). *Cosmopolitan Democracy: An Agenda for a New World Order*. Cambridge et al.: Polity Press.

Ahvenharju, S., Minkkinen, M., & Lalot, F. (2018). The Five Dimensions of Futures Consciousness. *Futures, 103*, 1–13.

Barilla Center for Food and Nutrition (2016). *Eating planet. Food and Sustainability: Building our Future*. Parma, Italy (https://www.barillacfn/com/media/pdf/Barilla_Eating_planet2016_Eng_S_abstract.pdf)

Beck, U. (1992). *Risk Society: Towards a New Modernity*. London: Sage.

Beck, U., Giddens, A., & Lash, S. (1994). *Reflexive Modernization. Politics, Tradition and Aesthetics in the Modern Social Order*. Cambridge: Polity.

Berkes, F., Colding, J., & Folke, C. (2003). *Navigating Social-Ecological Systems: Building Resilience for Complexity and Change*. Cambridge et al.: Cambridge University Press.

Calvelli, J. (2015). *The Future Is an Image: Unsustainability, Plasticity and the Design of Time*. New York and Dresden: Atropos Press.

Carson, R. (1991 [1951]). *The Sea Around Us*. Oxford: Oxford University Press.

Cash, D. W., Clark, W. C., Alcock, F., Dickson, N. M., Eckley, N., Guston, D. H., et al. (2003). Knowledge Systems for Sustainable Development. *PNAS, 100*(14), 8089–8091.

de Senarclens, P., & Kazancigil, A. (Eds.). (2007). *Regulating Globalization: Critical Approaches to Global Governance*. Tokyo et al.: United Nations University Press.

Drucker, P. F. (1993). *Post-Capitalist Society*. New York: HarperCollins.

Edelman, M. (2001). Social Movements: Changing Paradigms and Forms of Politics. *Annual Review of Anthropology, 20*, 285–317.

Erb, K., Haberl, K., Krausmann, F., Lauk, C., Plutzar, C., Steinberger, J. K., et al. (2009). *Eating the Planet: Feeding and Fuelling the World Fairly, Sustainably and Humanely – A Scoping Study*. Social Ecology Working Paper 116. Vienna: Institute of Social Ecology.

Gerst, M.D., Raskin, P.D., Rockström, J. (2014). Contours of a resilient global future. *Sustainability, 6*(1), 123–135.

Haberl, H., Fischer-Kowalski, M., Krausmann, F., & Martinez-Alier, J. (2011). A Socio-Metabolic Transition to Sustainability? Challenges for Another Great Transformation. *Sustainable Development, 19*, 1–14.

Hagens, N. J. (2020). Economics for the Future – Beyond the Superorganism. *Ecological Economics, 169*. https://doi.org/10.1016/j.ecolecon.2019.106520.

Hahnel, R. (2012). *Of the People, By the People: The Case for a Participatory Economy*. AK Press Distribution.

Hartmann, T. (2012). Wicked Problems and Clumsy Solutions. *Planning Theory, 11*(3), 221–241.

Heilbroner, R. L. (1968 [1959]). *The Future as History*. New York et al.: Harper & Row.

Held, D. (2010). *Cosmopolitanism: Ideals and Realities*. Cambridge, UK and Malden, MA: Polity Press.

Howkins, J. (2001). *The Creative Economy. How People Make Money from Ideas*. London.

Hubert, B., Rosegrant, M., van Boekel, M., & Ortiz, R. (2010). The Future of Food: Scenarios for 2050. *Crop Science, 50*, 33–50.

Institute for the Future (2011). Four Futures of Food: Global outlook alternative scenarios briefing. Palo Alto, CA (https://www.iftf.org/uploads/media/IFTF_SR1388_GFOFuturesFood.pdf).

Keil, F., & Stieß, I. (2007). Wissen was wir nicht wissen: Umweltforschung als gesellschaftlicher Lernprozess. *GAIA, 16*(3), 193–199.

Korten, D. (2009). *Agenda for a New Economy. From Phantom Wealth to Real Wealth*. San Francisco: Berrett-Koehler Publishers.

Krausmann, F., Fischer-Kowalski, M., Haberl, H., & Eisenmenger, N. (2008). The Global Sociometabolic Transition: Past and Present Sociometabolic Profiles and Their Future Trajectories. *Journal of Industrial Ecology, 12*(5–6), 637–656.

Leopold, A. (1949). *A Sand County Almanac*. Oxford: Oxford University Press.

Lorey, I., & Neundlinger, K. (Eds.). (2012). *Kognitiver Kapitalismus*. Wien.

Luhmann, N. (1976). The Future Cannot Begin: Temporal Structures in Modern Society. *Social Research, 43*(1), 130–152.

Mason, P. (2015). *Post Capitalism: A Guide to Our Future*. London: Allen Lane.

Millennium Institute (2013). Global Food and Nutrition Scenarios: Final report. Washington DC: Millennium Institute (http://www.un.org/en/development/desa/wess/_bg_papers/bp_wess2013_millennium_inst.pdf).

Miller, R. (Ed.). (2018). *Transforming the Future: Anticipation in the 21st Century*. London: Routledge.

Moore, J.W. (Ed.). (2016). *Capitalism in the Web of Life: Ecology and the Accumulation of Capital*. Oakland: Verso.

Munoz-Erickson, T. A. (2014). Co-Production of Knowledge-Action Systems in Urban Sustainable Governance: The KASA-Approach. *Environmental Science and Policy, 37*, 182–191.

Popa, F., Guillermin, M., & Dedeurwaerdere, T. (2015). A Pragmatist Approach to Transdisciplinarity in Sustainability Research: From Complex Systems Theory to Reflexive Science. *Futures, 65*, 45–56.
Popper, K. (1957). *The Poverty of Historicism*. London: Routledge.
Ramirez, R., Mukherjee, M., Vezzoli, S., & Kramer, A. M. (2015). Scenarios as a Scholarly Methodology to Produce "Interesting Research". *Futures, 71*, 70–87.
Raskin, P., Electris, C., Rosen, R. (2010). The Century Ahead: Searching for Sustainability. *Sustainability, 2*(8). https://doi.org/10.3390/su2082626.
Reckwitz, A. (2020). *The Society of Singularities*. Cambridge: Polity.
Rieckmann, M. (2011). Schlüsselkompetenzen für eine nachhaltige Entwicklung der Weltgesellschaft: Ergebnisse einer europäisch-lateinamerikanischen Delphi-Studie. *GAIA, 20*(1), 48–56.
Robertson, T. (2012). *The Malthusian Moment: Global Population Growth and the Birth of American Environmentalism*. Rutgers University Press.
Robinson, W. I. (2009). Saskia Sassen and the Sociology of Globalization: A Critical Appraisal. *Sociological Analysis, 3*(1), 5–29.
Routledge Encyclopedia of Philosophy (Online). (2020). Article "Ignorance". Retrieved February 20, 2020, from https://www.rep.routledge.com/articles/thematic/ignorance/v-1.
Saritas, O., & Smith, J. (2011). The Big Picture – Trends, Drivers, Wild Cards, Discontinuities and Weak Signals. *Futures, 43*, 292–312.
Schaffartzik, A., Mayer, A., Gingrich, S., Eisenmenger, N., Loy, C., & Krausmann, F. (2014). The Global Metabolic Transition: Regional Patterns and Trends of Global Material Flows, 1950–2010. *Global Environmental Change, 26*, 87–97.
Schaubroeck, T., Baustert, P., Elorri Igos, E., & Benetto, E. (2020). Is a Sustainability Assessment a Shot in the Dark? How to Deal with Its Nonquantified Uncertainty? *Environmental Science & Technology, 54*, 2051–2053.
Schneidewind, U. (2013). Transformative Literacy. *GAIA, 22*(2), 82–86.
Schutz, A. (1967). *The Phenomenology of the Social World*. Ivanston, IL: Northwestern University Press.
Schutz, A., & Luckmann, T. (1973). *The Structure of the Life-World*. Evanston, IL: Northwestern University Press.
Sousa, W. (1984). The Role of Disturbance in Natural Communities. *Annual Review of Ecology and Systematics, 15*, 353–391.
Srnicek, N., & Williams, A. (2015). *Inventing the Future: Postcapitalism and a World Without Work*. London: Verso.
Taleb, N. (2007). *The Black Swan: The Impact of the Highly Improbable*. New York: Random House.
Thissen, W.A.H. & Walker, W. (Eds). (2013). Public Policy Analysis: New Developments. New York: Springer.
Tilly, C. (1997). *Roads from the Past to Future*. Lanham et al.: Rowman & Littlefield.
van Dorsser, C., Walker, W. E., Taneja, P., & Marchau, V. A. W. J. (2018). Improving the Link Between the Futures Field and Policymaking. *Futures, 104*, 75–84.
van Rij, V. (2010). Joint Horizon Scanning: Identifying Common Strategic Choices and Questions for Knowledge. *Science and Public Policy, 37*(1), 7–18.

van Wuren, D. P., Bijl, D. L., Bogaart, P., Stehfest, E., Biemans, H., Dekker, S., et al. (2019). Integrated Scenarios to Support Analysis of the Food-Water-Energy Nexus. *Nature Sustainability, 2*, 1132–1141.

Vogt, W. (1948). *Road to Survival*. New York: William Sloane Associates.

Voros, J. (2003). A Generic Foresight Process Framework. *Foresight, 5*(3), 10–21.

Voß, J.-P., Bauknecht, D., & Kemp, R. (Eds.). (2006). *Precaution, Foresight and Sustainability: Reflection and Reflexivity in the Governance of Science and Technology*. Cheltenham, UK: Edward Elgar.

Walker, W. E., Marchau, V. A. W. J., Kwakkel, J. H. (2013). Chapter 9: Uncertainty in the Framework of Policy Analysis. In: Thissen, W.A.H., Walker, W.E., eds. Public Policy Analysis: New Developments. New York: Springer.

World Economic Forum (2017). *Shaping the Future of Global Food Systems: A Scenarios Analysis*. Geneva, Switzerland (http://www3.weforum.org/docs/IP/2016/NVA/WEF_FSA_FutureofGlobalFoodSystems.pdf)

Recreating Sustainability: Conjectures and Conclusions

10

Global sustainability cannot be achieved in the coming decades. It will be a long process, in which new and continuously improving strategies of sustainability transformation will have to be developed. This book has discussed many potential multi-scale strategies that will have many components and will experiment with different approaches and development paths. *The theoretical concept for a renewed sustainability process is a social-ecological transformation. In this book that concept has been constructed as a strategy that emerged from a progressive interdisciplinary knowledge synthesis that crosses the borders of existing scientific disciplines and infuses all of them with knowledge from new interdisciplinary research fields that have developed as part of the sustainability discourse over the last three decades.* This strategy, as it has unfolded in the chapters of the book, will solve the economic and the ecological problems of knowledge application in sustainable development, if it is applied correctly,

10.1 The Guiding Questions of the Book

The role of economic knowledge in the sustainability process and its connection with other scientific and practical knowledge. This book's many analyses prompt the following conclusions.

1. Disciplinary traditions of specialisation within *mainstream economics and environmental economics* impede interdisciplinary knowledge integration. Merely translating ecological problems and concepts into economic ones does not

K. Bruckmeier, *Economics and Sustainability*,
https://doi.org/10.1007/978-3-030-56627-2_10

promote interdisciplinary communication and knowledge transfer. Environmental economics's chosen path to sustainability is based on the following belief: that economic growth must come in new forms of green, blue, or sustainable growth that support sustainable development and supersede the unsustainable growth of the industrial society. Economic valuation and commodification of nature have correctly become prominent themes in the growth debate. Sustainable development is illusory at the national levels, on account of the general emphasis on growth, cannot, therefore, be extended to global economy, which has a longstanding tendency to ignore the negative effects of unequal resource flows in the globalising economy. It has taken many years for thinkers to construct alternative ideas of degrowth and postgrowth and to include them in the sustainability discourse. These ideas gain acceptance via experimentation with different approaches to limiting and transforming growth. But this end can only be achieved, if economic growth is connected with growth-reducing factors and processes such, such as reducing regular hours for paid work. As long as we lack sufficient knowledge and experience with strategies of sustainable growth and (even) degrowth, it will necessary to experiment with both approaches.

2. *Interdisciplinary sciences and practices of sustainability* can integrate economic and ecological knowledge from different disciplines and apply the resulting who to solve problems in the sustainability process through different forms of action. This can be achieved by continually developing knowledge integration through pluralistic knowledge practices that insist on working with different ideas, concepts and approaches. Such practices develop, for example, in sustainability science and in ecology under the name of adaptive management. The search for solutions in terms of environmental sustainability can develop in a way that balances economic and ecological principles, so that the economy remains within the limits to growth and planetary boundaries.
3. The exponential growth of scientific knowledge and the development of interdisciplinary knowledge cultures demand that *science-practice relations* be included in the sustainability discourse, so that the two can be connected in new ways. This will happen as we develop new forms of transdisciplinary knowledge production and integration. Both the science and of policy spheres can be rethought through change in knowledge practices that has been wrought by criticism of ineffective knowledge transfers in resource management practices. The disciplinary practices of academic research have produced only selective and fragmented analyses of social and environmental problems. Knowledge practices in environmental policy and governance are too orientated toward economic thinking, embracing traditional approaches to monetary valuation and

the commodification of natural resources. As a result, problems in resource distribution and the nature of ecological limits on natural resources have been overlooked.

10.2 Conclusions from the Chapters Regarding the Future Sustainability Process

The discussion of sustainable development from different, political, economic, historical, and knowledge-related perspectives yielded specific approaches for reconceptualizing the sustainability process as a process of social-ecological transformation:

Chapter 1—the global policy process of sustainable development. The problems that the Brundtland report identified in the global sustainability remain unsolved. The problems include the environmental consequences of economic growth, unequal growth and asymmetric power relations in the global economy, growing social and economic inequality, and diminishing support for sustainable development. The transformation of the global economy and its systematic reorganisation remains paralysed thanks to conflicting interests, lack of knowledge about the necessity and the consequences of economic growth, and lack of experience and practical knowledge about forms of sustainable growth, degrowth and social-ecological transformation. The dilemma of sustainable development is that the "capitalisation of everything" makes it impossible to "de-commodify natural resources". There is, in short, no way to "economise ecology" or "ecologise economics". In response to the worsening global situation, due to global environmental change and the obvious limits on global growth, the Millennium Ecosystem Assessment developed in 2005 a critical scientific and political perspective on the preconditions, implications and temporal perspectives necessary for the transition to sustainability, bringing forth new ideas for a social-ecological transformation or "another great transformation". Nevertheless, the policy and the economics worlds remain wedded to an unsustainable growth paradigm (Büchs and Koch 2019).

Chapter 2—the history of sustainability thinking. The end of the eighteenth century witnessed the rise of both classical political economy and a critical discourse that attacked the stationary state, economic growth and growth's consequences for the environment, with the latter tradition now providing core themes for present debates about limits on growth and sustainable development. The arguments against modern industrial society varied, depending on the philosophical, religious, political, scientific and cultural discourses to which the critics responded. Hence, the discussion of alternatives to industrialisation id not advance very far. For example, the vision of an agricultural welfare state and the Chayanovian discussion of a peasant economy and simple commodity production as another mode of production within the modern capitalist economy did not yield true alternative development paths, but became part of dis-simultaneous industrialisation, where earlier forms of economic production were integrated into the industrial division of labour. Only present environmental and sustainability discourses have sought paths out of industrial society.

Chapter 3—inter- and transdisciplinary knowledge creation for sustainable development. In the knowledge chain that runs from production into integration, application and dissemination of knowledge, the core process on which the success of sustainability governance and global transition depends is inter- and transdisciplinary knowledge integration. Epistemologically and methodologically such knowledge integration is difficult. In the sustainability process knowledge integration and synthesis are often shifted over to the policy and practice of sustainable development, where everything is evaluated through power-based knowledge selection. The new modes of inter- and transdisciplinary knowledge production have developed in response to the failure early sustainability policies to produce more systematic and efficient forms of knowledge synthesis and application, especially with respect to practices of knowledge sharing between different actors.

Chapter 4—natural resource management as economic practice. Ecological and economic research (environmental economics and ecological economics) have discussed the core question of knowledge use as a means to combat environmental problems caused by natural resource overuse within natural resource management, whether pertaining to sector-specific forms (agriculture, fishery, forestry), or integrated forms of natural resource management (ecosystem-based management). The key questions include differences between the "economy of nature" and the "economy of society," in addition to problems with interdisciplinary concept transfer. They have also discussed forms of property rights and their relevance in natural resource use (especially common or private property: "tragedy of the commons" or "tragedy of enclosures") and the development of new forms of property rights for the commodification of nature and natural resources (intellectual property rights regrading genetic resources and economic use of ecosystems and biodiversity). Within the sector-specific management of natural resources and the integrated or ecosystem-based forms the conditions for integrating economic and ecological principles are clear: *when the overuse of resources or the pollution of the environment increase the number of ecological risks, political regulation and restrictions of use through enforcement of ecological principles become necessary, in order to reduce economic growth.*

Chapter 5—environmental economics as knowledge about economic growth. In the short run the paradigms of a) sustainable growth and b) degrowth or postgrowth are incompatible. Attempts to maintain growth and to defend it with new ideas and constructions as sustainable growth will continue, although ecological research shows that growth's use of natural resources has exceeded ecological limits. This situation is not to be understood as an impending breakdown of the sustainability process, but rather as a crisis that can help to renew it. The knowledge available for the structuring and governing of the long transformation process remains insufficient. That one must make growth sustainable by limiting growth and developing a non-growing economy is only gradually becoming clearer, with critical reviews and refinement of growth calculations and economic and ecological indicators now showing more clearly the negative consequences of economic growth. Specialised transition research (especially in the journal "Environmental Innovation and Societal Transitions": Johnstone and Newell 2018; Araposthatis and Pearson 2019; Kanger and Schot 2019; Hopkins et al. 2020; Turnheim et al. 2020) has discussed the possibility of achieving a sustainable future society and economy via technological change and innovation. This discussion has revealed the difficulty in connecting the research about sociotechnical systems and technological change with broader and more critical research about sustainability transformation, confirming thus that *locked-in growth and the "technical fix" are illusory engineering solutions, because they do not alter the political and economic power relations and the forms of resource use that are causing the problems.*

Chapter 6—ecological economics in the dilemma between academic science and movement. The concepts or models from ecological economics that are applicable to the interdisciplinary sustainability discourse are mainly those in ecological indicators and economics that critique economic growth, all of which suggests the need to separate natural capital from neoclassical economics. How the economic concepts and knowledge can be integrated in the broader sustainability discourse is learned mainly from inter- and transdisciplinary knowledge production and application. The new methods include: the integration and synthesis of knowledge across disciplinary boundaries; learning to work and act in situations where multiple and contradicting ideas and knowledge are used; and development of new knowledge practices where joint learning and knowledge sharing are as important as scientific knowledge. Collective learning is part of a broader strategy of creating agency, awareness and behavioural changes through environmental education and learning among the entire population, that is, all resource users, consumers, and citizens. It includes *educational initiatives to break the dominance of technology and growth-oriented education in the search for pathways to sustainable futures.* Among the key initiatives are: the development of collective transformative education that challenges education that supports the growth-based model and the unequal power relations between the Global North and South (Selby and Kagawa 2015); breaking with the progress, competition and innovation based STEM (science, technology, engineering, mathematics) education of the technological orthodoxy (Alonso Yanez et al. 2019); decolonizing ways of thinking and knowing through transformative sustainability learning with "epistemological stretching" or learning to accept different ways of knowing for deconstruction of power relations and for worldview bridging (Harmin et al. 2017); building of new learning practices for scientists to learn to participate in interdisciplinary sustainability research (Freeth and Caniglia 2020).

Chapter 7—conflicts, disasters, relapse. Assuming that environmental and resource use conflicts continue during the sustainability process, new perspectives and improved procedures of conflict resolution will need to be developed in the governance process that can bring about a social-ecological transformation. Knowledge about global social and environmental change, conflict mitigation and sustainability transformation comprise the core of the renewal of governance. Scientific and practical learning in the transformation is a two-step process: 1) to learn how to integrate empirical research, theories and practical action, and 2) to learn how to synthesise knowledge from social and natural scientific research with practical and local knowledge. Learning from the failures of prior forms of policy and research is crucial to preparing for more and intensifying conflicts that will be a consequence of global environmental change and the difficulties of conflict resolution. Beyond the transformation of conflicts into cooperation between the conflicting parties, we need to learn more about conflict resolution—that is, to advance from the solution of single and specific conflicts to strategies where conflicts become driving forces of sustainability transformation.

Chapter 8—changing relations between science, politics and civil society in the sustainability process. Changing relations between science, politics and civil society in the research and in the practice of sustainable development are part of the broader process of new knowledge production which began about at the same time that the sustainability discourse appeared in the early 1990s. With these changes came critical reflections on the processes of knowledge transfer, dissemination of knowledge, knowledge's sharing and application, which has been described with the terms "mode 2" or "transdisciplinarity". In the epistemological discourse and the research about science and technology studies, inter- and transdisciplinary forms of knowledge production, post-normal science, and sustainability science will develop new forms of knowledge transfer and communication that will allow us to apply scientific knowledge more systematically and actively (in discursive knowledge creation), more efficiently socially (through joint learning), and more effective practically (to achieve the goals of collective action).

Chapter 9—the future of sustainability as social-ecological transformation. The new sustainability process is one of social-ecological transformation of the global economy and society in which the problems to be dealt with are more consequently addressed, including the limits of natural resources, unequal development, climate change, and other forms of global environmental change. The adoption of new knowledge practices for imagining possible pathways and futures of sustainability, and the building of transformative action groups in the sustainability process is a component of the joint learning of scientists, practitioners and activists in the further transformation process towards sustainability. It began half a century ago with the environmental discourse in science and public policy, and the spreading of interdisciplinary practices of research in environmental research. It is now developing as a broader process of social change and transformation of economy and society. To develop long-term transformation perspectives, for example, through global scenarios, helps to identify more possibilities for transition paths to sustainability, more locally and regionally adapted forms of transition. Nevertheless, it also reveals potential risks, conflicts, and economic and environmental catastrophes. The different discourses about a transformation to the future post-industrial and sustainable society need to be brought together in a coherent picture of the transformation process that is informed by historical knowledge of modern society's development, including knowledge from future studies that can delineate the path toward a new future society. This is shown in an exemplary way by Wilenius (2017), who influenced the theory of social evolution of modern society by the future scientist Malaska (Wilenius 2014), and showed what changes are necessary for overcoming the dynamics of the long waves (Kondratiev cycles) that shaped industrial society from its inception. The first step in thinking the future is to distinguish between what can be maintained what needs to be changed in this dynamic of development through innovation. The book does not consider the long transformation, but it does bring the ideas of sustainable development up to the point where critical analyses of the system can foment change by developing degrowth-strategies (Chap. 6) that can produced a transformation. Fergnani (2019) discusses four scenario variants of "continued growth", "disarray", "steady state" and "transformation" to outline potential paths of change.

10.3 Sustainability Dilemmas: Complexity of Changes and the Limits of Science and Governance

The dilemmas of sustainable development are connected to the dilemmas of economic growth as exhibited in the controversy between orthodox and ecological economists (Chap. 6):

1. *The dilemma of short-term practices, interests and processes that undermine long-term perspectives of sustainable development.* This was a fundamental dilemma for all forms of economically-organised use of natural resources, for which no lasting solutions were found either. in earlier phases of human history or in modern society. With industrialisation, economic and population growth, the dilemma has, in turn, become global. The controversy about the short and the long rationality and perspective of survival, of satisfying the needs of the presently living generations or taking into account the needs of all future generations, has become a struggle between economics and ecology over the criteria for organising and limiting economic production, consumption and growth. The dilemma can no longer be solved through externalisation—the shifting of burdens of ecological damages to other countries or to future generations—when present forms of resource use and pollution already create risks for human survival and survival of other species on the earth.
2. *The mismatch between the quantity of scientific knowledge from environmental and sustainability research and its limited and selective application in the sustainability process.* The exponential growth of knowledge and publications in research on environment and sustainability has not been mirrored in knowledge practices within the sustainability process. Large parts of research and knowledge never reach policy and governance of sustainability, because they are not communicated and discussed adequately with practitioners and actors but remain archived in the many scientific journals and publications that are used by specialists. To communicate scientific knowledge successfully, however, requires new forms of active communication and transfer in new forms of knowledge production such as inter- and trans-disciplinary research in sustainability science, ecological economics, social and political ecology, where knowledge is generated in the contexts of knowledge application.
3. *The dilemma within social-ecological discourse is that of the building of a new sustainable society without knowing how to do it.* In the history of human societies all new societies or modes of production gestated during long processes of development. The social groups and actors participating in the transformation could not foresee the consequences of their actions and remain, thus, behind a

"veil of ignorance" about the future. Nevertheless, transformations occurred as a result of continued processes of trial and error, as experimenting and learning combined with growing experience and new practical and scientific knowledge. The people, actors and institutions that worked to solve the social and environmental problems of early modernity did not know (and could not foresee) that they would build a new industrial society with a new mode of production. This dilemma reappears in a different form in the transformation from the industrial to a post-industrial, sustainable society.

4. *A contested concept to deal with complex problems of environmental change.* The critique of the notion of sustainable development has not yielded alternative concepts that can integrate analyses of complex systems and processes. The alternatives suggested--resilience and environmental justice—resist knowledge integration and efforts to initiate transformation. They became adaptive reactions to disturbance (as resilience) or reduce the problems of intra- and intergenerational solidarity to norms of environmental justice. It is possible to integrate the concepts of resilience and environmental justice in the sustainability concept, as specific parts within a larger process of change, but not the other way around—that is, to reduce sustainability to resilience or environmental justice. To note the complexity that. Is necessary for dealing with sustainable development is not an argument for giving up on the idea, but a call to deal with the problem more consequently and via new knowledge practices. The limits of science and governance have not yet been reached. Solutions must be sought within the longer process and through improvements in knowledge development and transformative capacity.
5. *The limited success at regional and national levels in some countries of experiments with green growth cannot conceal the failures and resulting environmental deterioration at global levels*, as is shown clearly in the global assessments that have appeared since the Millennium Ecosystem Assessment of 2005. *Roadblocks on the way to global sustainability are growing judging by natural resource use, material and energy, and the concomitant forms of environmental pollution and emissions (especially of greenhouse gases) at global levels*. These continuing processes reveal the limited nature of success in combatting global change processes—climate change, biodiversity loss, land use change through deforestation, urbanisation and urban sprawl. Economic discourse has not kept up with ecological debates. From Hotelling's (1931) rule for the management of non-renewable resources to Hartwick's (1977) rule for observing intergenerational equity conventional economic knowledge practice has been to introduce specific conditions for the use of non-renewable resources into mathematical models of growth. The expectation that losses of natural capital can be

substituted through other forms of capital have, however, turned out to be illusory, as the discussion in critical economics and ecology showed. The new strategies for green and blue growth adopted by the OECD and the EU have kept the discussion of possibilities of a non-growing economy off of policy agendas, because they expect continued growth to come through new technologies. The transition to a new sustainable society is still thought as new phase of industrialisation or re-industrialisation of Europe (Sautter 2016), even when sustainability and growth-critical questions are discussed (Foxon 2018). The economics of weak sustainability, ecological modernisation of the industry and new economic growth as green growth create, however, new illusions that one can solve global change problems through further growth. Green growth does not happen where industrial growth and pollution happens (in the newly industrialising countries in the Global South) but in small, affluent and developed national economies in Europe and only under exceptional conditions. To renew the sustainability discourse and to recreate the governance process via more effective and efficient forms of a global transformation to sustainability *the actors need to bring about a series of connected changes in the knowledge- and power-related practices of governance, that will be difficult to realise because of manifold resistance and vested interests of powerful actors and institutions.*

10.4 The Way Beyond "Agenda 2030"

Early policy and governance processes that embraced sustainable development first appeared with the UNCED Conference in 1992 and the global "Agenda 21". These achieved little more than to introduce the idea of sustainable development to national and global policy makers. They did not clarify the pathways towards sustainability, with the consequence that the notion of sustainable development was subjected to various political and economic interests and, in turn, was reduced to simple visions in the practice of sustainability governance. The reasons for the lack of success became evident from the way sustainable development was dealt with in public and private policies, in governmental policies and in the economic practices of development. The societal project of transition to global sustainability fell apart in piecemeal engineering of social and environmental damage. The policies formulated under the influence of powerful global players, the national and international political and business elites, did not address the key problems of global limits to growth and intra- and inter-generational solidarity in sharing the earth's limited resources. In fact, the "Agenda 2030" is an impediment to transformation.

The "Agenda 2030" does not opt for a strategy of social-ecological transformation, although it adopts the term transformation for the changes to be achieved with the sustainable development goals. The strategy adopted by the UN and the global actors, the green growth and green economy strategy, is open to question with regard to its environmental goals. The impossibility to achieve in the foreseeable future a solution in terms of limiting growth or matching growth with the global limits (mitigating climate change and reducing the global pollution and the emissions of greenhouse gases) requires critique and renewal of the global sustainability policy.

The environmental and social problems that should be solved through sustainable development are formulated in the seventeen sustainable development goals of the UN. The appear in normative and abstract forms, giving only a vague idea of the social, economic and environmental components of sustainability that must be realised. The duration of the sustainability process remains unclear, as the transformation process and the changes necessary in economy, politics, and society as a whole to achieve sustainability have not been formulated. That the goals can be achieved by 2030, the goal of the global policy programme "Agenda 2030", is highly unlikely, as the UN admits in its Global Sustainable Development Report from 2019, which indexes the achievement of goals. The slowest progress happens with regard to the environmental goals. Only very few (and small) countries—Sweden, Denmark, Finland—have achieved three quarters of the UN goals (Arora and Mishra 2019: 339). The exceptional situation of these countries has been discussed in the sustainability debate (see Chap. 6), and many have questioned whether the "green growth" they achieved can be replicated in all counties. Frankly, this seems impossible during the twenty first century.

Steps towards more realistic strategies for sustainable development cannot develop through policy review and single successful projects and programmes in forms of resource management projects or planning at local, regional or national levels. These are necessary to maintain the sustainability process and to experiment with new ideas. They can create some synergy effects through "bottom up"- strategies of networking and integrating local projects, national development programmes and global policies. Additionally, to develop transformation strategies, requires new forms of knowledge work with two main components: interdisciplinary knowledge integration and the use of theoretical knowledge from interdisciplinary sustainability research (social ecology, political ecology, sustainability science, transformation science, ecological economics). To work with theoretical concepts and knowledge in policy and governance processes seems a strange element in policy processes. *The theories for the sustainability transformation are not specialised theories and theoretical models from economics or ecology, but interdisciplinary theories for the analysis and explanation of large-scale social and ecological systems. Many of these have developed in social ecology, including theories for the analysis of complex systems in society and nature, their interaction, the social metabolism and the resource flows between ecosystems and social systems in modern society, the modes of production in ecological and economic perspectives, the global process of social and environmental change. Little of this knowledge is discussed in economics, although more has penetrated into ecological economics.* The political governance process that will strengthen the sustainability process is not to be found in the sustainability diplomacy of the UN and UNEP, but in newly developing processes of global environmental governance or earth system governance.

Neither economics nor economic knowledge as discussed in this book is at the forefront of efforts to renew sustainable development through building strategies of social-ecological transformation. Also, ecological economics with its inter- and transdisciplinary opening for knowledge exchange and integration has only limited knowledge of a new sustainability strategy, although some important ideas such as the concept of sustainable degrowth can help to create a breakthrough in the growth discourse and to develop alternatives to traditional growth strategies. When alternatives to growth strategies develop, the economic valuation and valorisation of nature and natural resources and the methods for this purpose can be reduced. At that point, the ideas and knowledge from ecological economics will become important. However, in the projects for developing sustainable development—that of the TEEB, for example—ecological economists are often caught in a dilemma, because they are participating in projects that promote growth, as opposed to calling for finding alternative ecological practices of resource management. Moreover, other ideas discussed in ecological economics are exclusive of the discourse of ecological economics. Ideas taken from other critical discourses, for example, that of transformation, were created in institutional economics by Polanyi. In the melding of ecological economics with other heterodox approaches, the adoption and integration of ideas and knowledge from other approaches makes the profile of the subject less clear. Social-ecological economics has borrowed ideas that have been discussed in other interdisciplinary fields, such as social and political ecology, and sustainability and transformation science, before they became part of the discourse of ecological economics. Ecological economics as an open and pluralistic subject supports, as do other subjects, the development of sustainable development strategies with ideas, concepts and theories that have been recreated several times.

Recreating sustainability by tackling the great challenges of transforming the modern economy and society means: (a) building a new global economy about which not much more is known presently than those components that have been formulated in interdisciplinary theoretical discourses, in critical economics and social and political ecology and that will include new, non-commercial forms of a sharing and solidarity economy, an economy with redistributive institutions, with markets and global exchange not determined by growth-driven forms; (b) building the new sustainable economy and society through new forms of knowledge work in the sustainability discourse and process, for which the theoretical components include the ones summarised above; (c) creating new empirical knowledge not only through continued specialised research, but through a pluralisation of knowledge strategies that has been described as trans-disciplinarity. This will comprise, in turn, recovering practical and local knowledge and making use of the experience in the transformation process through experimenting and joint learning and action strategies. The new transdisciplinary transformation research is important here—not the established discourses of economics and ecology. The global process supporting sustainability transformation is that of global environmental governance as discussed in the chapters. The concept of "peer-to-peer learning" seems useful to organise the future sustainability discourse and process, since much of the joint learning will happen among the actors that cooperate in the process. This can also help to dissolve the hierarchies and asymmetric power relation in the sustainability policies.

10.5 Sustainability Related Themes Not Discussed in the Book

This book has not discussed systematically all of the potentially important themes and questions in the broader sustainability discourse. One possible theme is a more detailed discussion of *environmental and sustainability ethics* in the sustainability discourse and reasons for the difficulty in assessing ethical questions in sustainability practice (as described in Chap. 8 where the textbook by Kibert et al. (2011) was recommended as additional source). Ethical norms and principles need to be translated into another form of discourse that is not well-developed in philosophy and in the environmental and sustainability discourses. This is known as applied ethics. But here the debates are often casuistic, concerned with single ethical problems. The few forms of discursive ethics for use in political and public discourse, for example the discourse ethic of Habermas, failed to find their way into environmental and sustainability discourses.

Another possible theme of deeper discussion is also from the philosophical discourse, *ontology*. Beyond the epistemological questions that come up within. the interdisciplinary knowledge work that several chapters discussed, ontological questions have some significance in interdisciplinary discourses. However, the subject dissolves amid different interpretations and applications of the term, which remain abstract as the ontological reflections in philosophy (reflections about "being", "existence" or "reality"). Ontological questions touching knowledge production and use are discussed in combination with epistemological questions, also sustainability and sustainability-related knowledge. Ontology in information and computer science comprises the definition of categories and of the relations between concepts and data used in a discourse. This can be seen as a component of knowledge work and conceptual reflection in interdisciplinary knowledge practices. The examples of Martins (2011), Buttigieg et al. (2016) and Kuster et al. (2019) show, that ontological discussions in the environmental and sustainability discourses are connected with epostemological and methodological debates and normative questions of the values, worldviews, paradigms and visions of and its application.

Other themes excluded are of more practical nature. The book does not provide systematic information about *single policy fields that connect with the sustainability process* and become part of its practice under the broad set of sustainable development goals, including climate policy, energy policy, food policy, marine policy, land use policy, urbanisation, and others. Not much can be learned when these fields are illustrated through case studies and examples of single policy programmes. What would be necessary with such examples is critical analysis and discussion of their development, as it is done in one of the learning exercises in the book (in Chap. 4 as discussion of bioenergy, a controversial discussion in the development of sustainable energy sources). These policies require more systematic analysis, and for that purpose a series of textbooks and handbook have been mentioned in the chapters for deepening and specialising studies. Furthermore, *environmental planning* has not been discussed, nor has the development of *environmental movements*. These themes are touched upon in some chapters, but they are not core parts of the integration of the economic and sustainability discourses.

Finally, one controversial theme was discussed, albeit without resolving the controversy: whether the global North-South divide that brought the sustainability theme onto policy agendas is growing or dissolving. This controversy is not resolved in the book, but arguments are presented about why the divide seems to continue, and sustainable development has not yet brought any solutions. This is not a geographical, but a social and economic divide that has not been overcome, because some countries in the Global South began to industrialise rather late. The divide requires us then to consider the nature of the modern society as global society and the division of labour within the global economy, including analyses of extracting and producing economies and the "financialisation" of capital in the globalising economy that has created new development centres and hinterlands, global cities and villages, and rural areas decoupled from the growth areas. Also, the discussion of a "new world order" to be built with sustainable development is a theme for further clarification. The discussion has not advanced in the scientific debates beyond general discussions in philosophy, sociology and political science of new normative orders and a cosmopolitan society. Finally, the question of a "Eurocentric" perspective comes up in the discussion of the modern society that began in Europe, since the global economy remains under control of the Western and Northern countries in the OECD that dominate global policy and development. There is much more to discuss in the sustainability discourse and process. Nevertheless, the many and varied possible topics are outside this book's bounds.

References

Alonso Yanez, G., Thumlert, K., de Castell, S., & Jenson, J. (2019). Pathways to Sustainable Futures: A "Production Pedagogy" Model for STEM Education. *Futures, 108*, 27–36.

Araposthatis, S., & Pearson, P. J. G. (2019). How History Matters for the Governance of Sociotechnical Transitions: An Introduction to the Special Issue. *Environmental Innovation and Societal Transitions, 32*, 1–6.

Arora, N. K., & Mishra, I. (2019). United Nations Sustainable Development Goals 2030 and Environmental Sustainability: Race Against Time. *Environmental Sustainability, 2*, 339–342.

Büchs, M., & Koch, M. (2019). Challenges for the Degrowth Transition: The Debate About Wellbeing. *Futures, 105*, 155–165.

Buttigieg, P.L., Pafilis, E., Lewis, S., Schildhauer, M.P., Walls, S.E., & Mungall, C. (2016). The environment ontology in 2016: Bridging domains with increased scope, semantic density, and interoperation. *Journal of Biomedical Semantics 7*, 57. https://doi.org/10.1186/s13326-016-0097-6.

Fergnani, A. (2019). Scenario Archetypes of the Futures of Capitalism: The Conflict Between the Psychological Attachment to Capitalism and the Prospect of Its Dissolution. *Futures, 105*, 1–16.

Foxon, T. J. (2018). *Energy and Economic Growth: Why We Need a New Pathway to Prosperity*. Abingdon, UK: Earthscan for Routledge.

Freeth, R., & Caniglia, G. (2020). Learning to Collaborate While Collaborating: Advancing Interdisciplinary Sustainability Research. *Sustainability Science, 15*, 247–261.

Harmin, M., Barretta, M. J., & Hoessler, C. (2017). Stretching the Boundaries of Transformative Sustainability Learning: On the Importance of Decolonizing Ways of Knowing and Relations with the More-Than-Human. *Environmental Education Research, 23*(10), 1489–1500.

Hartwick, J. (1977). Intergenerational Equity and the Investing of Rents from Exhaustible Resources. *American Economic Review, 66*, 972–997.

Hopkins, D., Kester, J., Meelen, T., & Schwanen, T. (2020). Not More but Different: A Comment on the Transitions Research Agenda. *Environmental Innovation and Societal Transitions, 34*, 4–6.

Hotelling, H. (1931). The Economics of Exhaustible Resources. *Journal of Political Economy, 39*(2), 137–175.

Johnstone, P., & Newell, P. (2018). Sustainability Transitions and the State. *Environmental Innovation and Societal Transitions, 27*, 72–82.

Kanger, L., & Schot, J. (2019). Deep Transitions: Theorizing the Long-Term Patterns of Socio-Technical Change. *Environmental Innovation and Societal Transitions, 32*, 7–21.

Kibert, C. J., Monroe, M. C., Peterson, A. L., Plate, R. R., & Thiele, L. P. (2011). *Working Towards Sustainability: Ethical Decision-Making in a Technological World.* Hoboken, NJ et al.: Wiley.

Kuster, C., Hippolyte, J-L., Rezgui, Y. (2019). The UDSA ontology: An ontology to support real time urban sustainbility assessment. *Advances in Engineering Software*. https://doi.10.1016/j.advvengsoft.2019.102731.

Martins, N. (2011). Sustainability economics, ontology, and the capability approach. *Ecological Economics, 72*, 1–4.

Sautter, B. (2016). Futuring European Industry: Assessing the ManuFuture Road Towards EU Re-Industrialization. *European Journal of Futures Research, 4*, 25. https://doi.org/10.1007/s40309-016-0100-6.

Selby, D., & Kagawa, F. (Eds.). (2015). *Sustainability Frontiers: Critical and Transformative Voices from the Borderlands of Sustainability Education.* Toronto: Barbara Budrich Publishers.

Turnheim, B., Asquith, M., & Geels, F. W. (2020). Making Sustainability Transitions Research Policy-Relevant: Challenges at the Science-Policy Interface. *Environmental Innovation and Societal Transitions, 32*, 7–21.

Wilenius, M. (2014). Society, Consciousness and Change – An Inquiry into Pentti Malaska's Futures Thinking. *Futures, 61*, 58–67.

Wilenius, M. (2017). *Patterns of the Future: Understanding the Next Wave of Global Change*. New Jersey: World Scientific Publishers.

Glossary

Biodiversity The variety and variation of life on earth, described at different levels (genetic, species, and ecosystem) for terrestrial and marine ecosystems. Biodiversity loss or species reduction through human modification of ecosystems is one aspect of present global environmental change that creates new risks for life's future evolution.

Carrying capacity The maximum level of population of different species that an ecosystem can support without suffering negative impacts (conventional definition in ecology). Because human consumption differs strongly between individuals, the carrying capacity for humans has been re-defined as the maximum level of consumption, not of people (critical definition in human ecology: Rees).

Columbian exchange The environmental historian Crosby identified the decisive mechanism for the building of the modern economic world system since the Portuguese and Spanish colonial conquests in the sixteenth century: the exchange of colonial goods between Europe and the rest of the world, that brought exotic fruits and products to Europe and created a continuous flow of wealth from the poor to the rich parts of the world.

Commons and common pool resources Commons describes a form of local common property (of land, water, forests) that developed historically in agricultural societies. Common pool resources describe resources as they exist in their natural state (for example, fish in a lake or in the sea). These resources can be part of commons but can also become subject to other forms of property rights.

Conflicts In specialised conflict research conflicts are vaguely defined as all forms of interest or value conflicts between two or more persons that can be violent or non-violent. In this book the term is limited to environmental and sustain-

K. Bruckmeier, *Economics and Sustainability*,
https://doi.org/10.1007/978-3-030-56627-2

ability related conflicts in form of natural resource use conflicts, environmental conflicts, and ecological distribution conflicts.

Degrowth A concept for a non-growing economy that includes forms of creating wealth and welfare beyond monetary income, material consumption and economic growth, including especially new forms of material and immaterial consumption for creating better conditions of life (for example through reduction of the work time). Degrowth cannot be realised under the conditions of the capitalist growth-based economy where every form of non-growth results in crisis or collapse. The social forms of degrowth need to be found in the transformation to sustainability.

Development A concept with different meanings that can be related to personality development (psychology) or social and economic development (used in this book). Development can be differentiated from other forms of change as a slow, continuous change that does not alter the state of the social or economic system (for example, through growth of the economy or of capital). Structural change is a longer process that implies gradual change of a social system, such as changing the economy through technical innovations or other forms of modernization. Transformation differs from development and structural change in the sense that it bring such fundamental change that a new form of economy or society emerges.

Ecological distribution conflicts Social conflicts resulting from unequal access to natural resources and unjust distribution of benefits and burdens (i.e., pollution) between economic actors (resource users, producers, and consumers). The conflicts can be classified according to the phases of extraction, production and processing, transport, waste and pollution (Martinez-Alier).

Ecological modernisation A variant of growth- and industry-based development through non-polluting technologies and the reduction of environmental pollution in production and consumption, based on combinations of economic and ecological rationality and behaviour.

Ecology (a) the science of ecology is a natural science that studies the relationship between organisms and their environment, as part of biology; (b) further meanings refer to interdisciplinary scientific subjects as human, social, cultural, political ecology, and (c) to the relations between society or humans and nature.

Environmental justice An environmental movement (in the USA) with the goal of fair and equal distribution of environmental benefits and burdens. In a more social-scientific sense, the term refers to theories of the environment and justice with principles of fairness and participation of all people in the development, implementation, and enforcement of environmental laws and policies.

Global change A general term that comprises two kinds of global processes—social change (economic globalisation, global environmental and sustainability governance) and environmental change (anthropogenic climate change, biodiversity loss, land use change and urbanization). Both kinds of processes are connected and interacting.

Globalisation The forms of deregulation of markets and international capital flow since 1970 that intensified the global flows of goods, capital and money, information and people in the international economy. The world market and the modern economic world system in which the present globalisation happens were built in the sixteenth century during the period of European colonial conquest.

Great transformation The economist Polanyi called the building of an industrial and market-based society in England during the eighteenth century the "great transformation", because it produced dis-embedded markets, The process continues today with the economic globalisation. In the discourse of sustainable development the term "great transformation" describes a transformation of the whole economy and society to a post-industrial or post-capitalist society.

Growth Includes within the environmental and sustainability discourse different forms of growth connected with each other: population growth, economic growth, growth in the use of natural resources, growth of environmental damages and disasters.

Inter-disciplinarity Forms of producing, communicating, integrating and using scientific knowledge across the boundaries of scientific disciplines, for purposes of better understanding, explaining, influencing or managing complex processes in the social or physical systems where scientific knowledge is applied.

Kondratjev cycles An idea invented by the Russian economist Kondratiev, which holds that "long waves" (about fifty to seventy years) of innovation have driven the growth-based industrialisation process since its beginnings in late eighteenth-century Europe. Each cycle begins with a new technical innovation, as in the contemporary examples of the internet and information technologies.

Kuznets curve A hypothesis formulated by Simon Kuznets, which holds that in any country's economic development and industrialisation market forces first increase, then decrease economic inequality. An additional hypothesis holds that with respect to the environment the environmental Kuznets curve implies that industrialisation first increases, then decreases environmental pollution. (The latter hypothesis is disputed.)

Limits to growth The term used in the Report of the Club of Rome of 1972 to describe the connected forms of rapid, exponential growth of the global population, of natural resource use, of environmental pollution and ecosystem and of the damages that can result in a potential collapse of the global economy or society. (No system can exponentially grow for unlimited time.)

Maladaptive change Changes in social groups, organisations, social systems and coupled social-ecological systems with negative, destabilising consequences for the social or ecological systems or the environment. Examples are exponential growth, overuse of natural resources, degradation of the environment, climate change, and biodiversity loss.

Metabolic rift A concept used by Marx to describe the interruption of nutrient cycles in the modern, capitalist agriculture(based on the knowledge from the agricultural chemistry and theory of plant nutrition by Liebig). The term can be re-interpreted with today's ecological knowledge and connected with the manipulation and disturbance of material cycles of water, nitrate, phosphorous and so on in the earth system through humans.

"Mode One" and "Mode Two" Mode One signifies the traditional form of academic science and research (disciplinary specialisation). Mode two signifies new forms of inter- and transdisciplinary knowledge production, where scientists, decision makers, economic actors and civil society actors cooperate in knowledge's production and application.

Modernisation and modern society Inexact concepts for the development of a global society. They include the following: (a) a variety of cultural, social, political economic, and technical forms of development, change and innovation, such as secularisation (diminishing religious authority and building of non-religious institutions), modern democracy, the nation state, economic transformation through new technologies and modes of production in agriculture and industry, building of a global economy and world market; (b) the institutional core of modern society that developed after the middle ages in Europe through the colonial conquest of the Global South includes three closely coupled and overlapping macro-systems: the modern economic world system, the capitalist and the industrial society, and capitalist mode of production. Human modification of nature, environmental destruction and overuse of natural resources are the negative and neglected consequences of modernisation and modern society that are today discussed in the sustainability discourse.

Post-normal science A new form of scientific knowledge production (described by Silvio Funtowicz and Jerome Ravetz) where facts are uncertain, values disputed, stakes high and decisions urgent.

Resilience The capacity of social or ecological systems or of persons to survive disturbance or catastrophes without collapsing, through changes into another state of functioning, for example, a lower state of development.

Social-ecological system Ecosystems and social systems are not separate, but interact with each other and are coupled in many, geographically varying and historically differing forms. It is difficult to describe general structures and processes of social-ecological systems. In empirical ecological research the concept is simplified as a territorially limited ecological system with biological and physical resources and the social actors connected with it.

Social subjects of sustainability transformation The global transformation to sustainability is a process without a unified social or political subject (social class or state) that is guiding and directing the process. The transformation happens through the activities of many different actors, groups, movements, organisations and institutions with each cultivating different ideas and interests. The process becomes one of negotiation, finding consensus amid common interests, goals and priorities of action at a given time in the long transformation.

Societal (or social) metabolism A theoretical concept (in social ecology) to describe the complex forms of the social organisation of natural resource use in a specific form of economy or mode of production. The societal metabolism of the industrial society is based on the use of fossil energy resources (coal, oil, gas) that decoupled the modern economy from locally available energy sources and local land use. Every historical form of societal metabolism has specific social and natural limits to growth. In the industrial society these are the global limits of natural resources and the social forms of economic crises and resource-use problems that result from the economic mode of production (capitalist, based on private property, capital accumulation, market economy).

Socio-metabolic regimes (or socio-metabolic systems) The systemic forms of interaction of society and nature in forms of material and energy flows, necessary for the reproduction and the maintenance of biophysical structures of human societies (including also flows between different societies and within a society and country). In simpler forms the regimes can be described as the different types of material and energy use in early societies of hunters and gatherers, agricultural societies, and modern industrial society.

Stationary state economy A term from classical political economy (Smith, Ricardo, Mill) to describe the situation of natural limits to growth and resource use on Earth. In this perspective the economic growth that began with industrialisation was expected to result in a scarcity of resources and the accumulation of capital. A stationary state of the economy implies a that population and total capital remain constant.

Steady-state economy A renewal of the concept of the stationary state, it came back in the 1970s, thanks to the work of the ecological economist Daly. Daly defined the steady state as an economic system with a constant stock of capital and population that is maintained through the flow of natural resources (material and energetic resources) through the system.

Sustainability science The inter- and trans-disciplinary science of complex systems and of the interaction of social and ecological systems with the aim of generating knowledge for the transition to sustainability. Sustainability science differs from traditional academic research through the close connection between research and action (cooperation of scientists, practitioners and the public, joint learning of researchers and actors) and specific methods (semi-quantitative models, qualitative data, learning from case studies).

Sustainability and sustainable development In the general and abstract sense of the term (that encompasses a variety of definitions and interpretations) sustainable development is social or economic development that can last for an unlimited time without destroying the social, economic and ecological conditions of future development. Sustainability is the state or goal to be achieved through sustainable development. Example: a steady state economy in the sense of ecological economics.

Technical fix Technological shortcuts produced in an attempt to solve social, economic, and environmental problems created by earlier technologies. Nevertheless, these shortcuts often create new, unanticipated problems.

Tragedy of the commons and tragedy of enclosures Two different social forms of overuse of natural resources and subsequent ruin of the ecosystem or the resource base for human subsistence—either through overuse under conditions of common property (Hardin), or through overuse under conditions of private property or enclosures (Martinez-Alier). The controversy about overuse cannot be resolved as long as the commons is reduced to a consequence of property rights, because this legal situation makes it impossible to apply concrete forms of management to the use of the resources.

Trans-disciplinarity A variant of knowledge creation and integration from various sources that differs from conventional forms of interdisciplinary knowledge production through the inclusion of non-scientific knowledge and the creation of knowledge in the contexts of practical application of knowledge.

Transformation and social ecological transformation The specific form of change of economic and societal systems in the sense of transformation to sustainability is a large-scale, global transformation in a long process whose endpoint cannot be foreseen. It is a social-ecological transformation insofar as it includes transformations of modern society, economy, and of the network of ecological systems from which the society and economy are dependent for their functioning.

Transformation science A new form of trans-disciplinary research that developed with the discussion of sustainable development as a social, economic and ecological transformation. An important issue in transformation research is to find ways to build capacity and agency to transform an economy or society into sustainable forms.

Transition The two terms transition and transformation can be used interchangeably. Transition is mainly used to describe local and less complex forms of change, transformation for large-scale and complex changes of social systems and coupled social-ecological systems.

Vested interests The reasons and motives that drive specific actors to protect their (short-term) interests, including their professional, political, economic or scientific position, or influence and power for the purposes of securing material advantages from the given situation. With regard to sustainable development vested interests include interests in obtaining income, monetary gains and profit from the market and a growth-based economy, even though such a system is socially and environmentally unsustainable.

World system and modern world system According to Wallerstein who created world system analysis, a world system is any historical social system covering large territories (empires) with interdependent parts (core and periphery) in a common structure, with specific (political, economic) rules, or a single division of labour (or mode of production), with a variety of cultural systems. Before the modern world system (the capitalist world economy) there were other world systems in human history (political world systems in the form of empires, such as the Chinese or Roman Empire). The modern world system comprises a world economy, larger than any state or country. The different parts of centre and periphery are linked economically through market mechanisms in a capitalist world economy (accumulation of private capital, exploitation of human labour and natural resources for profit in a market and growth-based economy).

Index

K. Bruckmeier, *Economics and Sustainability*,
https://doi.org/10.1007/978-3-030-56627-2

GPSR Compliance
The European Union's (EU) General Product Safety Regulation (GPSR) is a set of rules that requires consumer products to be safe and our obligations to ensure this.

If you have any concerns about our products, you can contact us on

ProductSafety@springernature.com

In case Publisher is established outside the EU, the EU authorized representative is:

Springer Nature Customer Service Center GmbH
Europaplatz 3
69115 Heidelberg, Germany

www.ingramcontent.com/pod-product-compliance
Ingram Content Group UK Ltd.
Pitfield, Milton Keynes, MK11 3LW, UK
UKHW021942040925
462611UK00004B/279
* 9 7 8 3 0 3 0 5 6 6 2 6 5 *